History, Humanity and Evolution

History, Humanity and Evolution

Essays for John C. Greene

EDITED BY
JAMES R. MOORE
Lecturer in History of Science and Technology, The Open University

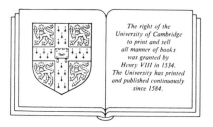

The right of the
University of Cambridge
to print and sell
all manner of books
was granted by
Henry VIII in 1534.
The University has printed
and published continuously
since 1584.

CAMBRIDGE UNIVERSITY PRESS
Cambridge
New York   Port Chester   Melbourne   Sydney

Published by the Press Syndicate of the University of Cambridge
The Pitt Building, Trumpington Street, Cambridge CB2 1RP
40 West 20th Street, New York, NY 10011, USA
10 Stamford Road, Oakleigh, Melbourne 3166, Australia

First published 1989

Printed in the United States of America

Library of Congress Cataloging-in-Publication Data
History, humanity, and evolution: essays for John C. Greene /edited
by James R. Moore.
p.   cm.
Includes index.
ISBN 0-521-33511-6
1. Biology – Philosophy – History.   2. Evolution – History.
3. Science – Philosophy – History.   I. Greene, John C.
II. Moore, James R. (James Richard), 1947–
QH331.H54   1989
575′.009 – dc20

89-32583
CIP

British Library Cataloguing in Publication Data

History, humanity and evolution : essays for
John C. Greene.
1. Man. Evolution. Theories, history
I. Moore, James R. (James Richard)
II. Greene, John C.
573.2′01

ISBN 0-521-33511-6 hard covers

CONTENTS

v

PREFACE

This collection of essays is intended to serve as a benchmark and a monument. On the broad terrain of the history of evolutionary thought, it offers a point of reference from which the current generation of scholarship can be assessed. At the same time it pays tribute to the one scholar of a generation past who has most influenced today's leading practitioners in the field. John Greene has, indeed, inspired the contributors to this volume not only by the sweep and subtlety of his historical writings, but by their broad accessibility as well. The essays assembled here in his honour are thus written for a general academic readership and with a view towards a wider contemporary audience concerned with the religious, moral and political aspects of evolution.

Thirty years since scholars last joined forces to survey the history of evolutionary thought, the interests and the interpretations of leading practitioners in the field have undergone a fundamental shift. The historians who wrote in *Forerunners of Darwin, 1745–1859* and the biologists who contributed to *Evolution after Darwin* were more or less agreed (to use the words of the books' respective editors, Bentley Glass and Sol Tax), that a 'great, steadily enlarging current of biological thought' eventuated in a man named Charles Darwin who 'broke through a tremendous fog', liberated 'scientific conclusions' that had been 'stubbornly and blindly repressed', and 'gave us a new understanding and perspective on the basis of which we have done a hundred years of fruitful research'. Today's historians are more likely to fault than to flatter biologists' triumphal periodization of their disciplinary past. Darwin, for them, is not the revolutionary figure he once appeared to be; evolutionary ideas are not simply the rational outcome of a self-correcting science. In the nineteenth century, when disciplinary boundaries were being redrawn, or in some cases drawn for the first time, the sources of evolutionary ideas were diverse, their ramifications universal, and their scientific status constantly negotiated under pressure from a variety of theological, philosophical and ideological standpoints. Biology was not the touchstone, nor geology, or astronomy, or anything but all these emerging disciplines together in their complex cultural relationships. The human significance of evolution was paramount, and so it has remained ever since. For this insight, historians today owe more directly or indirectly to another title published thirty years ago, *The Death of Adam*, and to subsequent works by the same author, than to the writings of any other historian of his generation.

John Greene would of course be the first to acknowledge that the generation of scholars represented in these pages have pursued the history of evolutionary thought to a much greater depth than he was able, as well as in different historiographic directions, some of which remain controversial. Indeed, he says as much below in his Introductory Conversation with the editor and in his Afterword, where he responds to the essayists. These remarks, so indicative of change and continuity in the field, emerge with a biographical freshness that not only is fascinating in its own right, but may also incidentally shed light on the development of intellectual history as an academic specialism in mid-twentieth-century America. Their import will be evident to all who understand that historians write history under personal and epochal constraints every bit as real as those under which biologists practise biology.

The essayists themselves are among the chief interpreters of the history of evolutionary thought today. Most of them cut their academic teeth on John Greene's works; some have also benefited from his personal encouragement in their careers. Taking the past two centuries as their domain, they offer a series of new perspectives on people, institutions and societies that authorized and sponsored developmental conceptions in biology, geology, astronomy, anthropology and religion. Medical history and contemporary literature are shown to elucidate the ideas of Erasmus Darwin and Jean-Baptiste de Lamarck. Early Victorian professional and political struggles become the context for understanding the deployment of transformist physiology and the nebular hypothesis. Fresh manuscript sources serve to relate the private and public dimensions of the lives of Robert Chambers and Charles Darwin. Visual materials are shown to be richly relevant for understanding nineteenth-century beliefs about the prehistoric past. A whole range of wider cultural considerations evoked by feminist scholarship, popularization studies, and the histories of politics and art are brought to bear on post-Darwinian evolutionary theory to explain its extraordinary malleability and the diversity of its applications to human beings. Finally, two essays address perennial concerns in John Greene's work by offering historical reflections on the problem of constructing an evolutionary world view that does not cede the realm of human values to scientific expertise.

As editor, I only wish to add that this book would not exist except for the steady enthusiasm, encouragement and patience of the contributors, especially Simon Schaffer, who made its completion possible. Garland Allen, Gillian Beer, John Brooke, Richard Burkhardt, Pietro Corsi, Sandra Herbert, David Kohn, Maureen McNeil, Steven Shapin and Edward Yoxen also wished the project every success as a testimonial to one who they, too, regard as an outstanding scholar and friend. Above all, though, it has been John Greene's cooperation, and the warmth and hospitality he and Ellen have extended to me over the past five years, which has made this Festschrift a reality. I may use my editorial prerogative, speaking no doubt on behalf of the contributors, those named above, and a great many others, to thank him, and them, for enriching life in countless ways and for investing the historical enterprise with fresh significance.

Cambridge J.R.M.
31 March 1989

CONTRIBUTORS

ROY PORTER (b.1946) is Senior Lecturer in the Social History of Medicine at the Wellcome Institute for the History of Medicine in London. His publications include *English Society in the Eighteenth Century* (1982), *Mind Forg'd Manacles* (1987) and *A Social History of Madness* (1987). He has also edited (with Mikuláš Teich) *The Enlightenment in National Context* (1981) and *Revolution in History* (1986)

LUDMILLA JORDANOVA (b. 1949) is Senior Lecturer in History at the University of Essex. She has published *Lamarck* (1984) and *Sexual Visions* (1988) and has edited and contributed to *Languages of Nature* (1986). She has also edited (with Roy Porter) and contributed to *Images of the Earth* (1979). Her research interests include the relations between the visual arts and the biomedical sciences, and the cultural history of science and medicine

ADRIAN DESMOND (b. 1947) is in the Department of Biology, University College London. He specializes in the social history of Victorian zoology, palaeontology and evolution, and is the author of *Archetypes and Ancestors* (1982) and *The Politics of Evolution* (1989). His popular science books include *The Hot-Blooded Dinosaurs* (1975) and *The Ape's Reflexion* (1979)

SIMON SCHAFFER (b. 1955) is University Lecturer in the History and Philosophy of Science at the University of Cambridge. He has published widely on the history of eighteenth- and nineteenth-century astronomy and experimental philosophy, and is co-author (with Steven Shapin) of *Leviathan and the Air Pump* (1985) and co-editor (with David Gooding and Trevor Pinch) of *The Uses of Experiment* (1988). He is also an editor of *Science in Context*, a journal of science studies

JAMES A. SECORD (b.1953) is Lecturer in the History of Science at Imperial College of Science and Technology, London. He has published *Controversy in Victorian Geology* (1986) and numerous articles on the history of Victorian science. His research interests include the social history of science, the life and earth sciences since 1750, and the relations between science and its audiences

JAMES R. MOORE (b. 1947) is Lecturer in History of Science and Technology at the Open University in England. He has published *The Post-Darwinian Controversies* (1979) and many articles on science and religion in Victorian culture. He has also edited *Religion in Victorian Britain: Sources* (1988) and made historical documentaries for BBC radio and television. His research is concerned generally with the social history of natural theology, and in

xi

particular with a contextualist interpretation of the life and career of Charles Darwin

MARTIN RUDWICK (b. 1932) is Professor of History of Science at the University of California, San Diego. His publications include *Living and Fossil Brachiopods* (1970), *The Meaning of Fossils* (1972) and *The Great Devonian Controversy* (1985). He has also published widely on palaeobiology, nineteenth-century geology, and the historical relations of scientific and religious practices

EVELLEEN RICHARDS is Senior Lecturer in the Department of Science and Technology Studies at the University of Wollongong in Australia. She has made many contributions to the social history of evolutionary biology, the sociology of medical knowledge, and gender and science, including *Vitamin C and Cancer: Medicine or Politics?*

BERNARD LIGHTMAN (b. 1950) is Assistant Professor of Humanities at York University in Toronto. He has published *The Origins of Agnosticism* (1987) and (with Sydney Eisen) *Victorian Science and Religion* (1984). He has also edited (with Richard Helmstadter) *Victorian Faith in Crisis* (1989). His research interests include unbelief, science and religion in Victorian Britain

PAUL WEINDLING (b. 1953) is interim Director of the Wellcome Unit for the History of Medicine at the University of Oxford. He has published *Health, Race and German Politics* (1989) as well as numerous articles on German biology and medicine in their social contexts. He has also edited *The Social History of Occupational Health* (1985) and co-edited (with Pietro Corsi) *Information Sources for the History of Science and Medicine* (1983)

PETER J. BOWLER (b. 1944) is Reader in History and Philosophy of Science at the Queen's University of Belfast. He is the author of *Fossils and Progress* (1976), *The Eclipse of Darwinism* (1983), *Evolution: The History of an Idea* (1984), *Theories of Human Evolution* (1986) and *The Non-Darwinian Revolution* (1988)

JOHN R. DURANT (b. 1950) is Assistant Director (Research and Information Services) at the Science Museum, London, and Visiting Professor of the History and Public Understanding of Science at Imperial College of Science and Technology, London. He has published widely on the history of the biological and human sciences and is the editor of *Darwinism and Divinity* (1985) and the co-author and editor (under the pseudonym John Klama) of *Aggression* (1988)

ROBERT M. YOUNG (b. 1935) is a London-based writer, editor, publisher and psychoanalytic psychotherapist. He is the author of *Mind, Brain and Adaptation in the Nineteenth Century* (1970) and *Darwin's Metaphor* (1985), and editor of two journals, *Free Associations* and *Science as Culture*. His work is concerned with various aspects of the limits and potential of human nature, considered politically, morally, culturally and psychobiologically

INTRODUCTORY CONVERSATION

JM: On the occasion of your seventieth birthday, John, I thought it would be fitting to honour your contribution to the history of evolutionary ideas by assembling a collection of essays written by your younger friends and colleagues under the title *History, Humanity and Evolution*. This title was chosen with some care. Its terms not only mark out the general locus of your own interests over many years; they also indicate an order of cognitive priority that appears in your own work and reappears in the essays contained in this volume: *history* first, then *humanity* as its offspring, and finally *evolution* as the discovery and artefact of human genius working under particular historical conditions. In this introductory conversation I would like to discuss your contribution to our understanding of evolutionary ideas in the context of your own personal and professional history. I want to focus both on the historicity of evolutionary ideas and on the humanity that has evolved – and, according to some interpreters, is evolving still. In the course of our conversation I'm sure your own particular humanity as an interpreter of evolution will emerge, and this should prove instructive to those of us who take up the historical task where you leave it after a lifetime's work. Perhaps I may begin with a quotation. You once commended Benjamin Silliman, the nineteenth-century Christian geologist, for his 'resolute effort to see human life and the place of science in it steadily, and to see it whole'. Would you accept this as a statement of your own intentions in writing about the history of evolutionary thought?

JG: Yes, I would. I feel strongly that in our own time science has lost its historical bearings. As I see it, the history of Western civilization has three main intellectual components. The first is the classical component, from Greece and Rome; the second is the Judaeo-Christian component; the third is what we call modern science, dating from about 1500. Modern science is the offspring of the first

1

two: in the sixteenth and seventeenth centuries, by some kind of alchemy we do not entirely understand, the fusing of the classical and the Judaeo-Christian components produced it. But then the child began to react very powerfully on both its parents. What I like about the eighteenth century, on which I have spent a considerable amount of time, is that the three components subsisted in a marvellous balance. Take somebody like Thomas Jefferson: he was reared as a Christian, and his deism was shot through with Judaeo-Christian ideas; he read Homer and Epictetus in the original with great pleasure and he was enthusiastic about modern science. Not only in Jefferson, but in many people of the period, one sees this sort of balance. In the nineteenth century the balance began to break down. Science, with its technical applications, began to undermine both the Judaeo-Christian tradition and the classical humanistic tradition. In the twentieth century things have been completely torn apart. Whirl is king, having deposed both Zeus and Jehovah. But while science, with its technical applications, has shown a great power to influence and undermine the other two components, it has shown no power whatever to provide an alternative basis for values or a satisfactory replacement for traditional views of the nature and destiny of man. It seems to me that the problem of our time is to reconstitute the cosmos in such a way as to leave a reasonable place for modern science *without* sacrificing traditional values and meanings.

JM: To what extent do you think this reconstitution would involve a recovery of the particular 'balance' that you commend the eighteenth century for?

JG: I'm not sure we can predict what form it might take. I have been much influenced by Whitehead's attempt to construct a general picture of reality that would accommodate both natural science and aesthetic and moral sentiments. On the other hand, I don't think Whitehead ever took full account of those aspects of human nature that St Paul talks about – the good which I would I do not, and the evil which I would not, that I do. For the same reason, I have always viewed the eighteenth century with ambivalence. Jefferson, I think, was overly optimistic about human nature, not sufficiently realizing that if knowledge is power, as he believed, and, if power corrupts, as Lord Acton would later say, then knowledge can corrupt too, unless it is developed in an adequate moral and metaphysical framework. Of course I share many of Jefferson's values – freedom of speech, freedom of thought – and I think that the Enlightenment was in many ways a great movement for the liberation of people from institutional restraints. But at the same time I believe there is a wisdom in both the classical and the Judaeo-Christian traditions which has now been largely lost. In the quest to command nature we have forgotten how to command ourselves.

JM: I'd like to pursue the theme of fragmentation a bit further. In 1959, your first and best known book was published, *The Death of Adam*. Then in 1961 came your lectures, *Darwin and the Modern World View*, and exactly twenty years later a collection of your essays appeared as *Science, Ideology, and World View*. These titles reflect your concern with the wider intellectual context of evolutionary ideas; the latest one names three things – science, ideology and world view – which, in the postscript to the book, you say will be forever intertwined and interacting. Now how would you distinguish among science, ideology and world view in the history of evolutionary thought if these three are so closely related?

JG: Well, I admit 'ideology' is the odd term here. In my writings you will not find a great deal connecting the history of evolutionary ideas with ideology. There are some hints, but on the whole the connections are not well developed.

JM: The term 'world view' occurs in the titles of both your latter books on evolution, 'ideology' appears only in the latest. I wonder whether this has anything to do with the state of political debate in the 1980s. Perhaps it would not have been acceptable to use the word 'ideology' in the title of a book published in 1961. Wasn't 'the end of ideology' being proclaimed in America then?

JG: I don't think that would be a correct historical explanation of the difference in the titles. My dominant interest has always been in the interaction of science and world view. The appearance of 'ideology' in the title of my collection of essays, two of which had been published in the 1950s, probably reflects my reading of people like yourself. But I have always been aware of the ideological aspect of evolutionary ideas, and I am perhaps a little more attuned to it now than I was earlier.

JM: What do you mean by 'ideology'?

JG: To me ideology means primarily ideas that justify programmes of social action. Marxism is an ideology; Freudianism, broadly speaking, is an ideology too. There was an ideology of the American Revolution. Many writers describe as ideology what I would call 'world view', and this may be because world view is a less familiar, and possibly more difficult concept. By world view I mean explicit or implicit presuppositions of thought in a period. A world view is not necessarily any particular individual's outlook. It is a certain inter-related set of presuppositions that dominates the thought of the age, such as the static view of nature in the eighteenth century. And there are competing world views too: a subdominant view inherited from the past – the decline from original perfection – as well as an incipient view – the idea of nature as a law-bound system of matter in motion. Among these a kind of tension exists. As a student of the history of ideas, my problem is to find out what brings about transitions from the dominance of one world view to the dominance of another.

JM: But just consider the advocates of the static view of nature in the eighteenth century. Weren't the 'world view' utterances of Boyle lecturers and physico-theologians also ideological statements – statements about social order? Natural theology was so ideological that Basil Willey could call it 'Cosmic Toryism'.

JG: The trouble with that view is that Cosmic Toryism is not found just among the Tories. Whigs were natural theologians too – Tom Paine is full of it.

JM: That may be so, but my point is that eighteenth-century people, whoever they were, expressed a static world view in terms that bore the distinguishing marks of what you understand as ideology. It is pretty generally admitted now that classical natural theology, which presupposed the fixity of created kinds, was pursued in such a way as to uphold a providentially ordered and relatively static social world, and was therefore ideological. What I am questioning is how far the distinction between ideology and world view can be sustained.

JG: I would not regard static physico-theology as part of a programme to defend or attack some social movement. Certainly, it could be *used* in that way. When as a graduate student I first developed my notion of the dominant view of nature in the eighteenth century, I also had a parallel conception, the dominant view of society, in which the family, the State, and so on, were regarded as given, either by God or by nature, and relatively unchanging. In other words, I have always distinguished between presuppositions of social thought and presuppositions about nature.

JM: We agree that there can be these two sorts of presuppositions, about human relationships on the one hand, and about the non-human natural world on the other. What seems to be at issue is whether there may be some inner historical connection between the two, and if so, whether this connection has to be made explicitly within a period such as the eighteenth century for the historian to be justified in recognizing it and referring to a world view as ideological. Perhaps you would concede that ideology can be either explicit *or* as implicit as the assumptions you refer to as constituting a world view. If so, there would be no need to limit ideology to the conscious advocacy of social programmes, would there?

JG: I don't think I would object to that. Speaking historically and biographically, though, I have always understood ideology as being quite explicit. I would have to consider whether the more ideology, in your sense, becomes implicit, the more I would want to call it world view.

JM: Let's move on to that other key word from your titles – 'science'. If it is possible for a world view to be ideological, or an ideology to serve as a world view, is it possible also for an ideology to be scientific, or for science to become an ideology?

JG: To me science has always seemed distinct from both ideology and world view. Science attempts to describe or to discover relationships among phenomena in a value-free way, whereas ideology is connected with social action and so involves values of all sorts. Of course science is a form of action, but it is a form of action that stresses observation. The scientist as observer sets aside questions of religion and social values in order to understand nature as it is. At the same time, though, simply by being educated and practising as a scientist, the same individual is heavily conditioned by the prevailing world view: by conceptions of what science is, what nature is – ideas one takes in like mother's milk.

JM: The scientist as passive observer of phenomena is precisely what scientists have been taught about in textbooks, isn't it? It's part of the modern scientific world view. But when we reflect on how scientists are trained, and how their actual working practices are affected by basic presuppositions about the world, then we see science for what it is and are able, perhaps, to transcend its metaphysical constraints. It seems to me that in the early part of *Science, Ideology, and World View* you allow as much: you state that while 'general conceptions of nature, God, knowledge, man, society, and history' shape – or 'dictate', to use your word – every aspect of science, if the scientist has 'insight and intellectual integrity', then his or her findings may alter the general conceptions that shaped the scientific enterprise, rather than simply reinforce them. Now how does this 'insight and intellectual integrity' come in to preserve science from the taint of world view? Perhaps you can given an example from the history of evolutionary thought to illustrate what you mean.

JD: I don't know that I would say 'taint' – let's call it 'influence'. Also, I want to be clear that, as I say in the preface to *The Death of Adam*, there is no such being as a 'pure scientist'. But I do believe the best scientists are driven by intellectual curiosity, a desire to understand and explain how things work. And the 'insight' I referred to is of the kind that led Darwin and Wallace to the theory of natural selection. I know that natural selection has acquired a broad penumbra of sociological connotations, but, nevertheless, at the core of the theory is a genuine insight into a problem that Aristotle stated long ago: namely, how do you account for the general fact of adaptation in nature? The atomists had talked about concourses of atoms, but Aristotle did not believe adaptation could be explained in that way. Intellectually it was not satisfying. He required an explanation in terms of immanent purposiveness. Darwin and Wallace, however, hit on the idea of natural selection.

JM: Later, in the same context, where you mention a scientist's 'insight and intellectual integrity', you note the 'curious fact' that nearly all the proponents of natural selection in one form or another in the first half of the nineteenth century were British. It is 'strange that nature should divulge one of her profoundest secrets only to inhabitants of Great Britain', you say. The fact is explicable only by 'assuming' that political economy and the competitive ethos in the first industrial nation 'predisposed' British naturalists to think of competitive struggle in theorizing about living things. Do you mean by this that the capitalist economy of the eighteenth and nineteenth centuries was required for people to have a genuinely true insight about adaptation in nature?

JG: It's hard to say what would have happened if things had happened differently. A similar question is whether the Judaeo-Christian tradition was an essential ingredient in the rise of the mechanical world view and the science of mechanics. All that can be said empirically is that there was a predisposition in British thought in the nineteenth century, but the fact that someone was predisposed to see the world in a certain way does not detract in the least from the genuineness of his insight into nature. To me, the more curious fact is that Darwin chose to express his insight through the metaphor 'natural selection', which is a highly anthropomorphic idea. In reality, there is no selection of any kind; selection necessarily implies an intelligent choice. So why did Darwin use this metaphor? He said that he wanted to mark the analogy to the methods of plant and animal breeders, but I think his evolutionary deism must also be considered. Natural selection to Darwin was a set of processes ordained by God to bring about adaptation and improvement. And soon the metaphor took on a life of its own.

JM: Wasn't it very difficult for Darwin *not* to look at living things as purposefully contrived – naturally selected in the anthropomorphic sense? Why shouldn't his 'integrity' have consisted in his choice of the metaphor to describe what he perceived?

JG: No, I would regard that as extra-scientific. The intellectual integrity appears when, having got hold of natural selection – after all, Patrick Matthew came up with the idea a few years earlier –

Darwin exercised the imagination and the firmness of purpose to think it out, to test it, to explore all its ramifications, even though he realized the tremendous consequences for the whole structure of Western thought. And the curious thing about a genuine scientific insight like this is that, not only for the discoverer but for his successors, all kinds of new interpretations can grow up around it. If natural selection did not have some actual connection with nature, rather than just social reality, then it would not have led to further scientific developments.

JM: Is this what you mean by a 'genuine scientific insight': that it is no longer in touch with ideology or world view, but transcends the conditions of its origin?

JG: I think what you say is true of Darwin. The developing core of the theory of natural selection has been applied in many fields and tested in various ways by scientists. Like Newton's laws of motion, the theory has been modified. Newton's laws do not apply to the extreme macroscopic or the extreme microscopic worlds, and Darwin's theory has had to be modified to take account of discoveries in genetics. Nevertheless, Darwin, like Newton, got hold of something that is true about nature.

JM: And this truth no longer entails a world view or has ideological implications?

JG: Newtonian mechanics doesn't any longer. In Newton's day, a lot was built around it, but the modern physicist ...

JM: That's rather odd to hear from someone who has always pointed up the Newtonian view of nature as a law-bound system of matter in motion – a metaphysical world view if ever there was one. Yet Newton's particular metaphysics – his ideas of absolute space and time, for example – have not been found adequate.

JG: Right. The dominant world view has changed enormously since Newton. But force equals mass times acceleration. That's as true today as it was in the eighteenth century.

JM: Depending on what one means by mass.

JG: Yes. And, as I said, at the extreme macroscopic level and the extreme microscopic level Newton's laws of motion have to be modified. But in the ordinary world they hold. They are not absolute truths, only part of the developing body of science.

JM: Let me ask you another question about the history of evolutionary thought. How far are you willing to press your assumption that predisposing social and intellectual conditions made possible the discovery of scientific truths? You've made that assumption in the case of natural selection. Do you think it is generally helpful to make?

JG: I would understand the rise of the mechanical world view in terms of a fusing of the atomic idea, the Judaeo-Christian tradition, and Pythagorean-Platonic concepts. It was this world view, with or without a Creator-God, that informed numerous attempts to explain the natural development of life. In the nineteenth century, German *Naturphilosophie*, with contrasting antecedents, also had an important bearing on the pursuit of natural history.

JM: Would you be prepared to look as readily for the predisposing social and intellectual conditions under which the neo-Darwinian synthesis of the twentieth century emerged?

JG: There I have difficulties. Darwin and the others who conceived the theory of natural selection in the first place were all British; they shared in the British intellectual tradition, the British competitive ethos, and so on. But those who formulated the modern synthesis were not only Britons such as Julian Huxley and J.B.S. Haldane, but Theodosius Dobzhansky from Russia, George Gaylord Simpson and G. Ledyard Stebbins in the United States, and Bernhard Rensch in Germany. I don't know of any common social or intellectual tradition in which all these men can be embedded.

JM: But the point is, you wouldn't look on the search for common predisposing social and intellectual conditions as an illegitimate enterprise?

JG: No, on the contrary, I think it would be as challenging a task as explaining why only the British came up with natural selection a century earlier. I am struck, for instance, that French thought in the 1830s and 1840s was obsessed with the problem of social order. Solving this problem was Auguste Comte's *raison d'être*. Now it seems inconceivable to me that anyone obsessed in that way should hit on the idea of natural selection. Maybe it's possible, but the French did not do so as a matter of fact. And it seems equally far-fetched to me that German *Naturphilosophie*, with its doctrine that the whole of reality is a manifestation of Idea with a capital 'I', should have yielded a theory similar to Darwin's.

JM: Anyone who has perused your works must realize that you have argued for the embeddedness of evolutionary ideas in wider cultural, or at least intellectual, contexts for a very long time – certainly since the publication in 1957 of what was in many respects your keynote essay, 'Objectives and Methods in Intellectual History'. Were you at first conscious of maintaining unfashionable views on the subject, particularly in the face of claims by a resurgent neo-Darwinism to a detached scientific status?

JG: Not at all. My views on intellectual history were developed without reference to biology. When I was a graduate student at Harvard in 1947, I did audit a course on evolution taught by Alfred

Romer. We read Ernst Mayr's *Systematics and the Origin of Species*, Simpson's *Tempo and Mode in Evolution*, and Dobzhansky's *Genetics and the Origin of Species*. So I was not unaware of the modern synthesis, although I do not recall it's being referred to as such. But when I thought about intellectual history and later wrote the 1957 essay, my ideas derived from earlier work at Harvard and from a background of general reading. The essay was basically an exposition of views that I had adopted in the course of preparing my doctoral dissertation.

JM: Nevertheless, according to your views, many of the ideas considered by your contemporaries to be purely scientific were in fact historically indebted to that *bête noire*, metaphysics. After the appearance of your 1957 essay, followed by *The Death of Adam* and *Darwin and the Modern World View*, did you ever face opposition from scientists?

JG: On the contrary – though my contacts with scientists were then few and far between. I submitted the draft chapters of *The Death of Adam* to a series of specialists to make sure I had not written scientific nonsense. Bentley Glass, the geneticist from Johns Hopkins, gave the book quite a favourable review. In 1959, I was invited to present a paper on religion and science at the Darwin centennial celebrations sponsored by the American Philosophical Society in Philadelphia. It was there that I first met Dobzhansky. We were at a reception in Philosophical Hall, before the papers were delivered. I said, 'Professor Dobzhansky, I'm going to take a few pot shots at you and Julian Huxley for some of your metaphorical language.' Dobzhansky replied, 'Pray sir, do not bracket me with Julian Huxley: he is an atheist, I am a Christian.' We talked a little further and I agreed to leave him out of the paper. Afterwards Dobzhansky came up and was very complimentary about it. Stebbins also spoke to me briefly. I had referred to 'ultimate reality' in the paper and he objected to the adjective 'ultimate'. 'There's only reality', he said. So I encountered no hostility to my views from the scientific community.

JM: Have your views on the intellectual 'impurity' of the natural sciences – you yourself use this metaphor occasionally, so I'll use it too – been modified in the past thirty years, and if so how?

JG: I'm not aware of any great change in my views, though I must say that I am more convinced than ever that evolutionary biology is deeply enmeshed in telling figures of speech. And I have also become somewhat depressed at the inveterate tendency of evolutionary biologists to draw all kinds of religious, social and moral inferences from what I would regard as the scientific basis of evolution by natural selection. Getting 'sermons from stones', as people say.

JM: But you don't believe the stones *contain* the sermons, do you? That was the upshot of our discussion of natural selection. So how can sermons be got out of them?

JG: You bring the sermons *to* the stones. You find in nature what you expect or want to find. If you are a devout Christian and believe that nature is the work of God, it is not surprising that in studying nature you find evidence of design and providence. If, like Julian Huxley, you approach nature with a totally different set of preconceptions, then you find something else.

JM: We are talking in figures, aren't we? The 'stones' are the theories from which evolutionary biologists draw their precepts. I was questioning whether you might believe, after all, that natural selection has some inherent sermonic value.

JG: I don't think the 'stones' are theories. But the basic idea of natural selection – if there is random variation and population pressure, and consequently differential survival and differential reproduction, then the character of populations will change – does seem to me almost common sense. Who can quarrel with it? I can't see any sermons in it either.

JM: In *Darwin and the Modern World View*, you say that 'ideas and ideals can be social forces in their own right, ... they can be something more than expressions of class interest or libidinal drives'. Would you still maintain this view on the grounds you did thirty years ago?

JG: Very much so. I still believe that in some sense human beings transcend nature. If ideas are only manifestations of class interest or libidinal drives, then the whole intellectual enterprise is reduced to absurdity. Freud's theory, Marx's theory, Darwin's theory – the notion that any of these could be true goes out the window. All are merely manifestations of something essentially non-ideational, and I certainly do not believe that. There is a life of the mind, and it is very real and very important.

JM: Although you still hold, don't you, that ideas may owe a great deal to the non-intellectual conditions that authorized them?

JG: How can a non-intellectual thing authorize an idea? I don't understand. Ideas are real but non-corporeal.

JM: Well, it seems to me that you have done a good job explaining why a certain idea, natural selection, occurred to particular people at a particular place and time. But if ideas are really non-corporeal, then place and time could be irrelevant to their appearance. Natural selection could have occurred to anyone, anywhere, like a bolt from the blue.

JG: I don't think so. Yes, Darwin was an Englishman, but natural history also had to be in a coherent state of readiness for his theory to be conceived. Before you can talk about the origin of species you have to know what a species is and you have to develop methods of classification. The work in natural history accomplished on the basis of a static view of nature was an essential prerequisite for the emergence of Darwin's theory. The circumstances we talked about earlier were not a sufficient cause.

JM: The coherent state of natural history in Darwin's time was institutional and concrete as well as intellectual and abstract.

JG: That's right. But when my students tell me that the French Revolution brought about tremendous changes in society and politics, hence Lamarck's theory of evolution and Erasmus Darwin's were the reflex of these changes – well, that is all very nice, but how do you go about establishing it?

JM: I'm not sure that many of the younger scholars represented in this book would want to talk about ideas being a 'reflex' of changing social conditions. But what I and others like me would want to maintain is, I think, that the conditions under which ideas appear, flourish and become dominant or recede are not purely, or even predominantly, intellectual; there are crucial social, economic and political conditions as well. Tracing the pathways from society – from the concrete power relations of human beings – to the dominant and subdominant ideas in that society may appear to be more difficult for the historian than trading in pure ideas alone, but many of us would regard it as a worthwhile enterprise. Viewing scientific ideas as embedded in social relationships, in particular economic and political contexts, seems to be more faithful to the texture of history as it was actually lived than dealing with the same ideas as free-floating entities that may or may not in some transhistorical sense be 'true'.

JG: Fine. I would agree as long as you do not neglect the ideas themselves. The notion of the internal development of scientific ideas still seems to me a very sound one. If you want to understand Newton's ideas about mathematics, the answer is not to study the social conditions in England in his time, but to examine Newton's ideas in relation to what Descartes had done, what Apollonius had done, and so on. There is an internal evolution of ideas that both influences and is influenced by other things not so internal. There is a history of thoughts. I regard the history of science as a broad enterprise that needs people approaching it from many different angles.

JM: Perhaps on this note, John, we can turn from the more rarified form of intellectual history to the personal side of your achievements. I'd like to find out what it was about your upbringing and education that enabled you to see and defend the cultural conditioning of evolutionary ideas.

JG: Well, let's start at the beginning. I grew up in Vermillion, South Dakota. We moved there in 1919 when I was two years old. My father taught French at the state university. He was not a highly intellectual person. He was a good French teacher and a very fine man, but not theoretically inclined at all, so I had no close intellectual relationship with him. My mother was a highly educated woman, a Barnard graduate, with strong views about lots of things, but her mind did not run to the subjects I eventually became interested in. When I went off to graduate school in history at Harvard her hope was that I would do a dissertation on one of those fine old New England towns like Wethersfield. She never really had any feeling for intellectual history or philosophy. I went to high school in Vermillion – I enjoyed it very much, particularly debating and singing. And I had some good courses in English literature, memorized a good deal of poetry, wrote some editorials for the high school newspaper. Then in 1934, while still living at home, I went on to the University of South Dakota, which had about eight hundred students. There again I enjoyed debating very much, working up a case. The main intellectual stimulus, apart from debating, came from a couple of courses in philosophy – one in the history of philosophy, the other in ethics – where we read and discussed Plato's *Republic*. Also, there was a little group that met in one of the churches for weekly discussions. It included a philosophy professor and the former dean of the engineering school – a brother of Carl Akeley, the African explorer – a tremendously energetic man, full of ideas. One of his prize students was Ernest O. Lawrence, who invented the cyclotron.

JM: Did the group meet at the church you attended regularly?

JG: No, the group met in the Baptist church at that time. I went to the Congregational church, although there was nothing in my religious upbringing to predispose me strongly one way or the other. The Congregational church in Vermillion was rather liberal and not at all theological. Very little in the way of Bible study. In Sunday school we had little readers that would give us cases of John and Betty doing things and would ask us whether these were right or wrong. Moral stories for discussion. I sang in the choir. I can still sing a lot of those hymns and enjoy them. And I was always a sermon taster. I always listened to the sermon. I found that the next thing to a really good sermon to make me think was a really bad sermon, and I heard quite a few really bad sermons.

JM: What became of the discussion group?

JG: The discussion group went on all the time I was at the

university. I remember presenting a talk called 'An Esthetic Approach to Problems of Morality and Religion', which represented my view that moral decisions are somewhat like esthetic choices. I was now reading quite a bit of philosophy and some books about religion. Through the writings of a liberal theologian at the University of Chicago named Wieman, I learned about Alfred North Whitehead's works. I started to read them, beginning with *Science and the Modern World*.

JM: Was this extracurricular reading?

JG: Yes. I started out at the university and actually finished as a political science major. The political science department was heavily practical at that time. But I had a theoretical bent. I can remember my professor suggesting topics about public utilities for my senior thesis. I said that I would like to write on the idea of sovereignty. He replied, 'Well, really, that stuff doesn't get you anywhere', but he let me do it. All I can remember about the thesis is my conclusion that I still did not know what sovereignty was and I didn't think my authorites did either. But I had a good deal of intellectual interchange in writing it, not so much with Dean Akeley but with Professor Josey, the philosophy professor, who also taught psychology.

JM: You dedicated *The Death of Adam* to Dean Akeley and Professor Josey.

JG: That's right. Looking back, I don't think Professor Josey was a very profound philosopher but he was always willing to talk. He was probably my main intellectual stimulus apart from reading. Mostly I was on my own as far as ideas were concerned.

JM: What about your social life and politics?

JG: Socially, I was not very active. I lived at home. I was not highly gregarious nor have I ever been. I have always been perfectly able to amuse myself and enjoy life without much company. But those were the Depression years. No one could ignore the economic reforms. I was aware of the New Deal, and, as far as I can remember, generally sympathetic to it. My parents, I think, were brought up Republicans, but they liked Franklin Roosevelt. How they voted I really don't know. The convention in our family – and conventions can be very important – was that we hardly ever talked about politics or religion. There wasn't even much general discussion at meals. I remember my father saying to me once, not long before I left home, something about going to church being the 'social thing' to do, but what his own religious beliefs were I have no idea. I just assumed he was some kind of Christian.

JM: It sounds like a fairly liberal upbringing.

JG: It was. I was not indoctrinated in any way and I had no impulse, either external or internal, to engage with social issues. In the latter part of my college career a young sociologist named John Useem came to the university. This was about the time the Lynds produced their study of Middletown, and Useem would go down to the local pool hall and hang out. He was making a little study of Vermillion, you see. I found this behaviour strange. The pool hall was not quite respectable among the faculty, who were gentle folk mostly – my father and his friends. There was nothing at all hurly burly or *avant garde* about them. They were just nice people, whereas Useem, who was much younger, wanted to find out what made Vermillion tick. I should also mention Paul Parker, who taught in the art department for a while. He had rather unorthodox ideas, though not political ones especially. I usually managed to search out Useem and Parker for a talk. But my intellectual development, with the exception of Professor Josey's influence, was more or less on my own.

JM: Vermillion had a university but it was still a small town. Did you spend time in the countryside?

JD: Summers are very hot in South Dakota – it gets hot pretty early in the day. I used to get up at 4.30 in the morning, before sunrise, and go down by the bluff. Vermillion is situated on a bluff over-looking the broad valley of the Missouri River. The Vermillion River flows right under the bluff and joins the Missouri in the distance. The Missouri River runs along the bluffs on the other side of the valley, six or seven miles away, where the Nebraska hills begin. It is a beautiful view. I would go over behind a house where nobody lived in the summer and sit on the bluff there and watch the sunrise and read Milton's *Paradise Lost*. Then I would go down below the bluff by the Vermillion River and watch the birds. There was nothing scientific about it – these were just 'nature rambles'. From time to time I would also go up to the natural history museum – a small museum, but not bad for a small museum – and describe for Mr Over, the curator, some bird I had seen. Sometimes he would say, 'There ain't no such bird', my descriptions not being very good, and other times he would say, 'That's a Bell's vireo', or something like that. At this time my general temperament was romantic. In high school I read and memorized poetry – Wordsworth, Shelley, Keats, and others. But I read Milton on my own.

JM: You haven't mentioned the man to whom you dedicated your second book, *Darwin and the Modern World View*. Bert Loewenberg had completed his Harvard dissertation in 1934 and was publishing his well-known articles on evolutionary thought in America. How did he influence your life?

JG: Loewenberg's first teaching job was at the University of South

Dakota. He came out there in my senior year, 1937–8. He was an excellent teacher, extremely energetic, roving up and down like a caged animal as he lectured, with an explosive mode of speech. We took to each other immediately because he was interested in ideas and the history of ideas, while I tended to speak up in class, which he liked. Since it was a very small university, I of course had a chance to get to know him outside class too. I wrote a paper for him on Bergson, which he did not like at all, but he recommended me for scholarships at six different eastern universities. I got several offers, including one from Harvard; and since my father had a Harvard degree – in 1903 I believe – I chose to go there. So I owe a great deal to Bert Loewenberg. I don't know where I might have ended up or what I might be doing now except for him. As an undergraduate I didn't know what I wanted to do. I was interested in everything. I had some notion of going into law, but because Loewenberg's connections were all in history, and the scholarships were favourable, I never did. My being in history was more or less an accident.

JM: You were at Harvard from 1938 until 1942, and again after the war, from 1946 to 1948. This was Harvard under President James Conant – a pretty exciting place to be. Tell me about your graduate studies there and some of the people who influenced you.

JG: I went to Harvard in American history because Loewenberg had studied with Arthur Schlesinger, Sr. The first year I took Schlesinger's course in social and intellectual history, which was primarily about social history. He was, I must say, a tremendously impressive

At home in Vermillion on graduation day, June 1938.

and loyal person with his graduate students; he was also a major scholar for whom I have the greatest respect. But he did not excite me intellectually. I took British constitutional history with McIlwain, who was an excellent teacher. And I took American constitutional law with Professor Yeomans; we just discussed cases in class, which was very stimulating. Yeomans himself was not intellectually exciting, although he was very kind to me personally. But there were students in the political science class whom I got to know well. They were interested in ideas of all sorts – political ideas especially.

JM: What were the political discussions about in your early years at Harvard?

JG: There were the Loyalists in Spain and the coming war of course. One friend in economics, Arthur Billings, was a socialist – a follower of Norman Thomas – and very much against intervention in the war. I had a good deal of conversation with him on general subjects and at one point, more to please him than for any other reason, I joined the Socialist Party. But as the war issue became pressing, I came out on the interventionist side. Billings and I remained good friends but I resigned my Party membership. I continued to get their literature for a long time afterwards, and I am probably somewhere in the FBI files to this day.

JM: I take it that Billings was untypical of your friends.

JG: Almost all my political friends were New Deal enthusiasts – New Deal liberals, you might say – but certainly not to the left of Roosevelt. On the whole, that was true of the economists and the political scientists rather than the historians, who were more conservative. And it was about this time, in my second year I think, that I began to audit Talcott Parsons' graduate seminar in sociological theory. The next summer I read his book *The Structure of Social Action.* I also took Crane Brinton's course in European intellectual history from 1750 to 1850.

JM: Parsons and Brinton had been members of the Harvard seminar formed by Lawrence Henderson in the early 1930s to consider the ideas of the Italian sociologist Vilfredo Pareto as a bulwark against Marxism. Didn't you acknowledge Brinton's influence by dedicating *Science, Ideology, and World View* to him?

JG: Brinton's course is where I myself got interested in Pareto. The first week Brinton talked about nothing else. This was red meat to me because I was just interested in ideas and the course was apparently going to be theoretical. So I signed up. Well, Brinton never mentioned Pareto again, and the course was not after all what I would call theoretical. There was no general schema of any kind. Brinton was essentially of a sceptical temperament, and certainly not given to systematic thinking, whereas Parsons was just the opposite. He could not tell you the state of the weather without constructing a conceptual framework first. And I guess it was from reading Pareto that I got the idea of auditing Parsons' course. Pitirim Sorokin was there then. I read some of his work. I read quite a bit of sociological theory. I read Durkheim the first summer and Parsons' book the next. I also read Perry Miller's *The New England Mind*. That was another great source of intellectual stimulation. I was interested in ideas of all sorts, and I took the lead in forming a little discussion group. We were all history people and we invited McIlwain to come in and talk with us about Karl Mannheim's *Ideology and Utopia*. I was probably atypical of the history students at that time, not atypical in respect to someone like Stuart Hughes, who was there, but atypical in reaching out in all directions.

JM: How practical were your concerns? Were you conscious that this was a time of ferment, not only for yourself, but for the university in American society?

JG: I did not think much about the university in American society. I was pretty well wrapped up in my own interests.

JM: And the Spanish Civil War, the war in Europe, the bombing of Britain, then the engagement of the United States with the attack on pacifism and socialism – these were peripheral to your academic experience?

JG: I was not deeply engaged. Graduate life, particularly in those days when graduate students generally remained unmarried, was rather difficult psychologically because it was very competitive. I did feel that I ought to be doing something about world affairs. I started trying to read the *New York Times* every day, for instance, but I never succeeded. I remember having lunch with Professor Schlesinger – he was a very strong liberal, much engaged with current events – and I told him that I felt guilty about not *doing* something. It was pretty vague. And he said to me, 'Well, John, you'll have plenty of time to do something later on.' So he put my mind at rest.

JM: Did you feel socially disadvantaged, coming to the east coast and Harvard from South Dakota?

JG: No. My parents were both brought up and educated in the east, and probably half of my fellow students were from the midwest. I noticed in class the first day that when the roll was called the people with prep school training would say, 'Present, sir', and the rest of us would say, 'Here.' But after the first semester, when the grades were out, it didn't seem to make any difference where you came from.

JM: It wasn't a shock then?

JG: No, it was not a shock. I was rather maladroit when I first went

to the Schlesingers' Sunday afternoon teas. It took me a while to learn
ordinary drawing room manners, because my folks were not highly
social and my mother was partially deaf. But apart from that, there
was no difficulty in adjusting. The only real, honest-to-God Beacon
Hill gentleman I ran into was Professor Samuel Eliot Morison. I took
his course in colonial history. Because I knew Morison was famous
for his style, I went to him and said I would like to write a couple of
extra papers and get his criticisms. He was a bit leary at first, but he
agreed and asked what I wanted to write on. I had developed a
theory about Puritanism that I wanted to work out. He was pretty
sceptical of that. He said, 'Why don't you do something a little more
down to earth.' I ended up writing an essay on 'The Puritan Way of
Life and Thought', which he liked. But he was the only Boston blue-
blood I ran into.

JM: In 1941 you were elected to the Society of Fellows. Among the
senior members then were Henderson and Brinton. How did your
election come about?

JG: It was primarily through Brinton. After the first week in his
course on intellectual history, when he talked about Pareto non-stop,
I went up and said, 'Professor Brinton, if you don't mind I'd like to
write on Pareto.' And he said to go ahead. I'd taken three years of
French in college and I thought this would be a good time to brush
it up. So I read through the two-volume French translation, the
Sociologie générale, and wrote an essay. Brinton was rather pleased to
learn that I was reading Pareto in French, and when I turned in the
paper he liked it. I also wrote a paper on some eighteenth-century
figures to see to what extent they were aware of what Pareto calls the
non-logical aspect of human behaviour. I read Rousseau, Godwin
and Burke. Brinton was pleased with my quotations from Rousseau
and Burke, which showed that some of the ideas Pareto wrote about
were present in the eighteenth century. It was these papers, apparently, that gave him the idea that I might be timber for the Society
of Fellows.

JM: How did you emerge from reading Pareto? Plenty of influential people at Harvard put great store on him before the war, others
dismissed him as a sort of bourgeois Marx. Stuart Hughes, whom
you mentioned, has since called him 'the great rationalizer of autho-
ritarian conservatism in our time'. Did you feel that Pareto answered
any pressing questions for you as a historian?

JG: Generally speaking, when I read any thinker my first effort is
to understand his point of view. And I found Pareto's point of view
extremely interesting. I was not converted to it; I had criticisms.
It seemed to me at the time that what he called 'residues' were

really what I was beginning to think of as implicit major premises – presuppositions of thought. This wasn't yet formulated though. But Pareto also made an attempt to reduce thought to some kind of non-ideational basis – 'ultimate residues'. He talked about residues and 'derivations', the derivations being rationalizations of underlying residues. And I did not like that idea.

JM: Did you come away from Pareto with any firmer notion of a personal political philosophy or about the role of intellectuals in society?

JG: I paid no attention to his political philosophy. There are a lot of things I might have gotten out of Pareto but didn't. I knew he was the subject of some discussion. More about that when I get to Henderson.

JM: Let's go on to the Society of Fellows then.

JG: All right. Before I left Harvard on a travelling fellowship to research the thesis topic Professor Schlesinger had proposed – 'Geology and Religion, 1820–1860' – Brinton called me in and asked whether I would like to be considered for membership in the Society. I said I would. He suggested I submit some papers like those I had written for him. So I did that before I left. Then when I got to Philadelphia I received a notice that I should come back for an interview. The interview turned out to be very interesting. It took place at a horseshoe-shaped table in Eliot House, where the Society met. I sat at one end of the horseshoe. Harvard's former president, A. Lawrence Lowell, with his electric hearing aid – he was in his eighties then – was on my immediate right, and Brinton, Morison, Henderson, A.N. Whitehead, John Livingston Lowes and Arthur Darby Nock were ranged on the opposite sides of the table. They were the Senior Fellows. By this time I had formulated a project focused on the doctrine of the plenary inspiration of the Bible. I had come to believe that everything in the relations of science and religion turned on this idea. So when asked to explain what I would work on if I were elected to the Society of Fellows, I started expounding it. I had gone a little way when Mr Lowell interrupted: 'Mr Greene, what do you know about Unitarianism.' He had memories from his youth in this regard. I explained that I had worked on a lot of people but I just hadn't got to the Unitarians. He looked disappointed and lapsed into silence. Then I went on, and about every ten minutes he would interrupt and say, 'Mr Greene, what do you know about Unitarianism?' Well, this was a little disconcerting, but everyone else smiled as if they knew what to expect. Lowell was over the hill by then. Anyhow, somewhere in my exposition I happened to mention Pareto. Henderson sat up straight. 'Oh, well,' he said, 'have you *read* Pareto, Mr Greene?'

And I replied that I had read the *Sociology*. 'Well, what do you think of it?' he demanded. All the other Senior Fellows were very sympathetic toward me on this question. I said that, as far as I could see, what Pareto called 'residues' were just implicit presuppositions of thought in a period. Whitehead burst in, 'Exactly!' He obviously was not bowled over by Pareto, although I don't think there was any antagonism between him and Henderson. But Henderson was a very imposing figure. He sat erect with his beard and moustache. He looked very much the *savant*. And those blazing eyes that pierced right through you, as if to intimidate – I don't know that I ever saw him smile. Either you had to be his disciple or go at it hammer and tongs with him. There wasn't much of a middle ground. And so he continued to ask me about Pareto; everybody else was looking sympathetic. It didn't go on for long. Then the others asked me questions, and in due course I heard that I had been elected. The Society allowed me three years of independent study with all expenses paid, one year before the war and two after.

JM: Who were the other Junior Fellows?

JG: Of course some of them are now well known, such as Arthur Schlesinger, Jr, Carl Kaysen and James Tobin. John Moore was in classics. John Sawyer later became President of Williams College and after that President of the Sloan Foundation. Frank Sutton was in sociology with Parsons and later was high up in the Ford Foundation. Those are some of the people I think of first. John Kelleher, in Irish studies, became a professor at Harvard; so did Schlesinger, and Carroll Williams, a biologist. With John Brown and Saul Levin I discussed religious and philosophical issues that still interest me.

JM: What do you remember about the Society's gatherings? You got together every week, didn't you?

JG: On Monday night for dinner. The original idea for the Society of Fellows was conceived by Whitehead, Henderson and Lowell. They wanted to foster something like the 'high table' ethos of an Oxbridge college. They wanted to get around the Ph.D. fetish and to emphasize conversation and independent study. You would not need to have the union ticket. Lowell provided the money for the project, though no one said so. Anyhow, we would have sherry and talk a little before dinner, then we would go in and sit around the horseshoe table.

JM: You had been reading Whitehead since your undergraduate days in Vermillion. Did you talk philosophy with him?

JG: Well, you've probably heard the story, which is now widespread, but I was there the night it happened. When two or three of us asked about his relations with Bertrand Russell, Whitehead, with his little eggshell head you could almost see through and those wisps of white hair on top like a halo, replied in his high squeaky voice, 'Bertie thinks I'm muddle-headed, but I think Bertie is simple-minded.' It was a perfect portrayal of the two intellectual temperaments. Whitehead was a marvellous person and a marvellous conversationalist. You could bring up any subject and, without trying to dominate the conversation, he always entered in and had something fascinating to say. He had a tremendous sense of humour too.

JM: And Henderson – did you get to know him before he died in 1942?

JG: Henderson sat bolt upright in the centre of the horseshoe. The Junior Fellows who were his disciples usually sat across from him, but not always. Occasionally, I sat across from him. And, as I said, it was a little hard to carry on a conversation with Henderson because he always had a strong point of view. Sometimes you would bring up an argument and he would interject, 'Aha! Now you are manifesting residue number fourteen', or something similar. He felt there should always be serious discussion after dinner, and he seemed to monitor the proceedings. I did not seek him out much.

JM: After the war Brinton presided, didn't he?

JG: Yes, but Brinton did not want anything prescribed. He was completely opposed to that. We would have lots of little groups, some at one end of the horseshoe, some at the other, some in the middle, some out in the other room. And, personally, Brinton was not the kind of person you could have an argument with either. He did not like argument. I am the kind of person who wants to come out and say, 'This is my position – what's yours?' He was not. He would dance around a topic. He was no doctrinaire Paretan by any means. He was not doctrinaire in anything. He described himself as a posthumous son of the Enlightenment. He felt at home in the eighteenth century. What he liked about Voltaire and others was their lack of dogmatism. If the Fellows got into a serious discussion and someone made a complicated or pretentious remark, Brinton would simply say, 'Bullshit', or something like that. If someone tried putting intellectuals down, Brinton would ask, 'What are *you*, a blacksmith?'

JM: Why, then, did you dedicate *Science, Ideology, and World View* to Brinton? Was it just his historiographic style?

JG: Brinton's style is not my style. My view of the Enlightenment was much more qualified than his. He had read everything – literature, philosophy, you name it – but he did not impress me as a great intellect. As a person I liked him, but he was very shy. In conversations he would be constantly looking around. I never felt close to him. The reason I dedicated the book to him was that he encouraged me, we shared a lot of interests, including intellectual history, and my career owes him much. I feel I also owe a great deal to Arthur Schlesinger, Sr, but in a very different and more practical way.

JM: How did you get involved in the war?

JG: My first year in the Society of Fellows, in the spring, somebody came around recruiting people to do code work, which would involve learning Japanese. Several of us volunteered, and that summer we took an intensive language course with Edwin Reischauer. In the middle of the course I got my draft notice. I went out to Vermillion to be inducted into the Army; they sent me to the signal corps at Camp Crowder, Missouri, for the regular basic training. I spent my leisure hours in a town library nearby reading St Augustine's *Confessions*.

JM: Where did this training lead?

JG: I was just about to finish my message-centre training when a chance came to get officer training in the signal corps. I took the opportunity, but my eyes turned out not to meet the standards for vision.

JM: Was this the first you knew that your eyes weren't working properly?

JG: No, I discovered that at Harvard. My eyes would fatigue after just a few hours of reading. If I went on I would get eye aches, and if I persisted a bit further I would get nausea also. I went to all kinds of specialists, but to no avail. Since then I have never been able to work solidly for more than four or, at most, five hours a day, which has meant that I don't read newspapers or watch TV or go to movies very often. Anyhow, the immediate result of this was that I wound up in the administrative officer training corps and from there went overseas.

JM: Did you get to Japan?

JG: Never. But I did have the most marvellous world-tour at the expense of the Army and the US taxpayers. They sent two of us to an unknown destination in journalism. We were a long time getting there. Our first stop was Hobart, Tasmania; then we spent six weeks in India, where I kicked around in the Bombay province. Finally, we arrived in southern Iran, but we could not go up to the capital. Something very hush-hush was going on – the Tehran Conference. We didn't discover this until it was over. By the time we got to Tehran they needed only one person in journalism, and it wasn't me. So I sat around reading up Persian history until I was sent to help establish a sightseeing camp in Isfahan for GIs coming up out of the desert for rest and recreation. Then they decided to send the GIs over to Palestine, and I went there as a liaison officer. I got interested in the history of Palestine and began to organize sightseeing tours. But seven months later I was back in Tehran – that's when I first met Ellen at a Christmas dance – to serve as aide-de-camp to Brigadier-General Booth. I was based in Iran until the end of 1945.

JM: Tell me about Ellen.

JG: Ellen Wiemann. She had been stationed in Tehran with the Red Cross for several months, and we saw each other very steadily at the General's billet. Then she was transferred to Cairo and I realized that I missed her. In November 1945 I obtained a leave of absence to go over and propose to her, and another to return and get married. After our honeymoon in Cyprus, Ellen finished her work in Cairo and went back to her family in New York. I stayed on in Tehran before going home via Italy. We rented an apartment near Harvard.

JM: Did anything come of your historical studies while in the Near East?

JG: My very first two publications were on Iranian history. The Army needed something to give the GIs when they came to Isfahan. An expert on Iranian history was there at the time, Don Wilbur by name. I suspect now that he was with the OSS. He helped me write *A Sketch of Iranian History* and *A Guide to Isfahan*.

JM: Back at Harvard in 1946, how had the atmosphere changed – after Yalta, after the Bomb?

JG: So far as I can remember, the atmosphere was still fairly hopeful as to what might come out of the United Nations and atomic energy. But I cannot recall being deeply involved. Harvard itself had been somewhat democratized in the houses. Before the war you sat at table and were waited on hand and foot. Not afterwards. In the Society of Fellows there were some new Senior Fellows, including the zoologist Hysau. But the main changes came from having Brinton instead of Henderson as the chairman. And by this time I realized that my major interest was not science and the Bible; it was the downfall of the static view of nature.

JM: How did this realization come to you – was it during the war?

JG: It was during my year as a Fellow before the war, when I was planning the background chapter of my thesis on geology and religion, which Professor Schlesinger told me would have to be written first. That was fateful advice. I already knew that it wasn't just geology causing the trouble – there was the nebular hypothesis, the so-called 'development hypothesis', and archaeological discoveries too. But when I went back and read the debate between deists and Christians in the eighteenth century, what struck me almost immediately was that, whereas the two sides were totally at loggerheads about the Bible, they were in perfect agreement about nature. No two men differed more about the Bible than Tom Paine and Timothy Dwight, but they both believed in a static natural world that testified to the wisdom and goodness of the Creator. What also struck me at the same time was how different this view of nature was from the one accepted in the 1940s, and I began to wonder how the change came about. I ceased to focus on the relations of science and the Bible and, instead, set out to trace the gradual breakdown of the static view of nature in astronomy, in geology, in palaeontology, in biology, in physical anthropology, and if possible in philosophy also. Until then I had been reading American writers, but they repeatedly cited Linnaeus, Ray, Buffon, and others. So after the war I began to read the Europeans, and it was out of these researches that I distilled the dominant view of nature and the two subdominant views I spoke of earlier.

JM: Meanwhile, how did you and Ellen support yourselves?

JG: The Society of Fellows gave an extra allowance to married people. As a matter of fact, I was doing better then, even after the birth of our daughter in March 1947, than I did in my first job.

JM: Which was at the University of Chicago.

JG: Yes. I didn't have my Ph.D. when I left the Society of Fellows in 1948 – we weren't supposed to get a Ph.D. About that time the

Ellen, John and Ruth, June 1947.

dean of Hutchins's experimental college at Chicago came through looking for teachers, and I got hired. The college was the most intellectually stimulating place I have ever been. David Riesman was the presiding genius in the social sciences. For each course there was a staff of seven or eight; we taught entirely from original sources chosen by ourselves. In Social Sciences I, we read Locke, Hobbes, Jefferson, Calhoun, Supreme Court decisions, and so on; in Social Sciences II, we read Freud and Fromm, and Davis and Dollard, and Max Weber, and so forth. Riesman was on the staff, as well as Lewis Coser, Philip Rieff and Morton Grodzins. All those people eventually wound up doing rather interesting things. Discussions with them were invaluable. Apart from Parsons' classes, I owe my knowledge of the history of sociology entirely to Social Sciences II.

JM: Meanwhile, you finished your dissertation, then you had to leave Chicago.

JG: It was the other way around. I had to go because Hutchins left Chicago after I had been at the college two years, and there were a lot of changes. I worked hard in my spare moments, including summers, on what I had planned to be a book, because I realized now that if I were going to get another job, I would have to obtain a Ph.D. I simply submitted the completed chapters of the book as a dissertation. Schlesinger, Brinton and Oscar Handlin examined me. They asked questions about American history that I couldn't answer. But they knew I needed the degree and that I was no longer an American historian. So I passed.

JM: That was in 1952, the year you left Chicago, the beginning of your Wanderjahre. You went from Chicago to Wisconsin for four years, then to Iowa for six years and to Berkeley for a year, then to Kansas for four more years. Your family was growing at the same time. How did you and Ellen manage?

JG: During the eight years at Chicago and the University of Wisconsin we were as poor as churchmice. The universities did not pay a nickel more than they had to. And we now had two children. Ellen tried working a couple times. She was an expert secretary but she just could not take it physically – the young children and the job. Finally, when they told me I wasn't going to be kept on at Wisconsin, I seriously considered leaving the academic world. I talked to some acquaintances in the State Department, but fortunately I heard that Iowa State University was looking for someone in the history of science. And by now it had become clear that this really was my field. I had published a couple of articles on racial ideas from my chapter on physical anthropology. So I was interviewed, and Iowa offered me the job as an associate professor. A year later I got tenure, and by 1956 I was a full professor.

JM: What happened to your book on the breakdown of the static view of nature?

JG: It was still unfinished. But then in 1958 there was a competition for the best manuscript from any source submitted to Iowa State University Press. I completed my manuscript, sent it in, and it won. *The Death of Adam* came out a year later. Within a relatively short time, I began to get invitations to speak, and I became secretary of the History of Science Society.

JM: 1959 was the Darwin centenary year. Yet the book as you first conceived it seems to have had little to do with Darwin.

JG: The whole Darwin business was put in at the last minute. When Iowa accepted the manuscript, it stopped at the deaths of Cuvier and Lamarck. My plan had been that there would eventually be two volumes: the first would extend to about 1832, the second would deal with Darwin. But the Press wanted Darwin in the first volume for the centenary. And in six months I did all the research and writing for the last two chapters. People sometimes talk about me as if I were a Darwin expert. I am not. That was my first introduction to Darwin.

JM: During this period, how did you feel about your work? You had published only the two articles; the book won a prize only at the last minute; it had been ten years since you left Harvard; you had recently turned forty. Was this a difficult time?

JG: The work on *The Death of Adam* was always intensely interesting to me. I never tired of it. But from a professional point of view, and a family point of view, it *was* a difficult time because I had no grants to relieve me of teaching responsibilities. The only way I could find time for research was to get up in the morning and say, 'Today I am going to do research; if that means I go to class half-baked, well then I go to class half-baked.' I was working up courses in the history of science, and I had never taught history of science before. The first year I went to Iowa State I made a deal with them: I said, 'I will teach a course eventually that begins with the Greeks, but I don't know anything about that now; I am going to start with the sixteenth and seventeenth centuries, then I will add the other later on.' And that's what I did. The result was that I felt guilty doing research if it took time away from preparation of lectures. On the other hand, if I wasn't doing research and writing I felt guilty, as I knew damn well my family was not going to prosper if I didn't.

JM: It was a frustrating time.

JG: It was a very difficult time. But it was much harder on Ellen than on me. I am a pretty self-centred character anyway, and I enjoyed my work.

JM: I suppose it was a stage when your relationship was bound to come under strain.

JG: It came under strain from the day we were married, largely because I get all wrapped up in what I'm doing. Having weak eyes didn't help matters any. I was like a one-armed paper hanger. I was always in a hurry, feeling under pressure. When I was not actually working, I wished I could be working.

JM: Would you work until your eyes made you sick?

JG: I couldn't. I had eye aches when I pushed too hard. If I'd had a good pair of eyes, it is at least conceivable that I would have been able to do what I did and have more of a family life too. But it is also possible that I would have spent all of my time in the study, which would have been even worse.

JM: Before we move on to your period of professional recognition, I want to ask one more question about the earlier post-war era. You mentioned your two articles on the history of racial ideas. That reminded me of a number of events which tore and divided American society during the first half of your career, most notably the Mc-Carthyite inquisition and the civil rights movement in the 1950s, and the student anti-war movement a decade later. You left Iowa for the University of California, Berkeley, at the time of the 'free speech' struggle there. Then as the Vietnam conflict escalated you were at the Universities of Kansas and Connecticut. How did the events of those years influence your historical perception of science in society, bearing in mind the effort to 'see things whole' that we discussed at the start?

JG: I think, personally, they did not have much effect. Of course I read the newspapers and I was a longstanding liberal in my attitude toward racial discrimination. I remember when we lived in Wisconsin we seriously considered selling our house to a black family. We wondered, though, whether it would be fair to our working-class neighbours.

JM: And at the very time you were publishing articles about racial ideas in history.

JG: No connection.

JM: You made no connection between the articles and what was going on? Was it coincidental that they appeared at the time of the Supreme Court ruling on segregation?

JG: It was accidental. I had to publish, and the physical anthropology chapter from my dissertation seemed like something I could get in shape.

JM: Why did you choose that chapter rather than another?

JG: I don't recall that it had anything to do with current events.

JM: And the McCarthyite witch-hunt earlier – did that put the fear of God in you as a liberal? Joe McCarthy was a senator from Wisconsin, after all, and you were an untenured professor in the leading university of the state.

JG: We were upset by McCarthy. Ellen telephoned people in our neighbourhood to ask them to vote against him, but she was not very successful. We lived on the wrong side of the tracks politically speaking. I was very anti-McCarthy but never demonstrated my opposition.

JM: What about the influence of current events on other historians of science?

JG: As far as history of science was concerned, this was its period of professionalization. The whole idea was to be – to get – professional. It was not my idea – to enrol graduate students trained in science, to stop building bridges between science and the humanities ...

JM: To do it for its own sake.

JG: ... yes, make a real McCoy out of it. This wasn't my attitude, but Marshall Clagett and others felt that way. I saw the results especially in the History of Science Society, because at least once a year while I was secretary someone would write to me who had joined the Society in George Sarton's day. At that time members came from all sorts of disciplines – art, literature, banking, everything. They were essentially amateurs interested in the history of science. But in the 1960s they dropped out one by one. They told me, 'I can't read *Isis* any more – too technical.'

JM: Your Wanderjahre finally ended in 1967 when you moved to the University of Connecticut. How did that come about?

JG: I had a tenured post at Kansas, so I went there first as a visiting professor. One of the faculty had read *The Death of Adam* and the history department wanted a historian of science. On this basis they invited me to come for a year. Then in the middle of that year they offered to make it a permanent appointment. Ellen was dying to get back to the east coast, and I had already decided that if we moved from Kansas, it would be to the east coast where Ellen's family were. Also, we now had three children, and our daughter was at university, so the increased salary was attractive. When I told Ellen about the offer she said, 'Johnnie, you've got this programme going at Kansas – do you really want to give it up?' But as soon as she realized that I rather liked the idea of staying in Connecticut, she was delighted. And it all turned out very well. Until my retirement in 1987 I was the only historian of science in the university. But that has been more or less the case wherever I was, and it suited me. I was able to combine my Darwinian studies with research for my book *American Science in the Age of Jefferson*, which came out in 1984. That research had begun when I was a graduate student in American history.

JM: You were an office holder in the History of Science Society continuously from 1960 to 1977, and the last two years you served as President. You've also served as a consultant or adviser to numerous projects, and have refereed countless manuscripts for the press. What has happened to the way we *do* history of science since the death of Sarton?

JG: The big movements after Sarton centred on the name of Alexandre Koyré. Marshall Clagett, Thomas Kuhn, Charles Gillispie and others were great admirers of him. They emphasized the internal history of science. It was not strictly internal, of course, as it included the influence of philosophy, but it was at least intellectually internal. What they were really concerned with was the history of thought. This approach was dominant until about the mid-1960s, although there were some Marxist-influenced historians of science in England and elsewhere. I remember giving a talk at the time in which I predicted that the growing concern about the practical impact of science on the world – the anti-Vietnam spirit, the question of defence-related research, and so forth – would soon be reflected in the profession as a greater interest in the social history of science. I never published that paper, but now I wish I had because it was a good prediction. By 1970, or shortly thereafter, people like Clagett and Gillispie were lamenting the fact that *Isis* had begun to publish on all kinds of subjects that, to them, did not appear central to the history of science. And since then the social or sociological approach to the history of science has become the dominant one – the concern with so-called 'external' conditioning factors.

JM: Within little more than a decade, then, the newcomers were becoming outsiders. Does this suggest that the fortunes of the history of science as a discipline are linked inexorably with those of science and technology in the wider culture?

JG: I think it is a particular case of what has happened in the educational world generally. Certainly in the 1950s and 1960s money was readily available through the National Science Foundation to do research in the history of science, and those who served on the panels were mostly people like Gillispie and me. More recently, the funds have been harder to get, and the National Science Foundation itself has, partly through pressure from Congress and other sources, become more concerned with science and technology in their practical aspects. Also in the 1960s, we were mainly interested in graduate students – everybody wanted to teach graduate students who would go out and teach graduate students who would go out and teach graduate students. But that came to an end economically too, and we had to interest undergraduates. The result was courses on science and the occult, science and religion, and so forth. We have had to cater for the market.

JM: Under these circumstances, do you think that the history of science can any longer be regarded as a professional enterprise – something done 'for its own sake'? Have the days when it could be now come to an end? Should they end?

JG: I rather doubt it. I think scholarship will always be done for its own sake primarily. And I don't think that is likely to change unless we get a regime, dictatorial or otherwise, that prescribes a party line. But if you look at the projects now funded by the National Science Foundation in the history of science, they are not all that different from former ones, although the amount of sociological research is larger than it used to be. There are an awful lot of people still doing the origins of the oxygen revolution. There's an awful lot being done in a fairly traditional mode still.

JM: In 1975 you published a magisterial overview of Darwin studies – the 'Darwin Industry', as it's called. Your lifetime has seen the growth of this scholarly industry from its origins in the editorial work of Nora Barlow during the 1930s. And you of course made an important contribution to the first wave of interpretations stimulated by the centenary of 1959. Do you agree with some observers that Darwin studies are becoming too specialized for their own – or anyone else's – good?

JG: That's very hard to say. What about Newton studies or Galileo studies? My own feeling is that there will always be a need for – and certainly, there will *be*, whether a need exists or not – all kinds of specialized scholarly studies connected with Darwin. But there will also be broader interpretive studies. The pendulum will swing back and forth between very narrow specialization and broader interpretation. I don't see any watershed here.

JM: How far do trends in Darwin studies reflect theoretical trends in the study of biological evolution?

JG: Many of the leading evolutionary biologists have been interested in the history of their subject – Ernst Mayr is only one. This is a natural interest. The main danger is that they will all try to get Darwin on their side. I am more and more impressed by how hard it is for deeply engaged scientists to gain a genuine historical perspective. It is extremely difficult. The question of recruitment in the history of science also arises here. On the whole, graduate students have been recruited from the sciences on the theory that it is easier to give them some historical perspective than it is to give somebody from history or literature a knowledge of science. I think, however, that people underestimate how difficult it is to develop a genuine historical perspective. My own hope is that history of science will always recruit widely, among people with diverse backgrounds. Perhaps the major recruiting will be among those with science training, but it would be too bad if this became the only source of graduate students.

JM: You once wrote, referring again to the Christian geologist Benjamin Silliman, that 'every noble and useful life springs from some enduring vision of the world and man's place in it, from an ideal of individual and social existence'. Do you think it is legitimate, as well as possible, for the professional life of the historian of science today to be pursued in the interests of some such enduring vision of the world?

JG: I think it should be, very definitely. And I think in many cases it has been. I think Koyré had this orientation. I suppose in his own way Gillispie does too.

JM: Among younger historians whose historiographic concerns, as you pointed out, have become dominant today, have you found similar kinds of commitment, arising perhaps in some cases from the formative influence of current events?

JG: Yes, at Cambridge in 1974 I first met Bob Young and Roy Porter – some of the people who have contributed to this volume. Bob was very kind and had us to dinner. I don't remember having any particular arguments with him. I met Ludi Jordanova with Karl Figlio at a dinner given by Roy. It was extremely friendly and congenial. What struck me was that basically Bob and I – also Martin Rudwick, who was then at Cambridge – had a broad view of the history of science. We all believed it should include intellectual history and some social history too. I was generally aware that Bob may have taken a Marxist position, rather different from anything I held. But it was good to find like-minded people.

JM: Were you conscious that some of these people may have been doing history of science in a way that would have been considered unprofessional in the United States, though compatible with your own approach?

JG: I don't remember being particularly conscious of Bob's political orientation. I sensed that Martin had a religious interest somewhat like mine. There were people in the United States who had this kind of orientation too, but they were not mainstream figures like Gillispie and Kuhn and Clagett. So I certainly was aware that there was another dimension to the history of science at Cambridge, and that this set it apart from the dominant view in the United States, as well as from the views of older Cambridge academics such as Buchdahl and Hesse.

JM: Evidently your approach to intellectual history has ensured the continuing relevance of your work to younger generations of scholars. The present volume is intended as a witness to that fact. But isn't there a deeper affinity as well? Your work, like some of theirs, has not been done merely for its own sake, has it? It has been pursued in the interests of what you have called an 'ideal of individual and social existence', an 'enduring vision of the world'.

JG: Yes – at least I hope that I have always tried to see life whole. As I said earlier, one of our problems today with respect to science derives from the fragmentation of our conception of the world. So there is no reason for any of us to be apologetic about seeking a coherent view.

JM: For many years, John, you have been extremely critical of attempts to biologize human nature, to naturalize history, and in some sense to construct a science of society. You developed your critique historically in an essay on Auguste Comte and Herbert Spencer that appeared in 1959. And since then you have remarked critically on biologist-philosophers who follow in the Victorian naturalistic tradition – men such as Cyril Darlington and Julian Huxley. What view of human nature, history and society – what 'ideal of individual and social existence' – have you defended over against theirs?

JG: I have often said that sinking man and human nature entirely into nature leads to hopeless paradoxes and contradictions. My general method, beginning with *Darwin and the Modern World View*, has been to take what the biologist-philosophers say at face value, follow their line of argument and show where it leads. They wind up hoist with their own petard. That, in general, has been my method rather than advancing a position of my own. But it seems to me that in many ways human nature transcends nature. This is of course a premise of the notion that ideas have some real force in history. I think we can get a lot of light – a considerable amount of light – on human nature if we study the primates, but whether we can get an *adequate* conception in this way is quite another question. In considering human nature, we have to talk about it not only as it is, but as it might be – about a possible harmony of human nature as envisaged by Plato, for example, and likewise by the New Testament, where the 'full stature of man' appears in Jesus Christ. We have to consider human nature both as it seems to be from ordinary historical and scientific observations and as it appears in its highest manifestations. My viewpoint is pretty well spelled out in the last paragraphs of *Darwin and the Modern World View*.

JM: Yes, and as I recall, you make clear there that you believe human beings uniquely possess a spiritual nature which is both irreducibly transcendent and relatively free, a nature which participates in a divinely created ground as its source of value. If that is a fair statement of your view, don't your own critics among the biologist-philosophers have a point in saying that you are an 'essentialist'?

JG: Well, I'm not sure I know what they mean by essentialist. If they mean an essentialist is someone who believes in a real human nature, then, yes, I believe there is a real human nature.

JM: Is this human nature created and permanent, in the way biological species were once supposed to be, or is it a nature that has been *made* – a finite abstraction from the history of human becoming?

JG: We know about so-called hominid or hominoid creatures. And there is reasonable scientific evidence for tracing back some kind of hominidity two or three million years. So I suppose if you look over the very long haul, you might say that there have been changes in human nature. But within historic time – what we know anything about – I do not see much evidence of change. There is the possibility that by eugenics we might conceivably make some kind of change in the genetic structure of human beings that would amount to a revolution in human nature. But for practical purposes human nature is fairly stable.

JM: You said earlier that we have to consider not only what human nature is, but what it might become. Now genetic engineering has moved the eugenics debate precisely into that sphere. Do you have trouble reconciling the technological promise of modifying human nature with what you refer to as the New Testament revelation of what human nature might become?

JG: I am not only very sceptical but frankly alarmed at the possibility of human genetic engineering. I have no objection to getting rid of certain genetic diseases such as sickle cell anaemia; but the notion of restructuring human nature is an extremely perilous one. When I said human nature 'as it might be', I meant our nature now, not some future nature. The full stature of man as revealed in Christ Jesus is a vision of what human nature might be at present. So is Plato's idea of the life of reason.

JM: Does this imply that the idea of 'the full stature of man in Christ Jesus' can exercise the kind of force that you talk about ideas having in history?

JG: Yes. After all, everyone is moved by ideals of some sort, even the genetic engineers. And the ideal has a real force. I think a very interesting book could be written on the subject of 'The Idea of Reality in Modern Thought'.

JM: This reminds me of your first philosophy course at the University of South Dakota and that discussion group at the Baptist church.

JG: I read Plato then, yes.

JM: Your critics' other point, of course, is that you are not only an essentialist but a teleologist. You've conducted a dialogue at intervals with Ernst Mayr over the years, and he has charged you with this.

JG: Yes, and I admit to being a teleologist in some sense, although, again, it is one of those words that's hard to define. I simply do not see how one can account for the harmony and order that we perceive in the universe apart from some principle or agent of intelligence in nature, whether Aristotle's immanent purposiveness, Whitehead's principle of concretion, or the Judaeo-Christian Creator. The theory of natural selection deals with modifications of organisms, but organisms are 'going concerns' that have an internal harmony. You can say, 'Well, that organism came from other organisms', but they, too, were going concerns. And if all these organisms ultimately came from a macromolecule, that macromolecule was a going concern as well. You do not get around the question of the source of order by going back in time.

JM: So the point is that things *get* to where they are intended to, rather than *how* they get there.

JG: I do not much believe in the notion that evolution is going somewhere. If you think that God was trying to get from Eohippus to the modern horse, you must concede that He was a terrible bungler. But what reason have we to think that the modern horse is any more precious in the sight of God than Eohippus? I agree with Mary Midgley in *Beast and Man* that the whole idea of evolution going somewhere is a snare and a delusion.

JM: Except that once there were not human beings and now there are. That is important from your standpoint, isn't it?

JG: Sure, that's significant. But to say everything that happened before human beings appeared on the scene was on the way to producing human beings – I can't accept that.

JM: In *Darwin and the Modern World View* you say that 'science becomes pointless and even destructive unless it takes significance and direction from a religious affirmation concerning the meaning and value of human existence'. But where, in your view, is the 'meaning and value' of the process through which human beings evolved? If I understand you correctly, human existence cannot mean whatever we choose to make it mean. Nor does it mean anything that is ever likely to occur in the process of evolution.

JG: That's right.

JM: So it means something else quite beyond?

JG: In my view, the end of human existence is neither pure intellection nor command of nature for human purposes, but the harmony of life with life and with the Source of all life.

JM: And that includes the life of individuals in relationship to the divine life and the life of individuals in relationship to other life, including other individuals.

JG: Certainly – it means social harmony. We should seek harmony in all aspects of our lives. *— seeing evolution as seeking harmony —*

JM: Your emphasis on harmony sounds familiar to someone who *pacifist* has studied pre-Darwinian naturalists and natural theologians. You refer in *Science, Ideology, and World View* to books by Huxley, Hardin, Waddington, Simpson and E.O. Wilson as 'the Bridgewater Treatises of the twentieth century'. But by emphasizing harmony and transcendence, aren't you yourself also promoting a renascence of natural theology? Or, as one of your critics put it, isn't *Darwin and the Modern World View* a 'religious tract'?

JG: My reply to Michael Ghiselin when he accused me of that was to say that his *Triumph of the Darwinian Method* was a religious tract. I have a broad view of religion in this respect.

JM: Do you have as broad a view of natural theology?

JG: The point about Bridgewater Treatises, whether twentieth century or nineteenth century, is that they try to get sermons from stones. But it is one thing to say, as Aristotle or Aquinas said, that the human mind demands some explanation of the order and harmony of the universe; it is quite another thing to say that geology or palaeontology or some other science proves the existence of God. That is a very truncated thought.

JM: But Huxley and the others don't say anything remotely like, 'We can prove God's existence from the phenomena of evolution.' They simply derive meaning and value for human beings from the Darwinian vision of nature. In so far as your own view of the world incorporates Darwinian evolution and takes account of the order and purpose evident to creatures who have evolved, doesn't it amount to a natural theology in the same sense – not an argument from design for the existence of God, but an argument from what nature presents to some larger meaning and value? *— can see it a spiritual process not*

JG: My argument would not be constructed just on what nature *necessarily* presents. It is based on general knowledge *and* introspection. It is a *atheist* combination of philosophical and religious perspectives.

JM: So you differ from the biologist-philosophers, or latter-day natural theologians, in this respect: that while they look exclusively to nature, you look first within, to the spiritual essence of human nature.

JG: No, I would not say that. I would say that they look to science in a fairly modern sense to provide a knowledge of values and the meaning of life; my own position is much more traditional. As I see science, it seeks a very limited kind of intelligibility. It is enormously broad – the galaxies and the origin of species are its concerns. But it does not embrace our whole experience. It does not consider the existence of the scientist himself; it has nothing to say about the value of science. It is a restricted form of intelligibility – functional relationships among phenomena – whereas it seems to me that both in philosophy and in religious visions of life we try to find an intelligibility that makes sense of all these things.

JM: In your particular Christian vision, then, what order of priority do you assign to spiritual intuition and historical revelation as sources of knowledge of the larger intelligibility that science cannot find?

JG: I have not usually thought in those terms.

JM: Reason and revelation are the traditional Christian sources, aren't they?

JG: Well, I've never developed a systematic philosophy of nature, although Whitehead's philosophy has interested me. As far as the Christian revelation is concerned, I think it provides a very realistic view of human nature that is confirmed in our own experience and from observation of our fellow human beings. And I think it also discloses the possibility of harmonious human existence, both social and individual.

JM: 'The full stature of man in Christ Jesus', according to the New Testament?

JG: Right. You always have to think in terms of the alternatives. I do not think Plato's life of reason is as good an alternative by any means. And the notion held by some biologists of ever-increasing command of nature or ever-increasing knowledge of the phenomenal world is just not satisfying. What is ultimately satisfying is something that makes sense of our whole existence, not merely a part of it.

JM: Would I be right in thinking that in the last result, through all the varying fortunes of your career, you have maintained the same basic liberal Christian faith that you were brought up on in South Dakota?

JG: Yes, I have had a pretty consistent intellectual and spiritual development. But I would add that I do not have any final answers. Again, it comes down to a choice among alternatives. I haven't looked systematically at the Marxian alternative. In so far as I understand it, it seems to be shot through with scientism, and its conception of human nature does not seem to me very realistic.

JM: Almost twenty years ago, on first reading *The Death of Adam*, I was struck by the closing sentence, which had resonances with my own religious past: 'The historical Adam is dead, a casualty of scientific progress, but the Adam in whom all men die lives on, the creature and the creator of history, a moral being whose every intellectual triumph is at once a temptation to evil and a power for good.' I see now that on the last page of his most recent book, *Darwin's Metaphor*, one of the contributors to the present volume, whom you described earlier as 'likeminded', has written: 'The practitioners of science and the students of its history, philosophy, and social relations should recall the fate of its first advocate. Adam and Eve were innocent until tempted to hubris, to knowledge, wisdom, and good and evil. As punishment, their tempter was accursed.... If science is to take its place in the making of a better world, it is best to know that the problem of science in society begins with the Fall, with the serpent, and with Cain in the Land of Nod.' Do these biblical allusions represent a failure of nerve on both your parts, or has the history of evolutionary thought begun to speak with one voice to us all?

JG: I am doubtful that it will ever speak with one voice, but that's an interesting passage of Bob Young's, isn't it? I don't think that Adam and Eve failed because they desired wisdom. The knowledge of good and evil is a kind of wisdom, and I don't find anything wrong with that. But everything depends – I suspect Bob would agree – on where this wisdom is found. It is ironic that Darwin said he gave up Christianity in part because of the manifestly false historical account in Genesis, yet the account of man's fall from grace in Genesis, despite its historical inaccuracy, gives a better and truer picture of the human condition than Darwin's idea that, through the operation of natural selection and the inheritance of acquired characters, a more perfect man is being produced who will look back on Darwin as a mere barbarian. Adam and Eve fell from grace by partaking of the tree of the knowledge of good and evil, a tree of which science knows nothing. It is that knowledge which constitutes our transcendence, our danger, our humanity.

JM: If humanity is to be saved, will science have to acquire a world view that knows good and evil in this sense?

JG: Definitely. Religion apart from science tends to become obscurantist, dogmatic and bigoted; science apart from some general view of human nature in its total context becomes meaningless and destructive. Unless science is practised on the basis of a conception of human nature that does justice to our highest aspirations, the prospect for the future is bleak indeed.

JM: So we return to our opening theme – seeing the place of science in human life steadily and seeing it whole. Has that been your accomplishment these fifty years or more?

JG: I do not see much remarkable about my career. I have enjoyed tracing what Whitehead called the 'adventures of ideas', and I hope my writings and teaching have thrown some light on the problems facing humanity. I have not been much involved politically except in the sense that, as some see it, we are all politically involved. But I do find real satisfaction in the intellectual continuities we have explored in this conversation. Nothing whatever has happened to invalidate or relegate to the trash heap my belief that ideas and their history are important in their own right, and for understanding science. Of the history of ideas I would still say what Milton said of philosophy: '. . . not harsh and crabbed as dull fools suppose, but musical as Apollo's lute, and a perpetual feast of nectared sweets, where no crude surfeit reigns'.

At home in Storrs, Connecticut, August 1980.

1

Erasmus Darwin: Doctor of evolution?

ROY PORTER

Thirty years old now, John Greene's *The Death of Adam* remains the most powerful introduction to the history of evolutionary thought precisely because it regards its subject not as a narrow scientific doctrine but as a cast of mind, a temper. Greene offered a majestic vision of a protracted struggle between, on the one hand, a static, hierarchical and creationist orthodox way of seeing the living world and the natural order, and, on the other, a radical tradition, espousing dynamic process and directional change as the keys to Nature. Within this polar struggle Greene had no difficulty in placing Erasmus Darwin. His philosophy of Nature was clearly of the latter tendency. In Greene's words:

> Like Buffon and Lamarck, Darwin sought to explain the phenomena of life in terms of the operations of a system of matter in motion. The peculiarity of living matter, he observed, was its capacity to undergo progressive transformations resulting in the appearance of new structures, new needs, and new functions.[1*]

The 'history of ideas' framework within which Greene couched his interpretation has now become unfashionable. The Lovejoyan approach, which saw the history of scientific thought in terms of the destinies of Platonic unit ideas, is currently regarded by many historiographical schools as distortingly abstract.[2] And recent scholarship has been particularly loud in its warnings against the dangers of hypostasizing, with the dubious benefit of hindsight, a single hereditary line of 'evolutionary' thought, culminating in Charles Darwin and Alfred Russel Wallace, in which prominent 'forerunners of Darwin' would number such Enlightenment 'precursors' as Diderot, Robinet, Buffon, Maupertuis, Lamarck and, of course, Erasmus Darwin.[3]

Indeed, today's scholarship on Erasmus Darwin is inclined to soft-

---

* Superscript numerals refer to numbered notes at the end of each chapter.

39

pedal his 'contribution' to evolutionary thought, and of late his prosody and aesthetics have been receiving more attention than his science.[4] Moreover, the leading apologist for Darwin the scientist, who once argued that Erasmus Darwin's evolutionary theories were 'nearer the modern view' than those of his grandson,[5] has himself come to regard 'revolution' rather than 'evolution' as the metaphoric key to Darwin's thought-world.[6] And Erasmus Darwin's most recent interpreter has forcefully argued that it is misguidedly 'whiggish' for Darwin studies to focus on his evolutionism; instead he must be more sensitively contextualized within the ideologies of industrialization.[7]

Clearly, then, Darwin's evolutionism presents major historiographical dilemmas – problems that I have no illusions about resolving here. It would be a serious distortion, however, if discussion of Darwin's evolutionary theories simply fell victim to the vagaries of current interpretive fashion. No less, it would be quite wrong to assume that, merely because Erasmus Darwin's theories did not win acceptance, merely because they did not lead directly to Victorian evolution by natural selection, merely because Charles Darwin was tersely dismissive about his grandfather's 'speculations',[8] we may, for those reasons, conclude that Erasmus Darwin's evolutionism was flimsy, ill-conceived, or no more than a poetical *jeu d'esprit*. Judgements to the effect that Darwin 'indulged in some casual evolutionary speculations' in his *Zoonomia*, he 'never amplified them further', and 'there is no justification for a detailed presentation of his thoughts', are uninformed and condescending.[9] Further detailed consideration is called for, and here I wish to argue, with the necessary brevity, that we shall gain a better understanding of Darwin's evolutionism if we approach it from the viewpoint of a peculiarly neglected dimension of his scientific thought and practice: his medicine.

Resiting the doctor

There is a tradition of situating Darwin in the manufacturing Midlands, placing greatest stress upon his activities in the Lunar Society and his contributions to industrialization. Much has been made of his mechanical ingenuity; and, more recently, a very detailed case has been made for Darwin's role in generating a mystifying ideology of industrial capitalism. There is an element of truth in such interpretations that underline Darwin's lively involvement in the economic enterprise of his cronies, or of his region. But it is far from the whole truth, and can lead to gross distortions of Darwin's thought- and action-worlds.[10]

Darwin's social roots were professional and minor gentry. He attended Cambridge University, before studying medicine at Edinburgh. He passed the great bulk of his professional life living first in the small and tranquil cathedral city of Lichfield and then in a country house outside Derby. He was above all a provincial physician. He practised medicine for some forty years and delayed publishing his scientific poetry until his medical career was established beyond challenge. His largest book by far, the *Zoonomia* – the third edition runs to four fat volumes – was, as he says, essentially a work of 'medical theory'.[11] In the light of Darwin's preoccupation with medicine, it is odd that his biomedical ideas have never received sustained analysis (no extensive study of the *Zoonomia* exists). Certainly, Darwin's most assiduous biographer has never seen fit to advance Darwin's claims to fame in medicine in the way that he has in such other fields as meteorology, botany and technological invention. Despite noting that *Zoonomia* was indeed 'a great success, and was widely translated and republished', he adds, 'We cannot value the book so highly today, because its basic theory is wrong.' As a result he does not explore Darwin's biomedical ideas about the animal economy in any detail.[12]

This neglect, however, results in distorted accounts of the structure of Darwin's evolutionism. It is typically argued that the Achilles heel of Darwin's theory – perhaps even the factor that explains its neglect – was its transformism.[13] The underlying assumption is that Darwin's 'speculations' resemble a spectacular theatrical set, made up of an assemblage of striking props and painted flats, but held together largely by the *trompe-l'œil* of literary flair and some bombastic ideological rhetoric (what hostile critics called 'Darwinizing'). Thus, it is commonly noted, and not incorrectly, that Darwin drew on Buffon's ideas of the original spontaneous generation of life out of a primeval chemical soup in the infancy of a terraqueous globe destined for a progressive future.[14] He then called upon growing knowledge of the fossil record, and its indications of extinction,[15] to sketch in an emergent prehistory of life, similarly incorporating evidence of how ontology recapitulated phylogeny, and thus further supported organic transformism.[16] And, nailing his colours to the mast of analogy,[17] he argued from the apparent unity of the living world ('the whole is one family of one parent') to the idea of universal common descent (from 'a single filament'). Generous helpings of literary tropes and mythology (the visual and metaphorical figure, for example, of the hierophant in the *Temple of Nature*) were then piled on,[18] topped with a eupeptic faith in cosmic progress:

> Thus it would appear, that all nature exists in a state of perpetual improvement by laws impressed on the atoms of matter by the great CAUSE OF CAUSES; and that the world may still be in its infancy, and continue to improve FOR EVER AND EVER.[19]

All these elements, whirled together in the *Temple of Nature* through an inspired scientific showmanship on Darwin's part that explicitly appealed to his reader's imagination (enlisting 'imagination under the banner of science'),[20] certainly conjured up an impressive vista of unlimited evolutionary change. Yet, if we consider Darwin's evolutionism exclusively on this grand, synthetic scale, it is easy to see why even fairly liberal organs, such as the *Edinburgh Review*, recoiled in horror, condemning Darwin for abandoning the straight-and-narrow of induction for the siren seductions of speculation.[21]

But this vision of Darwin as a spinner of science fictions, or as a kind of poetic precursor of Herbert Spencer, is at best a half-truth, and it is important not to presume that Darwin's evolutionary theory was based on little more than analogy and poetic license. For, as a glimpse at the fine texture of any of his books, or indeed his letters, will show, Darwin was also acutely interested in Nature in microcosm, in the individual case. As a physician he spent a lifetime dealing with the complaints of his patients and their families – even his critics commented on his medical benevolence. As a naturalist his writings evince an unflagging curiosity towards the micro-economies of stamens and pistils, the techniques of grafting and breeding, the habits of wasps and ants, and the characters of domestic pets – an inquisitiveness bearing comparison with his contemporary, Gilbert White.[22] And as a practical psychologist he possessed a quite exceptional gift for penetrating insights into the mental processes of his familiars, his patients and, not least, himself.[23] Charles Darwin was later to be snooty about the lack of a factual foundation for his grandfather's evolutionary superstructure, and it is certainly true that the busy medical practitioner did not have the leisurely years to ponder barnacles and earthworms enjoyed by the squire of Down. Yet the wealth of detail contained in the *Zoonomia* or the *Phytologia*, or even in his *Plan for the Conduct of Female Education*,[24] attest to his powers of observation and his overriding empirical bent.

Darwin as naturalist

Wherever Darwin looked, he saw Nature in motion. He was fascinated by the changes integral to the life-cycles of individual creatures: 'as in the production of the butterfly with painted wings from the crawling caterpillar; or of the respiring frog from the subnatant tadpole; from the feminine boy to the bearded man, and from the infant girl to the lactescent woman'.[25] But Darwin also saw changes more challenging to the doctrine of a Nature created in pristine perfection. Creatures adapted themselves to climate and environment: 'the hares and partridges of the latitudes which are long buried in snow, become white during the winter months'.[26] Interbreeding between species produced mules, and abnormalities of feeding habits ('exuberance of nourishment') often created monstrosities that, Darwin noted, often proved constant in breeding.

Even more significantly, through 'artificial or accidental cultivation', beings underwent 'great changes' that were then progressively passed on from generation to generation:

> as in horses, which we have exercised for the different purposes of strength or swiftness, in carrying burthens or in running races, or in dogs, which have been cultivated for strength and courage, as the bull-dog; or for acuteness of his sense of smell, as the hound and spaniel; or for the swiftness of his foot, as the greyhound; or for his swimming in the water, or for drawing snow-sledges, as the rough-haired dogs of the north; or lastly, as a play-dog for children, as the lap-dog.[27]

In fact, the human capacity to produce artificial breeds seemed to be transforming the face of Nature itself:

> Many of these enormities of shape are propagated, and continued as a variety at least, if not as a new species of animal. I have seen a breed of cats with an additional claw on every foot; of poultry also with an additional claw, and with wings on their feet.[28]

Long before his grandson, Darwin perceived the importance of fancy breeding:

> Mr. Buffon mentions a breed of dogs without tails, which are common at Rome and at Naples, which he supposes to have been produced by a custom long established of cutting their tails close off. There are many kinds of pigeons, admired for their peculiarities, which are monsters thus produced and propagated.[29]

In human beings, too, disease and bad habits often produce the equivalents of these monstrosities, 'which become hereditary ... through many generations':

> Those who labour at the anvil, the oar, or the loom, as well as those who carry sedan-chairs, or who have been educated to dance upon the rope, are distinguishable by the shape of their limbs; and the diseases occasioned by intoxication deform the countenance with leprous eruptions, or the body with tumid viscera, or the joints with knots and distortions.[30]

The dire effects of alcoholism running in families was a subject close to Darwin's heart. An ardent teetotaller himself, he witnessed many of his patients and friends succumbing to the chronic diseases of drunkenness, not least (perhaps) his first wife and her male kin.[31]

In short, Darwin's biomedical outlook – the product of his Edinburgh training – was informed by the evidence of change, both in degree and in kind, running ubiquitously through Nature. But an equally sharply defined feature of his thinking lay in his eye for continuities. Where other traditions expected to find evidence of rigid boundaries and ontological divisions, compartmentalizing Nature, Darwin – following Buffon – perceived connections and interpenetration. His sympathies for the use of analogy in scientific reasoning, and his generous reliance upon the figure of personification in poetry, of course supported such a stance;[32] but as a personal observer and interpreter of Nature's order and workings, what habitually caught his attention were features indicative of unity, integration and interdependence. In his *Phytologia*, he explored, along lines suggested by Priestley and others, the intricate cycles of nutrition which rendered vegetable life dependent both on atmospheric gases and on the soil;[33] which enabled animals to live off vegetable food; and through which the debris of all kinds of organic life were ultimately recycled to provide fresh resources for the future (in the form of fossils proving 'monuments of past delight').[34] Nature was less a ladder than a laboratory, less a fixed order than an organism, perpetually regenerated:

> While Nature sinks in Time's destructive storms,
> The wrecks of Death are but a change of forms;
> Emerging matter from the grave returns,
> Feels new desires, with new sensations burns;
> With youth's first bloom a finer sense acquires,
> And Loves and Pleasures fan the rising fires.[35]

Resemblance also played a key part in Darwin's hylozoic vision of natural continuity. His botanical writings argue for the view that plants are, quite literally, lower forms of animals, possessed of identical faculties to the brutes (including in some sense, sensitivity and volition), only in less sophisticated forms.[36] But the same sense of unity applied equally to the relations between humans and other animals. Observation of the expressions, habits and behaviour of dogs, cats and monkeys convinced him that the distinctions routinely drawn between 'man and the brutes', ascribing to man alone the powers of reason, free will and the capacity for progressive learning, were fallacious and vain. In this he drew upon personal experience:

> One circumstance I shall relate which fell under my own eye,
> and shewed the power of reason in a wasp, as is exercised
> among men. A wasp, on a gravel walk, had caught a fly nearly

as large as himself; kneeling on the ground I observed him separate the tail and the head from the body part, to which the wings were attached. He then took the body part in his paws, and rose about two feet from the ground with it; but a gentle breeze wafting the wings of the fly turned him round in the air, and he settled again with his prey upon the gravel. I then distinctly observed him cut off with his mouth, first one of the wings, and then the other, after which he flew away with it unmolested by the wind.

Go, thou sluggard, learn arts and industry from the bee, and from the ant!

Go, proud reasoner, and call the worms thy sister![37]

In many other departments too – Darwin drew upon standard Enlightenment instances such as polyps, corals and hermaphrodites – Nature's products and processes insensibly shaded into each other, defying the fine metaphysical distinctions erected by proud system builders.

Physiology and evolution

Darwin's fundamental discussion of the elements of the animal economy (set out primarily in his *Zoonomia*, though all his publications revolve around these themes), expounds the physiological basis for his understanding of change and continuity in Nature, and, more largely, for his overarching evolutionism. Like many zoologists of the late Enlightenment, Darwin drew copiously upon Haller, while also incorporating Scottish medical theories – he was at least as sympathetic to John Brown as to Cullen.[38] Outgrowing the Boerhaavian 'hydraulic' legacy, Darwin declined to account for bodily operations merely through mechanical categories; nor did he – unlike many English Boerhaavians – interpret life in 'animistic' terms as a separate transcendental force, not integrally associated with the body.[39] Rather, for Darwin, the starting point for understanding the economy of life lay in the inherent motility possessed by all organized and animated entities, lodged fundamentally in their fibrous matter:

In every contraction of the fibre there is an expenditure of the sensorial power, or spirit of animation; and where the exertion of this sensorial power has been for some time increased, and the muscles or organs of sense have in consequence acted with greater energy, its propensity to activity is proportionately lessened; which is to be ascribed to the exhaustion or diminution of the quantity.[40]

Darwin insisted that his phrase, 'sensorial power, or spirit of animation', should be taken in an essentially nominalistic sense, as merely a heuristic term, with no higher ontological pretensions:

> By the words spirit of animation or sensorial power, I mean only that animal life, which mankind possesses in common with brutes, and in some degree even with vegetables, and leave the consideration of the immortal parts of us, which is the object of religion, to those who treat of revelation.[41]

Living beings, in other words, were those entities that did not simply react to environmental interference according to the Newtonian laws of mechanics. Partly through being able to draw nourishment from the environment, they possessed a quantum of animation of their own, sustained through nutrition and discharged through action, the ultimate diminution of which spelt death.[42] Living bodies, in other words, were those entities capable of entering into dialectical interplay with their external environment.

Darwin classified life into four separate, though temporally and organically interrelated, capacities:

> The sensorium possesses four distinct powers, or faculties, which are occasionally exerted, and produce all the motions of the fibrous parts of the body; these are the faculties of producing fibrous motions in consequence of irritation, which is excited by external bodies; in consequence of sensation, which is excited by pleasure or pain; in consequence of volition, which is excited by desire or aversion; and in consequence of association, which is excited by other fibrous motion . . . [in other words, those powers] of irritation, those of sensation, those of volition, and those of association.[43]

'This is about as clear as mud', comments Darwin's chief biographer, but he surely misses the point.[44] For the combination of powers had a deep significance, both in the wider development of 'physiology', and within Darwin's analysis of life. In its wider historical dimension the formulation represents an extremely ambitious attempt to unite the Hallerian physiology of nervous stimulus and response with the utilitarian associationism of Locke, Hartley and his friend Priestley.[45] At the same time, in a narrower organic focus, the formulation can be seen as Darwin's attempt to articulate a naturalistic understanding of the organism, which could derive the entire range of animal operations, both normal and morbid, from the simplest needs, stimuli and powers. By emphasizing organic integration, Darwin aimed to avoid on the one hand collapsing into implausible mechanistic reductionism, and on the other floating off into the transcendental. Within his naturalistic physiological psychology, understanding the functioning of an organism would entail a grasp of the internal development of its own faculties.[46]

Given that the key to life lay in the contractility of fibres, and that fibrous motion was a function of the nerves, the most elementary power was clearly that of irritation. Darwin defined it as 'an exertion or change of some extreme part of the sensorium residing in the muscles or organs of sense, in consequence of the appulses of external bodies'.[47] Irritation encompassed all those immediate reflex responses of the organism to stimuli, both internal and external, required for the moment-to-moment sustenance of life. It governed such functions as digestion and excretion:

> Many of our muscular motions are excited by perpetual irritations, as those of the heart and arterial system by the circumfluent blood. Many others of them are excited by intermittent irritations, as those of the stomach and bowels by the aliments we swallow; of the bileducts by the bile; of the kidneys, pancreas, and many other glands, by the peculiar fluids they separate from the blood; and those of the lacteal and other absorbent vessels by the chyle, lymph, and moisture of the atmosphere.[48]

Even irritation, lowly though it was in the hierarchy of complexity of organic functions, manifestly involved more than mere mechanical or chemical cause and effect. To exemplify this, Darwin often referred to his favourite experiments with ocular spectra, which demonstrated that a constant quantity of light stimulus, bombarding the optic nerves, produced greatly varying degrees of response, as the eye adjusted or grew fatigued.[49] Similar experiments proved the same point:

> Thus when a circular coin, as a shilling, is pressed on the palm of the hand, the sense of touch is mechanically compressed; but it is the stimulus of this pressure that excites the organ of touch into animal motion, which constitutes the perception of hardness and of figure: for in some minutes the perception ceases, though the mechanical pressure of the object remains.[50]

Such tests showed that nervous response was conditioned not merely or essentially by the power of the stimulus, but by an adaptational response of the 'sensorial power' in the organism: irritation could cease though the irritant remained. To that degree the organism was far from mechanistically passive but was self-regulating.[51]

Another sign of these internal, 'vital' powers lay in the fact that irritation further stimulated sensation. In other words, certain irritants called into action higher planes of nervous organization, the capacity to experience pains and pleasures.[52] If many stimuli, like those exciting the circulation of the blood, were unattended by feelings, others were. Environmental pressures such as the experience of heat or cold could produce a variety of sensations that would dispose the organism to counter or cope with the environmental stimuli.[53] In their turn, the sensations of pleasure and pain generated feelings of desire and aversion,[54] leading organically to a further level of bodily operation: 'A certain quantity of sensation produces desire or aversion; these constitute volition.'[55]

Volition – 'an exertion or change of the central parts of the sensorium, or of the whole of it, terminating in some of those extreme parts of it, which reside in the muscles or organs of sense'[56] – constituted the capacity of a body to act in the light of the sensations of pleasure and pain created by sense impressions applying pressure upon the nerves:

> As the sensations of pleasure and pain are originally introduced by the irritations of external objects: so our desires and aversions are originally introduced by those sensations; for when the objects of our pleasures or pains are at a distance, and we cannot instantaneously possess the one, or avoid the other, then desire or aversion is produced, and a voluntary exertion of our ideas or muscles succeeds.[57]

Considered thus, as the centrally directed power of an organism acting upon the stimulus of pleasure or pain, Darwin explained that volition should not be mistaken for the old conception of free will, an undetermined act of mind or understanding, as commonly found in moral philosophy or theology – disciplines he clearly distrusted.[58] Darwin insisted that he was instead making a behavioural statement about real purposive action: he could on occasion be brusquely dismissive of metaphysical niceties. Traditional moral theory he judged erroneous, for volition, properly understood, must comprehend a whole range of actions that are not included under the rubric of the rational will. For one thing, the acts of madmen, though not conventionally seen as free or rational, must qualify as volitional. So must purposive actions performed without an immediate consciousness of them. Someone lost in thought brushes away an irritating fly without any awareness of performing the act. Yet it is an act of volition, even in the absence of interior consciousness.[59] Similarly, Darwin explored other modes of purposive action unattended by normal forms of consciousness, including sleepwalking, trance, reverie and kindred phenomena taken up a century later by scientific spiritualists.

In his investigations of the different levels of the mind – further examples were taken from states of intoxication – Darwin aimed to elucidate the links between volition and the rise of habit.[60] Frequent repetition of an action, often conferring the pleasures of security and familiarity, built up a familiar pattern, the subsequent performance of which required less conscious play of mind and will, or indeed less attention. The tyro at the piano must focus every nerve; expert pianists can allow their mind to wander because the succession of finger movements for a familiar tune has become habitual. Habit, however, does not supersede will; it merely pitches it on to a higher plane, better adapted to the complex needs of beings that must be capable of performing a range of actions simultaneously.[61]

The power of the will to advance from isolated acts to chains of settled behaviour, triggered by habitual sensory inputs, was a crucial step forward within Darwin's vision of organic capacities. On the one hand, it provided an exemplary explanation of the ubiquitous phenomena of change in individuals. Humans and other animals alike were not born inherently endowed with a fixed repertoire of dispositions, capacities, skills and propensities. Darwin was thus as fierce an opponent of the assumption of innateness as Locke.[62] On the other hand, his view of volitional power underpinned a vision of continual, indefinite change in nature. Particular actions were repeated; habits thereby formed, often on the basis of imitation;[63] and these habits, appropriately modified over time, adapted the behaviour of organisms to environmental pressures and opportunities. Thus the sanctions of the senses – pleasures and pains – enabled organisms to *learn*, and to advance through learning. Sense responses translated into habit; volition rendered creatures – *all* creatures, not just humankind – progressive.[64]

What enabled adaptive behaviour to assume such complex forms lay in a fourth power of the organism: association. Darwin's definition was rather abstract:

> All animal motions which have occurred at the same time, or in immediate succession, become so connected, that when one of them is reproduced, the other has a tendency to accompany or succeed it. When fibrous contractions succeed or accompany other fibrous contractions, the connexion is termed association; when fibrous contractions succeed sensorial motions, the connexion is termed causation; when fibrous and sensorial motions reciprocally introduce each other, it is termed catenation of animal motions.[65]

In reality, what Darwin had in mind was something close to the classic conception of the association of ideas as spelt out in empiricist epistemology from Locke through Hartley and Hume. (Darwin accepted Hume's three-fold subdivision of association into contiguity, causation and resemblance.) Precisely why the brain had such an associative capacity remained unclear, Darwin confessed, although he

argued it was best understood by analogy to gravitational attraction.[66]
Association held the key to the exceptionally complex interactions
of organic activity as a whole: that is, what were commonly, but
unhelpfully, distinguished as the mental and the physical components
of behaviour. As Darwin recognized, such facets of behaviour as the
expression of emotions constituted the learned product of a whole
series of responses, in practice largely adopted from creature to creature, from parents to offspring, over the generations by the power
of imitation. Thus he considered the manifestations of protective
behaviour under threat:

> So when the famish'd wolves at midnight howl,
> Fell serpents hiss, or fierce hyenas growl;
> Indignant Lions rear their bristling mail,
> And lash their sides with undulating tail.
> Or when the Savage-Man with clenched fist
> Parades, the scowling champion of the list;
> With brandish'd arms, and eyes that roll to know
> Where first to fix the meditated blow;
> Association's mystic power combines
> Internal passions with external signs.[67]

The suggestive logic of association thus offered insight into modes of
behaviour, bodies of belief, and concatenations of actions, words and
gestures, otherwise essentially unintelligible, or merely resolved, by
those eager to cut the Gordian knot, into mysterious givens.

Still more evocative was Darwin's classic account of the origins of
the sense of beauty:

> All these various kinds of pleasure at length become associated
> with the form of the mother's breast; which the infant embraces
> with its hands, presses with its lips, and watches with its eyes;
> and thus acquires more accurate ideas of the form of its mother's
> bosom, than of the odour and flavour or warmth, which it
> perceives by its other senses. And hence at our maturer years,
> when any object of vision is presented to us, which by its
> waving or spiral lines bears any similitude to the form of the
> female bosom, whether it be found in a landscape with soft
> gradations of rising and descending surface, or in the forms of
> some antique vases, or in other works of the pencil or the
> chissel, we feel a general glow of delight, which seems to
> influence all our senses; and, if the object be not too large, we
> experience an attraction to embrace it with our arms, and to
> salute it with our lips as we did in our early infancy the bosom of
> our mother.[68]

Such an explanation accorded well with Darwin's general interpretive strategy. It rescued the sense of beauty from being just a mystery by turning it into an acquired characteristic amenable to investigation. Moreover, it bridged the otherwise perplexing and frustrating gap between the mental and the physical by conjuring up instead an interplay between the two.

There were two special reasons why association loomed large in Darwin's account of the laws of life. First, from a medical point of view, it contained the key to disease propensities and patterns. For in each living being association produced life habits that issued in quite specific dispositions to disease. These Darwin styled 'temperaments', with more than a hint of a glance back to Classical medicine. People could be loosely grouped into classes of distinct temperaments (that is, those specially marked by irritability, sensibility, and so forth) and these in turn could be further subdivided into persons of decreased, or heightened, manifestation. Each such group was characterized by a particular susceptibility to certain clusters of complaints, within Darwin's highly physiological conception of disease. For example, those with increased capacity of association – people blessed with powerful memories or unusual powers of sympathy – would be more liable to hysterical afflictions.[69] Thus the doctrine of temperaments could be diagnostically invaluable, for association pointed the way to morbid clusterings.[70]

More specifically, a grasp of the association of ideas was vital to the physician for understanding specific kinds of disease. In many forms of mental illness, above all, consciousness translated sense impressions into ideas involving wild distortions of the realities of the objective world. Thus people would register natural, if unusual, occurrences, as marks of unique divine favour or wrath.[71] Darwin's explanation was that association had linked sense impressions with imaginative notions acquired from Bible-reading or from tales told as a child. Indeed, a glance at his day-to-day practice as a physician, as recorded in the *Zoonomia* and in his correspondence, shows just how far he believed sickness experiences were triggered, enhanced or complicated by association. And as his own therapeutic preferences show, he recognized that such 'psychosomatic' disorders required both sympathy and cunning, while deploring those belief systems such as hellfire Methodism that predisposed believers to morbid associations.[72]

In the second place, the power of association figured crucially in Darwin's concept of progress, and thus in his evolutionism. Through association, the operations of thought came into their own, producing behaviour at more complex levels. Association, as mentioned above, had generated the sense of beauty and, more broadly still, those feelings of sympathy facilitating mutual affection and society among humans and other animals. The brain stored up these capacities based on collective experience, and thus, through its power of imagination, became a key factor in hereditary change.[73]

Generation and evolution

For Darwin, the imagination played a crucial part in the mechanics of reproduction – such a fiercely disputed issue in eighteenth-century science. In the controversies over rival hypotheses in generation and embryology he repudiated the preformationist doctrine, arguing that experience showed both that foetal growth amounted to far more than the mechanical enlargement of microscopic parts all 'given' from the beginning, and that offspring did not in any simple way automatically remain identical down the generations.[74] To that extent he was an epigenesist. Generation consisted of interactions between reproductive matter endowed with certain propensities, and the forces acting upon it from without and within.[75]

Most important, Darwin held that mind determined hereditary transmission to the offspring. The general notion was not uncommon. Both popular folklore and traditional medicine through the early modern period and into the eighteenth century believed that the mother could impress the contents of her imagination ('mind') upon the embryo at the moment of conception or even subsequently, thereby producing monstrous or defective offspring.[76] This view (which underpinned the sad history of Laurence Sterne's hero in *Tristram Shandy*) was meeting increased medical and embryological hostility during Darwin's lifetime, and he dissociated himself from it.[77] Yet he did lend his authority to a parallel but much less common doctrine, the notion that the contents of the *male* imagination were impressed upon the 'form' of the foetus:

> I conclude, that the act of generation cannot exist without being accompanied with ideas, and that a man must have at that time either a general idea of his own male form, or of the form of his male organs; or an idea of the female form, or of her organs; and that this marks the sex, and the peculiar resemblances of the child to either parent. From whence it would appear, that the phalli, which were hung round the necks of the Roman ladies, or worn in their hair, might have effect in producing a greater proportion of male children; and that the calipaedia, or art of begetting beautiful children, and of procreating either males or females, may be taught by affecting the imagination of the male-parent; that is by the fine extremities of the seminal glands

imitating the action of the organs of sense, either of sight or touch.[78]

Thus in Darwin's theory of generation the female did little more than provide a seedbed within which the newly conceived foetus, a male 'conception', could grow. Her role was essentially nutritive. The male planted the basic 'filament' ('part of the father')[79] that became the next generation. To a very large extent it was the quality and contents of his mind that determined the 'form'[80] of the offspring. For example, the infants of a man infatuated by the beauty of another woman would resemble her rather than his wife, just as a man of diseased imagination – a habitual drunkard, for example – was likely to produce a diseased or defective child.[81]

Darwin regarded sexual reproduction as the happiest mode Nature possessed for securing the future of a species. Simpler, pre-sexual forms of reproduction led to degeneration over the course of time.[82] Sexual reproduction also provided the opportunity for 'joy'. But its greatest advantage pertained to evolution. By providing the means whereby the 'ideas' of the mind or imagination could be conveyed to the next generation, sexual reproduction ensured that life forms could actually be progressive: the learned wisdom, the adaptive behaviour of one generation, could be passed down to the next. That this happened was a simple fact of experience, Darwin believed:

> Some nations of Asia have small hands, as may be seen by the handles of their scymeters; which with their narrow shoulders shew, that they have not been accustomed to so great labour with their hands and arms, as the European nations in agriculture, and those on the coasts of Africa in swimming and rowing. Dr. Manningham, a popular accoucheur in the beginning of this century, observes in his aphorisms, that broad-shouldered men procreate broad-shouldered children. Now as labour strengthens the muscles employed, and increases their bulk, it would seem that a few generations of labour or of indolence may in this respect change the form and temperament of the body.[83]

Thus the biomedical theories of *Zoonomia* (and, less systematically, of Darwin's other zoological and botanical writings) could lead to a vision of change through learning, of boundless progress, and ultimately, the transformation of species. Yet *Zoonomia* was not primarily advancing a theory of evolution. Its explicit rationale lay in an understanding of the operations of life itself and of the organic economy of living entities, specifically with a view to adumbrating the laws of health, the causes and types of disease, and (not least) their remedies. Indeed, the work opens with a lengthy exposition of the different natures of irritation, sensation, volition and association; following which comes an account of their respective several motions, of the temperaments associated with them; and this culminates in an elaborate taxonomy showing how the gamut of diseases can be allotted to the four main kinds of power in the animal economy.[84]

Consistently with his thoroughgoing 'monist' approach, Darwin aimed to explore the complex interplay of mind and body in the generation of disease. Thus he viewed both sea sickness and many fever epidemics as attributable, in part at least, to a sort of suggestion, a contagiousness of perceiving others suffering from illness;[85] and he was particularly interested in the aetiology of handicaps, such as stammering, from which he himself suffered:

> That this impediment of pronunciation is altogether a disease of the mind, and not of the organs of speech, is shown by the stammerer being able to speak all words with perfect facility when alone, as in repeating a play; but begins to hesitate if anyone approaches, or even if he imagines that he is listened to. Those words also are most difficult to him to pronounce which he is conscious he cannot change for others, as when he is asked his own name, or the names of other persons, or of places; and the more so if he is aware that the hearer is impatient to be informed, and that he cannot conjecture the name before it is spoken.[86]

Equally consistently with Darwin's emphasis on the physical media of transmission and change was his desire to investigate the hereditary nature of certain diseases and their effects, not just on the health of individuals, but upon the health of the nation – indeed, the species at large. As he put it:

> Other similarities of the excitability, or of the form of the male parent, such as the broad or narrow shoulders, or such as constitute certain hereditary diseases, as scrofula, epilepsy, insanity, have their origin produced in one or perhaps two generations; as in the progeny of those who drink much vinous spirits; and those hereditary propensities cease again, as I have observed, if one or two sober generations succeed; otherwise the family becomes extinct.[87]

Fears about alcohol-based degeneration led Darwin to offer advice about choice of marriage partners: 'As many families become gradually extinct by hereditary diseases, ... it is often hazardous to marry an heiress, as she is not unfrequently the last of a diseased family.'[88]

Thus Darwin's biomedical theorizing culminated in practical understanding. Here comparison with Lamarck is apposite, for Darwin's primary aim was not to develop a view of organic evolution but to provide a comprehensive account of the economy of the living being through an identification of sensorial powers:

> There appears to be a power impressed on organized bodies by the great author of all things, by which they not only increase in size and strength from their embryon state to their maturity, and occasionally cure their accidental diseases, and repair their accidental injuries, and also a power of producing armour to prevent those more violent injuries, which would otherwise destroy them.[89]

An analysis of the multifarious powers of living beings led Darwin to the conviction that life contained within itself the capacity for repeated, continued, gradual modification, change transmissible to futurity:

> From their first rudiment, or primordium, to the termination of their lives, all animals undergo perpetual transformations; which are in part produced by their own exertions in consequence of their desires and aversions, of their pleasures and their pains, or of irritations, or of associations; and many of these acquired forms or propensities are transmitted to their posterity.[90]

And so an argument for evolution welled up from within the general discussion of life.

Within *Zoonomia* the statement of organic transformation was, as it were, a subtext, frequently present, but spliced into the argument. It arose out of asides about the powers of individual creatures. Thus Darwin aimed to show how animals were capable of progressive advances, 'just like mankind':

> The monkey has a hand well enough adapted for the sense of touch, which contributes to his great facility of imitation; but in taking objects with his hands, as a stick or an apple, he puts his thumb on the same side of them with his fingers, instead of counteracting the pressure of his fingers with it; from this neglect he is much slower in acquiring the figure of objects, as he is less able to determine the distances or diameters of their parts, or to distinguish their vis inertiae from their hardness. Helvétius adds, that the shortness of his life, his being fugitive before mankind, and his not inhabiting all climates, combine to prevent his improvement. . . . There is however at this time an old monkey shewn in Exeter Change, London, who having lost his teeth, when nuts are given him, takes a stone into his hand, and cracks them with it one by one; thus using tools to effect his purpose like mankind.[91]

Such individual cases led Darwin to more species-related views of change:

> Some animals have acquired wings, instead of legs, as the
> smaller birds, for the purpose of escape. Others great length
> of fin, or of membrane, as the flying fish, and the bat. Others
> great swiftness of foot, as the hare. Others have acquired hard
> or armed shells, as the tortoise and the echinus marinus.....
> The colours of many animals seem adapted to their purposes of
> concealing themselves either to avoid danger, or to spring upon
> their prey.[92]

And this could induce Darwin to occasional apostrophes to the evolutionary process as a whole:

> From thus meditating on the great similarity of the structure of
> the warm-blooded animals, and at the same time of the great
> changes they undergo both before and after their nativity; and
> by considering in how minute a portion of time many of the
> changes of animals above described have been produced; would
> it be too bold to imagine that in the great length of time since
> the earth began to exist, perhaps millions of ages before the
> commencement of the history of mankind, all warm-blooded
> animals have arisen from one living filament, which THE
> GREAT FIRST CAUSE endued with animality, with the power
> of acquiring new parts, attended with new propensities,
> directed by irritations, sensations, volitions, and associations;
> and thus possessing the faculty of continuing to improve by its
> own inherent activity, and of delivering down those
> improvements by generation to its posterity, world without
> end?[93]

But it was only in the *Temple of Nature*, posthumously published in 1803, that Darwin so organized his ideas as to make evolution the main plot of the whole work. There is no need here to analyse the poem's sublime panorama of change, ranging forward from the coagulation of nebulae to contemporary society, from mushrooms up to monarchs.[94] As noted earlier, many different sorts of evidence are integrated to add force to the evolutionary argument. What must be emphasized, however, is that the animal physiology of the *Zoonomia* is central to the evolutionary dynamo. As in *Zoonomia*, irritation is regarded as the initial trigger of the life forces, unlocking the potentialities of animated powers:

> FIRST the new actions of the excited sense,
> Urged by appulses from without, commence;
> With these exertions pain or pleasure springs,
> And forms perceptions of external things.

Thus, when illumined by the solar beams,
Yon waving woods, green lawns, and sparkling streams,
In one bright point by rays converging lie
Plann'd on the moving table of the eye;
The mind obeys the silver goads of light,
And IRRITATION moves the nerves of sight.[95]

The awakening of sensation follows:

Next the long nerves unite their silver train,
And young SENSATION permeates the brain;
Through each new sense the keen emotions dart,
Flush the young cheek, and swell the throbbing heart.[96]

Sensation in turn quickens the perceptions of pleasure and pain, so expanding consciousness into volition:

From pain and pleasure quick VOLITIONS rise,
Lift the strong arm, or point the inquiring eyes;
With Reason's light bewilder'd Man direct,
And right and wrong with balance nice detect.[97]

Whereupon sensation and volition produce association:

Last in thick swarms ASSOCIATIONS spring,
Thoughts join to thoughts, to motions motions cling;
Whence in long trains of catenation flow
Imagined joy, and voluntary woe.[98]

And with the association of ideas, come habit, imitation, imagination, and the higher organizing powers of the mind:

The impatient Senses, goaded to contract,
Forge new ideas, changing as they act;
And, in long streams dissever'd, or concrete
In countless tribes, the fleeting forms repeat.
Which rise excited in Volition's trains,
Or link the sparkling rings of Fancy's chains;
Or, as they flow from each translucent source,
Pursue Association's endless course.[99]

Most notably (but not exclusively) in human beings, this progression leads to language, the arts and sciences, the love of beauty, and the moral and social powers that sympathy engenders. As elementary physical sensations successively enrich the mind, so the pattern of progress assumes complementary physical and sociocultural directions. Thus, echoing the evolution of the sensation of taste, 'taste' itself develops as a cultural expression. Similarly, the organic sense of touch not only gives rise to the sense of beauty, but also, being so much more highly developed in bipedal humans, leads to their much enhanced 'grasp' of mental as well as physical reality. 'Man' has become the lord of creation not because he has been innately endowed with any superior power of intellect, but because the possession of highly sensitive hands has stimulated his superior powers of volition and understanding:

Immortal Guide! O, now with accents kind
Give to my ear the progress of the Mind.
How loves, and tastes, and sympathies commence
From evanescent notices of sense?
How from the yielding touch and rolling eyes
The Piles immense of human science rise? . . .[100]

But all living creatures have the potential for unlimited improvement even if parallel tendencies to degenerate also exist, and the endless competition of burgeoning life forms against each other within a finite terraqueous globe also results in death, destruction and even extinction.

A physician's vision

I have been arguing that theories of physiology and of reproduction lie at the core of Darwin's evolutionism, showing how the logic of life sets ordered adaptive change in motion, and how future generations inherit such acquired characteristics. These hylozoist ideas seem to give the lie to the common dismissal that Darwin lacked a mechanism for evolution, just as they also provide their own commentary on aspersions about the 'casual speculations' that inform Darwin's 'so-called evolutionary ideas'.[101] Clearly, Darwin's medico-physiological basis for evolutionism was not the one used by his grandson in the *Origin of Species*. Yet there is no denying how fascinated Charles Darwin was with problems concerning the development of expression, the inheritance of disease, the continuity of mind and body and its materialistic implications, problems that are so prominent in *Zoonomia*. Charles's 'M' Notebook, written in 1838, is crammed with references and allusions to Erasmus's *Zoonomia* – for instance, to the wasp that learned to bite off a fly's wings.[102] Via his own father, Robert Waring Darwin, and the patients in Robert's extensive medical practice, Charles Darwin was doubtless personally familiar with some of the very families whose abnormalities and hereditary diseases crop up in *Zoonomia* and in the 'M' Notebook. There were many plausible reasons why Charles chose not to follow up medico-physiological problems in his *Origin*. He did not utterly abandon them as dead ends, for many such problems form the core of his later *Expression of the Emotions in Man and Animals* (1872).[103] It is possible that Erasmus's concerns are so noticeably absent from the *Origin of Species* precisely because they were redolent of materialism, and because they placed human evolutionary history at centre stage.[104] Charles was well aware that he lived in less tolerant times than had his grandfather – he noted, for example, how remarkable it was that Erasmus had been able quite openly to father two bastards and bring them up at home without ruining his medicial respectability – and he was desperate to avoid any indiscretions that would rock the frail barque of evolution by natural selection.[105]

This is not the place to assess how far Erasmus Darwin's physiological vitalism triggered the general rejection of his evolutionism. It clearly was not unimportant. In his own refutation of evolutionism William Paley dismissed Darwin's attribution of 'appetencies' to matter;[106] Anna Seward rejected Darwin's vision of the unlimited capacity of living beings to respond to stimulus and change by learning (for her the suggestion violated the fundamental boundary between human beings and the animals);[107] and the *Edinburgh Review*, assessing *The Temple of Nature*, identified Darwin's basic heresy as the undermining of the distinction between mind and matter, thereby endowing matter with inherent vitality.[108] But, instead of pursuing the discussion further, I would like to turn briefly in conclusion to the question of the ideological implications of Erasmus Darwin's particular vision of evolution.

As his most recent interpreter has demonstrated, Erasmus Darwin's writings can and should be read as a vindication and celebration of the outlooks of the promoters of industrial society, outlooks naturalized through a social biology. And she has vividly explored the mystifications produced by that ideology.[109] Darwin was an ideologue, promoting a capitalist vision of heroic entrepreneurship, the benefits of competition, and eventual progress. His technique was to naturalize these socioeconomic attributes, and thereby legitimize them. I only wish to add that historians should not oversimplify Darwin's vision. It would be a mistake to impose upon Darwin a stilted and anachronistic rendering of what bourgeois ideology eventually amounted to. The same historian undoubtedly does this when she stresses how, to serve his wider apologetic purposes, Darwin legitimized war. In reality, like other bourgeois, he felt the reverse: 'I hate war.'[110] What in fact may be worth more attention than it has received is Darwin's vision of human nature and the human condition, and how these fitted into his wider ideological perspective. Here Darwin's repeated emphasis upon love, sympathy and social cooperation are crucial. Especially in the light of Josiah Wedgwood's ambition of turning his workforce into such machines 'as do not err', we might anticipate that Darwin's vision would be an adumbration of *l'homme machine*, a simple mechanistic behaviourism in which humans responded unerringly to external stimuli and pressures.[111] Such, after all, was the environmentalist ideology for industrial society soon to be advanced in Robert Owen's materialism, equally rooted in Enlightenment utilitarian meliorist associationism.

But not so with Erasmus Darwin. His world was not primarily that of the Lunar Society or of industrialization. His physician's vision was dominated by the living organisms he saw fighting disease, changing over time, involved in subtle interplay with the personalities they housed. Darwin was concerned to *rescue* 'man' from the aspersions of being just a machine.[112] His whole constitution of man constantly stressed the inner energy and drives, both the capacity and the need to learn, the inventiveness and adaptiveness of *homo faber*, the man who makes himself. It is a vision of man for the machine age, but it is not a vision of man the machine.

Notes

1. John C. Greene, *The Death of Adam: Evolution and Its Impact on Western Thought* (Ames: Iowa State University Press, 1959), p. 166. In my text the surname 'Darwin' always refers to Erasmus.

2. Arthur O. Lovejoy, *The Great Chain of Being: A Study in the History of an Idea* (Cambridge, Mass.: Harvard University Press, 1936); William F. Bynum, 'The Great Chain of Being after Forty Years: An Appraisal', *History of Science*, 13 (1975), 1–28; *idem*, 'Time's Noblest Offspring: The Problem of Man in the British Natural Historical Sciences, 1800–1863' (Ph.D. thesis, University of Cambridge, 1974). John Greene discusses the methodological implications in *Science, Ideology, and World View: Essays in the History of Evolutionary Ideas* (Berkeley: University of California Press, 1981).

3. Bentley Glass, Owsei Temkin and William L. Straus, Jr, eds., *Forerunners of Darwin: 1745–1859* (Baltimore, Md.: Johns Hopkins Press, 1959). For a balanced modern perspective, see Peter J. Bowler, 'Evolutionism in the Enlightenment', *History of Science*, 12 (1973), 95–114.

4. For example, see Irwin Primer, 'Erasmus Darwin's *Temple of Nature*: Progress, Evolution, and the Eleusinian Mysteries', *Journal of the History of Ideas*, 25 (1964), 58–76; Robert N. Ross, '"To charm thy curious eye": Erasmus Darwin's Poetry at the Vestibule of Knowledge', *Journal of the History of Ideas*, 32 (1971), 379–94; Donald M. Hassler, *The Comedian as the Letter D: Erasmus Darwin's Comic Materialism* (The Hague: Martinus Nijhoff, 1973); and *idem*, *Erasmus Darwin* (New York: Twayne, 1973).

5. Desmond King-Hele, *Erasmus Darwin* (London: Macmillan, 1963); *idem*, *The Essential Writings of Erasmus Darwin* (London: MacGibbon and Kee, 1968), p. 82 ('he was nearer the modern view than . . . his own grandson Charles').

6. Desmond King-Hele, *Doctor of Revolution: The Life and Genius of Erasmus Darwin* (London: Faber and Faber, 1977).

7. Maureen McNeil, *Under the Banner of Science: Erasmus Darwin and His Age*

(Manchester: Manchester University Press, 1987).

8. Charles Darwin, 'Preliminary Notice', in Ernst Krause, *Erasmus Darwin*, translated by W.S. Dallas (London: John Murray, 1879). See *idem, On the Origin of Species by Means of Natural Selection, or the Preservation of Favoured Races in the Struggle for Life*, 6th edn (London: John Murray, 1872), p. xiv: 'It is curious how largely my grandfather, Dr. Erasmus Darwin, anticipated the views and erroneous grounds of opinion of Lamarck in his "Zoonomia".' See also Michael T. Ghiselin, 'Two Darwins: History versus Criticism', *Journal of the History of Biology*, 9 (1976), 121–32.

9. Ernst Mayr, *The Growth of Biological Thought: Diversity, Evolution, and Inheritance* (Cambridge, Mass.: Harvard University Press, 1982), p. 340.

10. See King-Hele, *Erasmus Darwin; idem, Essential Writings; idem, Doctor of Revolution*; and McNeil, *Under the Banner of Science*. For the fullest record of Darwin's own preoccupations, see Desmond King-Hele, ed., *The Letters of Erasmus Darwin* (Cambridge: Cambridge University Press, 1981). For the Lunar Society, see Robert Schofield, *The Lunar Society of Birmingham: A Social History of Provincial Science and Industry in Eighteenth Century England* (Oxford: Clarendon Press, 1963).

11. Darwin's chief works are: *The Botanic Garden: A Poem in Two Parts. Part I: The Economy of Vegetation. Part II: The Loves of the Plants* (London, 1791) (henceforth *Botanic Garden*); *Zoonomia: or, The Laws of Organic Life*, 2 vols. (London, 1794); 3rd edn, 4 vols. (London, 1801) (henceforth *Zoonomia*; the 3rd edition will be cited); *Phytologia; or, The Philosophy of Agriculture and Gardening* (London, 1800) (henceforth *Phytologia*); and *The Temple of Nature; or, The Origin of Society* (London, 1803) (henceforth *Temple of Nature*). For the role of 'theory', see *Zoonomia*, I, viii.

12. King-Hele, *Doctor of Revolution*, p. 234.

13. Samuel Lilley, 'The Origin and Fate of Erasmus Darwin's Theory of Organic Evolution', *Actes du XIe Congrès International d'Histoire des Sciences* (1968), 5:70–5. See also James Harrison, 'Erasmus Darwin's View of Evolution', *Journal of the History of Ideas*, 32 (1971), 247–64 and Clark Emery, 'Scientific Theory in Erasmus Darwin's *The Botanic Garden* (1789–91)', *Isis*, 33 (1941), 315–25.

14. *Temple of Nature*, p. 54; 'Perhaps all the productions of nature are in their progress to greater perfection! an idea countenanced by modern discoveries and deductions concerning the progressive formation of the solid parts of the terraqueous globe, and consonant to the dignity of the Creator of all things.'

15. *Zoonomia*, II, 194: 'This idea is shown to our senses by contemplating the petrifactions of shells, and of vegetables, which may be said, like busts and medals, to record the history of remote times. Of the myriads of belemnites, cornua ammonis, and numerous other petrified shells, which are found in the masses of limestone which have been produced by them, none now are ever found in our seas'.

16. *Botanic Garden*, I, 5–6: 'There are likewise some apparently useless or incomplete appendages to plants and animals, which seem to show they have gradually undergone changes from their original state; such as the

stamens without anthers and styles without stigmas of several plants, . . . the halteres, or rudiments of wings of some two-winged insects; and the paps of male animals; thus swine have four toes, but two of them are imperfectly formed, and not long enough for use. . . . Perhaps all the supposed monstrous births of Nature are remains of their habits of production in their former less perfect state, or attempts towards greater perfection.'

17. Darwin also recognized the dangers of analogy. *Zoonomia*, I, vii: 'The great CREATOR of all things has infinitely diversified the works of his hands, but has at the same time stamped a certain similitude on the features of nature, that demonstrates to us, that the whole is one family of one parent. On this similitude is founded all rational analogy; which, so long as it is concerned in comparing the essential properties of bodies, leads us to many and important discoveries; but when with licentious activity it links together objects, otherwise discordant, by some fanciful similitude; it may indeed collect ornaments for wit and poetry, but philosophy and truth recoil from its combination.'

18. Primer, 'Erasmus Darwin's *Temple of Nature*'.

19. *Zoonomia*, II, 245–6.

20. *Zoonomia*, I, 76.

21. '*The Temple of Nature; or, The Origin of Society: A Poem with Philosophical Notes*. By Erasmus Darwin, M.D. F.R.S. Author of the *Botanic Garden*, of *Zoonomia*, and of *Phytologia*', *Edinburgh Review*, 2 (1803), 491–506.

22. Darwin, *Phytologia; idem, Botanic Garden*. On patient contacts, see E. Posner, 'Josiah Wedgwood's Doctors', *Pharmaceutical History*, 3 (1973), 2–5.

23. See King-Hele, *Letters of Erasmus Darwin*, pp. 180–2, 200–1 for Darwin's sympathetic treatment of the hypochondriacal James Watt. His comments on lunatics bear this out: *Zoonomia*, IV, 54–90.

24. Erasmus Darwin, *A Plan for the Conduct of Female Education in Boarding Schools* (London, 1797) (henceforth *Female Education*).

25. *Zoonomia*, II, 236: 'And all this exactly as is daily seen in the transmutation of the tadpole, which acquires legs and lungs, when he wants them; and loses his tail, when it is no longer of service to him.'

26. *Zoonomia*, II, 231.

27. *Zoonomia*, II, 233. The phrase 'exuberance of feeding' is interesting given the girth of Erasmus Darwin and his son, Robert Waring.

28. *Zoonomia*, I, 505.

29. *Zoonomia*, II, 234.

30. *Zoonomia*, I, 186.

31. King-Hele, *Doctor of Revolution*, pp. 40–2.

32. Philip C. Ritterbush, *Overtures to Biology: The Speculations of Eighteenth-Century Naturalists* (New Haven, Conn.: Yale University Press, 1964).

33. *Phytologia*, pp. 283ff.

34. *Temple of Nature*, p. 77.

35. *Temple of Nature*, p. 18.

36. *Botanic Garden*, pp. 79ff.

37. *Zoonomia*, I, 263.

38. Jacques Roger, *Les sciences de la vie dans la pensée française du XVIIIe siècle:*

La generation des animaux de Descartes à l'Encyclopédie, 1963; 2nd edn (Paris: Armand Colin, 1971); Christopher J. Lawrence, 'Early Edinburgh Medicine: Theory and Practice', in R.G.W. Anderson and A.D.C. Simpson, eds., *The Early Years of the Edinburgh Medical School* (Edinburgh: Royal Scottish Museum, 1976), pp. 81–94; *idem*, 'The Nervous System and Society in the Scottish Enlightenment', in Barry Barnes and Steven Shapin, eds., *Natural Order: Historical Studies of Scientific Culture* (Beverly Hills, Calif.: Sage Publications, 1979), pp. 19–40; Guenter B. Risse, 'The Brownian System of Medicine: Its Theoretical and Practical Implications', *Clio Medica*, 5 (1970), 45–51; Theodore M. Brown, 'From Mechanism to Vitalism in Eighteenth Century English Physiology', *Journal of the History of Biology*, 7 (1974), 179–216. Darwin had of course received most of his medical training in Edinburgh.

39. 'It would be curious to know (but he alone could have told us) the progress of your father's mind from the narrow Boerhaavian system, in which man was considered as an hydraulic machine whose pipes were filled with fluid susceptible of chemical fermentations, while the pipes themselves were liable to stoppages or obstructions (to which obstructions and fermentations all diseases were imputed), to the more enlarged consideration of man as a living being, which affects the phenomena of health and disease more than his merely mechanical and chemical properties. It is true that about the same time, Dr. Cullen and other physicians began to throw off the Boerhaavian yoke; but from the minute observations which Dr. Darwin has given of the laws of association, habits and phenomena of animal life, it is manifest that his system is the result of the operation of his own mind' (James Keir to Robert Waring Darwin, cited by C. Darwin, 'Preliminary Notice', in Krause, *Erasmus Darwin*, pp. 13–14).

40. *Zoonomia*, I, 92; cf. I, 37: 'I. The fibres, which constitute the muscles and organs of sense, possess a power of contraction. The circumstances attending the exertion of this power of contraction constitute the laws of animal motion, as the circumstances attending the exertion of the power of attraction constitute the laws of motion of inanimate matter. II. The spirit of animation is the immediate cause of the contraction of animal fibres, it resides in the brain and nerves, and is liable to general or partial diminution or accumulation. III. The stimulus of bodies external to the moving organ is the remote cause of the original contractions of animal fibres. IV. A certain quantity of stimulus produces irritation, which is an exertion of the spirit of animation exciting the fibres into contraction.' See Karl Figlio, 'The Metaphor of Organization: An Historiographical Perspective on the Bio-medical Sciences of the Early Nineteenth Century', *History of Science*, 14 (1976), 31–53 and *idem*, 'Theories of Perception and the Physiology of Mind in the Late Eighteenth Century', *History of Science*, 13 (1975), 177–212.

41. *Zoonomia*, I, 37.

42. *Zoonomia*, I, 96: 'Now as the sensorial power, or spirit of animation, is perpetually exhausted by the expenditure of it in fibrous contractions, and is perpetually renewed by the secretion or production of it in the brain and spinal marrow, the quantity of animal strength must be in a perpetual state of fluctation on this account.'

43. *Zoonomia*, II, v.

44. King-Hele, *Doctor of Revolution*, p. 238.

45. Darwin retained Haller's basic distinction between irritability and sensibility, which was overridden in the bridging concept of 'excitability' prominent in Edinburgh medicine. See E. Halévy, *The Growth of Philosophic Radicalism*, translated by Mary Morris, 1934 (Clifton, N.J.: Augustus M. Kelley, 1972), pp. 439–43 and Robert Hoeldtke, 'The History of Associationism and British Medical Psychology', *Medical History*, 11 (1967), 46–65.

46. Roger Smith, 'The Background of Physiological Psychology in Natural Philosophy', *History of Science*, 11 (1973), 75–123. Darwin saw himself advancing a secularized version of Hartley (*Zoonomia*, II, 199): 'The ingenious Dr. Hartley in his work on man, and some other philosophers, have been of the opinion, that our immortal part acquires during this life certain habits of action or of sentiment, which become forever indisoluble, continuing after death in a future state of existence; and add, that if these habits are of the malevolent kind, they must render the possessor miserable even in heaven. I would apply this ingenious idea to the generation or production of the embryon, or new animal, which partakes so much of the form and propensities of the parent.'

47. *Zoonomia*, I, 39. This, like similar definitions, is repeated verbatim in *Temple of Nature*.

48. *Zoonomia*, I, 45.

49. *Zoonomia*, I, 25.

50. *Zoonomia*, I, 18, 24.

51. *Zoonomia*, I, 25.

52. *Zoonomia*, I, 37: 'V. A certain quantity of contraction of animal fibres, if it be perceived at all, produces pleasure; a greater or less quantity of contraction, if it be perceived at all, produces pain; these constitute sensation. Sensation is an exertion or change of the central parts of sensorium, or of the whole of it, beginning at some of those extreme parts of it, which reside in the muscles or organs of sense.'

53. *Zoonomia*, I, 44: '2. Sensitive motions. That exertion or change of the sensorium, which constitutes pleasure or pain, either simply subsides, or is succeeded by volition, or it produces fibrous motions: it is termed sensation, and the sensitive motions are those contractions of the muscular fibres, or of the organs of sense, that are immediately consequent to this exertion or change of the sensorium.'

54. *Zoonomia*, I, 72: 'All our emotions and passions seem to arise out of the exertions of these two faculties of the animal sensorium. Pride, hope, joy, are the names of particular pleasures; shame, despair, sorrow, are the names of peculiar pains; and love, ambition, avarice, of particular desires; hatred, disgust, fear, anxiety, of particular aversions.'

55. *Temple of Nature*, note ii, p. 12.

56. *Zoonomia*, I, 44.

57. *Zoonomia*, II, 12; I, 67, 73.

58. *Zoonomia*, I, 40, 69.

59. *Zoonomia*, I, 272.

60. *Zoonomia*, I, 44.

61. *Zoonomia*, II, 153.

62. *Zoonomia*, I, 273; II, 11.

63. Darwin's views on imitation are spelt out most graphically in the *Temple of Nature*. Thus, canto 3, lines 285–8:

Whence the fine power of IMITATION springs,
And apes the outlines of external things;
With ceaseless action to the world imparts
All moral virtues, languages, and arts.

and canto 3, lines 331–4:

Hence to clear images of form belong
The sculptor's statue, and the poet's song,
The painter's landscape, and the builder's plan,
And IMITATION marks the mind of Man.

The education of the mind through imitation and habit is a key theme of Darwin's *Female Education*.

64. *Zoonomia*, I, 376: 'Man is termed by Aristotle an imitative animal; this propensity to imitation not only appears in the actions of children, but in all the customs and fashions of the world: many thousands tread in the beaten paths of others, for one who traverses regions of his own discovery. The origin of this propensity of imitation has not, that I recollect, been deduced from any known principle; when any action presents itself to the view of a child, as of whetting a knife, or threading a needle, the parts of this action in respect of time, motion, figure, are imitated by a part of the retina of his eye; to perform this action therefore with his hands is easier to him than to invent any new action, because it consists in repeating with another set of fibres, viz. with the moving muscles, what he has just performed by some part of the retina.'

65. *Zoonomia*, I, 38, 61, 76; cf. I, 13: 'Association is an exertion or change of some extreme part of the sensorium residing in the muscles or organs of sense, in consequence of some antecedent or attendant fibrous contractions.'

66. *Zoonomia*, I, 66.

67. *Temple of Nature*, p. 113.

68. *Zoonomia*, I, 201.

69. *Zoonomia*, IV, 66.

70. *Zoonomia*, II, 11f.

71. *Zoonomia*, IV, 71.

72. *Zoonomia*, IV, 84.

73. *Zoonomia*, II, 255.

74. *Zoonomia*, III, 6f. See Joseph Needham, *A History of Embryology*, 1934, 2nd edn (Cambridge: Cambridge University Press, 1959).

75. *Zoonomia*, II, 263.

76. See Bentley Glass, 'Heredity and Variation in the Eighteenth-Century Concept of Species', in *idem et al.*, *Forerunners of Darwin*, pp. 144–72 and James Blondel, *The Strength of Imagination in Pregnant Women Examin'd* (London, 1727). The 'Imaginationist' position is also denied in I. Bellet, *Lettres sur le pouvoir de l'imagination des femmes enceintes* (Paris, 1745). For discussion,

see P.G. Boucé, 'Imagination, Pregnant Women, and Monsters in Eighteenth Century England and France', in G.S. Rousseau and Roy Porter, eds., *Sexual Underworlds of the Enlightenment* (Manchester: Manchester University Press, 1987) and G.S. Rousseau, ed., *Tobias Smollett: Essays of Two Decades* (Edinburgh: T. and T. Clark, 1982), pp. 164ff.

77. *Zoonomia*, II, 264: 'It would appear, that the world has long been mistaken in ascribing great power to the imagination of the female.' Monsters could equally be products of the male imagination (*Zoonomia*, II, 266). See Roy Porter, 'Monsters and Madmen in Eighteenth Century France', in D. Fletcher, ed., *The Monstrous* (Durham: Durham French Colloquies, no. 1, 1987), pp. 83–103 and Valerie Grosvenor Myer, 'Tristram and the Animal Spirits', in *idem*, ed., *Laurence Sterne: Riddles and Mysteries* (London: Vision Press, 1984), pp. 99–114.

78. *Zoonomia*, II, 270.

79. *Zoonomia*, II, 273; cf. II, 210: 'The living filament is a part of the father and has therefore certain propensities or appentencies, which belong to him; which may have been gradually acquired during a million of generations.'

80. *Zoonomia*, II, 263. Darwin compares the male role to a sculptor's mind and chisel acting upon raw marble.

81. *Zoonomia*, II, 269: 'In respect to the power of the imagination of the male over the form, colour, and sex of the progeny, the following instances have fallen under my observation, and may perhaps be found not very unfrequent, if they were more attended to. I am acquainted with a gentleman, who has one child with dark hair and eyes, though his lady and himself have light hair and eyes; and their other four children are like their parents. On observing this dissimilarity of one child to the others he assured me, that he believed it was his own imagination, that produced the difference; and related to me the following story. He said, that when his lady lay in of her third child, he became attached to a daughter of one of his inferior tenants, and offered her a bribe for her favours in vain; and afterwards a greater bribe, and was equally unsuccessful; and the form of this girl dwelt much in his mind for some weeks, and that the next child, which was the dark-eyed young lady above mentioned, was exceedingly like, in both features and colour, to the young woman who refused his addresses.'

Darwin noted how, in mixed marriages, where the husband was white and the mother negroid, the offspring progressively took on the colour of the father over the generations (*Zoonomia*, II, 256). His doctrine of the power of imagination probably influenced Mary Godwin Shelley. See Anne Mellor, '*Frankenstein*: A Feminist Critique of Science', in George Levine, ed., *One Culture: Essays in Science and Literature* (Madison: University of Wisconsin Press, 1987).

82. *Phytologia*, p. 33.

83. *Zoonomia*, II, 13–14.

84. *Zoonomia*, II, 20ff.

85. *Zoonomia*, II, 138ff.

86. *Female Education*, p. 86.

87. *Zoonomia*, II, 274. See also *Phytologia*, p. 557: 'As all animal existence

must perish in process of time ... it is so ordered that as soon as any organized being became less irritable and less sensible, and in consequence feeble or sickly, that it is destroyed and eaten by other more irritable and more sensible, and in consequence more vigorous organized beings.'

88. *Temple of Nature*, note xi, p. 43.

89. *Phytologia*, p. 350. Cf., for Lamarck as not primarily an evolutionist, L.J. Jordanova, *Lamarck* (Oxford: Oxford University Press, 1984), pp.71–83.

90. *Zoonomia*, II, 325.

91. *Zoonomia* I, 188.

92. *Zoonomia*, I, 198.

93. *Zoonomia*, II, 240.

94. *Temple of Nature*, p. 147.

95. *Temple of Nature*, p. 87.

96. *Temple of Nature*, p. 24.

97. *Temple of Nature*, p. 25.

98. *Temple of Nature*, p. 15; cf. p. 107:

> Hence when the inquiring hands with contact fine
> Trace on hard forms the circumscribing line;
> Which then the language of the rolling eyes
> From distant scenes of earth and heaven supplies;
> Those clear ideas of the touch and sight
> Rouse the quick sense to anguish or delight;
> Whence the fine power of *IMITATION* springs,
> And the outlines of external things...

99. *Temple of Nature*, p. 107.

100. *Temple of Nature*, p. 86. Cf. *Zoonomia*, I, 263–4: 'It was before observed how much the superior accuracy of our sense of touch contributes to increase our knowledge; but it is the greater energy and activity of the power of volition ... that marks man, and has given him the empire of the world.'

101. Mayr, *Growth of Biological Thought*, p. 340.

102. Paul H. Barrett and Howard E. Gruber, eds., *Metaphysics, Materialism, and the Evolution of Mind: Early Writings of Charles Darwin* (Chicago: University of Chicago Press, 1980). For the wasp, see p. 17 (*M*, 63e): 'Newport says Dr. Darwin mistaken in saying common wasp cuts off wings of flies from intellect, but it does it always instinctively or habitually. – good Heavens it is disputed that a wasp has this much intellect, yet habit may make it act wrong, as I have done when taking lid of ⟨right⟩ side of tea chest, when no tea.'

For other direct references or allusions, see pp. 6, 8, 13, 21 and 25 (*M*, 104–6), where there appears this reflection, highly reminiscent of Erasmus Darwin: 'August 24th. As some *impressions* (Hume) become unconscious, so may some *ideas*, i.e., habits, which must require idea to order muscles to do the action. ? is it the impression becoming *very often* unconscious, which makes the *idea* unconscious, if so (think of this) study what impressions become unconscious those which are viewed with little interest, & those which are viewed very often. – former do not give rise to ideas so much as objects of interest. – do. I was much struck with observing how the Baboon / ⟨Macaco⟩ Cyanocephalus Sphynx Linnaeus / constantly moved the skin of

forehead over eyes, at every motion & look / turn / of the head. I could not
perceive/any/distinct *wrinkle*, but such movements in skin of eyebrow
important analogy with man.'

And consider the reference on p. 37 (*M*, 157) to inherited drunkenness:
'Has my father ever known ⟨intemperance⟩ disease in grandchild, when
father had not had it, but where grandfather was the cause of his intem-
perance. ⟨No⟩ Cannot say. –'

All this seems to give the lie to Charlies Darwin's statement, 'There is a
good deal of psychology in the "Zoonomia", but I fear that his speculations
on this subject cannot be held to have much value' ('Preliminary Notice', in
Krause, *Erasmus Darwin*, p. 104).

103. Janet Browne, 'Darwin and the Expression of the Emotions', in David
Kohn, ed., *The Darwinian Heritage* (Princeton, N.J.: Princeton University
Press, 1985), pp. 307–26.

104. Barrett and Gruber, *Metaphysics, Materialism, and the Evolution of Mind*,
p. 16 (*M*, 57): 'To avoid stating how far, I believe, in Materialism, say only
that emotions, instincts degrees of talent, which are hereditary are so because
brain of child resembles parent stock. – (& phrenologists state that brain
alters).'

105. C. Darwin, 'Preliminary Notice', in Krause, *Erasmus Darwin*, p. 46.

106. William Paley, *Natural Theology; or, Evidences of the Existence and
Attributes of the Deity, Collected from the Appearances of Nature* (London, 1802),
chs. 24–25.

107. McNeil, *Under the Banner of Science.*

108. 'Art. XX . . .', *Edinburgh Review*, 2 (1803), 499: 'Another error, nearly
akin to that we have been describing, but which deserves particular notice as
fatally characterizing many of the metaphysical speculations of Dr Darwin,
arises from constantly blending and confounding together the two distinct
sciences of matter and of mind. In this censure, we would not be understood
as referring directly to that hypothesis of materialism, which is everywhere
assumed by him with the utmost confidence. Ignorant as we are of the nature
of matter, beyond a few of its sensible qualities, it would be rash and idle to
limit dogmatically the modifications of which it may be susceptible. For
similar reasons, indeed, we cannot but regard it as still more rash and
unphilosophical, to assert the identity of substances between the known
qualities and attributes of which no sameness of analogy have yet been
recognized.' See Thomas Brown, *Observations on the 'Zoonomia' of Erasmus
Darwin, M.D.* (London, 1798) and Norton Garfinkle, 'Science and Religion in
England, 1790–1800: The Critical Response to the Work of Erasmus Darwin',
Journal of the History of Ideas, 16 (1955), 376–88.

109. McNeil, *Under the Banner of Science.*

110. *Temple of Nature*, p. 159:

> So human progenies, if unrestrain'd,
> By climate friended, and by food sustain'd,
> O'er seas and soils, prolific hordes would spread
> Erelong, and deluge their terraqueous bed;
> But war, and pestilence, disease, and dearth,

Sweep the superfluous myriads from the earth . . .
The births and deaths contend with equal strife,
And every pore of Nature teems with Life;
Which buds or breathes from Indus to the Poles,
And Earth's fast surface kindles as it rolls!

See King-Hele, *Letters of Erasmus Darwin*, p. 128; see also pp. 94, 166, etc.

111. Neil McKendrick, 'Josiah Wedgwood and Factory Discipline', *Historical Journal*, 4 (1961), 30–55.

112. *Zoonomia*, I, 187: 'But all those actions of men or animals, that are attended with consciousness, and seem neither to have been directed by their appetites, taught by their experience, nor deduced from observation or tradition, have been referred to the power of instinct. And this power has been explained to be a divine something, a kind of inspiration; whilst the poor animal, that possesses it, has been thought little better than a machine!'

2

Nature's powers: A reading of Lamarck's distinction between creation and production

LUDMILLA JORDANOVA

Every world view involves an emotional, as well as an intellectual apprehension of nature.

John C. Greene[1]

The distinction between creation and production lies at the centre of many world views, and above all of Christianity. In the Nicene Creed, which declares, 'I believe in one God, the Father, Maker of heaven and earth, ... And in the Lord Jesus Christ, ... Begotten not made ...', there is evidently considerable slippage in the way the verb 'to make' is used, as it can express both divine and non-divine actions.[2] None the less, for centuries Western intellectual traditions have meditated upon both the differences and the similarities between the kinds of making undertaken by people, nature and God(s), respectively. Natural philosophical traditions were no exception. At the same time, the distinction has a history. The eighteenth century has been seen as a decisive turning point in the novel, direct and sustained use of a language of creation in relation to human performance that it witnessed.[3] The century was also marked by fundamental new interpretations of the notion of production.

In setting Jean-Baptiste de Lamarck in the context of such questions, I intend to achieve two goals. The first is to offer a reading of certain key aspects of his natural philosophy in the light of contemporary texts of a rather different kind: natural theological writings on the one hand, and Mary Shelley's novel *Frankenstein* (1818) on the other. The second is to reveal the ways in which the elements of world views articulate with one another and the forms that 'an emotional ... apprehension of nature' takes on. Both goals entail showing the stakes invested in the distinction between creation and production and the historically specific resonances it generated. Accordingly, the rest of this essay is divided into four sections: the first offers an

71

exposition of Lamarck's ideas on the subject; the second contrasts these with the natural theological writings of William Paley and François-René de Chateaubriand; the third introduces *Frankenstein* in the context of the claim that it is useful to extend our analysis of creation/production by considering creation myths – the narratives that explain different kinds of making. In the final section, the threads are drawn together.

Lamarck employed a distinction, which is the subject of this essay, between the creativity of God and the productivity of nature.[4] The pair of terms, creation/production, took on a special significance in the late eighteenth and early nineteenth centuries. They evoked authority, order and hierarchy; they spoke about origins, sequences and the historical underpinnings of rulership; they described actions and processes as apparently diverse as work, reproduction, authorship and divine intervention. They served at once to link and to discriminate between human, natural and divine operations. The significance of these terms can be better understood if we place Lamarck in a wider cultural setting.

Reading Lamarck in the context of contemporary natural theology sharpens our sense of what he was trying to do. His natural philosophy and natural theology seem almost deliberate inversions of each other. An examination of *Frankenstein* also facilitates our understanding of Lamarck's project. The literary licence Shelley took, by contrast, for example, with Lamarck or Paley, is important because it enabled her to open up and make explicit issues that surely troubled them but were far harder for them to write about. Yet she offered no settled answers to the problems she raised. Her novel neatly reveals the facets of the creation/production distinction present in European culture. It does so, furthermore, in a highly condensed, potent form. Her work also shows that the distinction was important, not just at an intellectual level, but because it touched the nature of personal identity and cosmic meaning.

Thus this essay is not about influences, but about distinct contemporaneous styles of thinking, their wider implications and their associations with transformism. It is, I argue, of the utmost importance that we comprehend what Lamarck was arguing against, namely, natural theological traditions. Contrary to common opinion, these were well established in eighteenth-century France. L'Abbé Pluche's *Le spectacle de la nature* (1732–50) was, we know, widely read. The naturalist Réaumur, whose work Lamarck knew, employed the argument from design as did Bernardin de Saint-Pierre, briefly a colleague of Lamarck's and a best-selling author.[5] Above all, the notorious hostility between Cuvier and Lamarck included profound disagreements about religious matters. Although careful not to proclaim himself a natural theologian, Cuvier's ideas were widely understood by his contemporaries to have precisely this cast. And it has recently been claimed that 'Cuvier's functionalism ... gave support to the traditional argument from design'. Cuvier's protests against reducing 'Nature to a sort of slavery, into which ... her Author is far from having enchained her', on the grounds that 'the world itself would become an indecipherable enigma', articulate the central tenets of a natural theological world view in direct opposition to Lamarck and his sympathizers.[6] We can, therefore, confidently assert the historical appropriateness of juxtaposing Lamarck and natural theological writers.

There are additional reasons for setting Lamarck in the context of Paley, who was, after all, totally explicit about his metaphysical and epistemological commitments. Throughout his *Natural Theology*, which appeared in England in 1802 and in a French version published in Switzerland in 1804, Paley consistently argued against the power of habit – a lynchpin of Lamarck's transformism. Inadvertently, then, Paley took issue with Lamarck, just as Lamarck, deliberately, criticized natural theology when he redefined God and nature. Furthermore, we can usefully compare Lamarck with Chateaubriand whose *Génie du Christianisme*, also published in 1802, created a sensation in France, and who discussed the proof of the existence of God from the wonders of nature.[7] There is no evidence that Lamarck read either Paley or Chateaubriand, although he could hardly have been unaware of the latter's existence and ideas. He rarely cited other authors, few of his papers are extant, and only the meagre evidence of which books he owned at his death remains.[8] We know little of Lamarck's personal beliefs, but it is fair to assume that he was not a devout Catholic. It is not necessary to show direct links between these authors for the following arguments to work. I will be considering the intellectual possibilities available in a given culture, not seeking to establish lineages of ideas.

Lamarck's philosophy of nature

Lamarck's science always had a strongly reflective dimension to it, beginning with his earliest botanical work and continuing into his last publications. This aspect of his thought has received surprisingly little attention. His reputation among historians is built on his botany, his theories of organic change and his researches on invertebrates; other aspects of his natural philosophy have been neglected.[9] The meteorology, geology and 'psychology' remain little studied, as does Lamarck's last book the *Système analytique des connaissances positives de l'homme* (1820). This was not a new work in any significant sense. It contains long quotations from earlier published work, especially *Philosophie zoologique* (1809) and the introduction to *Histoire naturelle des animaux sans vertèbres* (1815). It also reproduced parts of a set of articles Lamarck wrote for Deterville's *Nouveau dictionnaire d'histoire naturelle* (1816–19). There he defined terms such as *Espèce, Faculté, Habitude, Homme, Idée, Intelligence, Jugement* and *Nature*.[10]

Lamarck expended considerable effort on elucidating key words. The results are revealing because they allow us to reconstruct the larger system of his natural philosophy. While his definitions may appear rather simplistic in retrospect, they fulfilled a significant function in clearing the ground, so to speak, in order that transformism could be planted in the right soil. Being sympathetic to the *idéologues*, Lamarck was aware that sciences were well-made languages, and he endeavoured to put this precept into practice.

Lamarck's ideas are generally structured into contrasting pairs, giving his thought a strongly dichotomous character. The universe was divided into matter and nature; and matter into living and inert; living things were of two kinds – animals and plants. The organization of the *Système analytique* to some extent reproduces such a structure. It centres on man and is divided into two principal parts devoted to phenomena inside and outside man respectively. The latter were then divided into phenomena, created and produced. There were two sets of created phenomena, matter and nature, and likewise two categories under production, consisting of inorganic and organic objects, and so on. Lamarck employed other important dichotomies, such as his distinctions between properties and faculties (the former are found in inert, the latter only in organic bodies), active and passive (only nature is active, matter remains passive), invention and imagination (the first is a useful skill for finding new relations between objects in contrast to the second, which results merely in the formation of images), and fiction and reality (ideas not rigorously grounded in observation as opposed to ideas rooted in nature).

Because Lamarck was evidently attempting to produce a consistent natural philosophy, it was necessary for him to deal with God, nature and man. In order to show what kind of a natural philosophy this was, and what its implications are for our understanding of his transformism, a theory of natural production, it will be convenient to examine Lamarck's definition of 'nature', published in Deterville's dictionary in 1818 and reprinted in the *Système analytique* in 1820. Lamarck began by establishing that nature was an 'order of things', a 'constantly active power' of which man was a part and which should therefore be of primary importance to him.[11] Like matter, nature was a created object that it was possible to know through observation. But Lamarck stressed the activity of nature, by which he meant that natural laws regulated all phenomena. Nature was, however, limited and – as he often put it – blind.

Other writers, according to Lamarck, used the word nature too liberally and vaguely – hence his attempts to clarify it. He had to purge the concept of unwelcome associations. It was not, he emphasized, an intelligence or an individual being – terms which personified nature in unacceptable ways – but something that acts out of necessity. Nature was 'a blind power, lacking intention or goal'.[12] It followed, for Lamarck, that God did not create natural objects directly, but did so through the initial creation of nature and matter. Nature, a system of dynamic laws, then produced all observable objects by acting upon passive matter.

God could have created everything directly – that is his prerogative – but evidence in nature indicated that direct creations have not occurred. The evidence that Lamarck had in mind was indications of change over time. While physical objects appeared permanent, signs of change were present for those who knew how to look for them. If natural objects were as old as nature itself, then it might be reasonable to assume them to have been created directly by God; whereas if they are subject to continual change, they must be understood as products of nature. Furthermore, if natural bodies were created directly, Lamarck pointed out, nature itself would have no real existence or significance. In fact, science must concern itself with 'the power of nature'. To argue this point, Lamarck chose the example of the 'specious idea . . . concerning the original creation and immutability of species'.[13] People espouse it because they cannot themselves directly perceive organic change, given the short life span of human beings. But close study of the 'monuments' on the earth's surface helps to dispel such a view. From these we learn of the continual activity of nature, evoked for the reader by geological and geographical examples.

For Lamarck, this ceaseless change was far from random. It could be resolved into two components: the destruction of bodies and their re-elaboration. This topic lay at the heart of his matter theory and had been explored at length in his earlier works *Recherches sur les causes des principaux faits physiques* (1794) and *Mémoires de physique et d'histoire naturelle* (1797). Such change was invoked in the service of Lamarck's assertion that there was a 'general power' – nature itself – constantly acting.[14] But while Lamarck stressed the perpetual destruction and renewal of nature, he was equally concerned to show that this 'great power' was limited and lawlike. He argued that characterizing nature in these terms is important both for its own sake and so that we can determine how to study it. Lamarck was quick to point out that his own work had benefited from such a broad understanding of nature. It had, for example, enabled him to discover and use a 'general plan' of nature to establish a natural order among invertebrates.

Lamarck's next move was to clarify the relationship between nature and matter. Nature can neither create nor destroy matter, only modify it continually. Again, the possibility that nature might be construed as a reasoning being was set aside by considering the factor of time. Nature, Lamarck said, does nothing except through time, which is unlimited and universal. These characteristics are in explicit contrast to acts of creation, which are without duration. Lamarck was particularly concerned about the danger of projecting human fantasies on to nature:

> It is to this blind power [i.e. nature], everywhere limited and subordinated, which, however great it may be, can do nothing but what it actually does; which exists, in short, only by the will of the supreme author of everything; it is to this power, I say, that we attribute intention, design, resolution in its acts.[15]

Such attributions were, for Lamarck, illegitimate.

So far we might summarize Lamarck's philosophy by saying that he argued against a natural theological perspective by carefully separating God from nature, creation from production. While natural theologians accepted these distinctions, they, unlike Lamarck, reconnected the dichotomies because they saw in nature the signs of God's creative power. Furthermore, Lamarck rigorously denied that nature can be seen in personal terms, although it was possible for him on occasion to refer to nature as 'this communal mother'. This was, I believe, a mere figure of speech, but it could well have been a rhetorical device knowingly used, a 'slip', or even a sign of the recalcitrance of language in being inevitably anthropomorphic.[16]

Lamarck's argument for the existence of limited 'orders of things' could be applied to specific natural phenomena such as 'life'. Like nature, life was not purposive or an intelligent being, just a set of laws giving rise to organic phenomena. Yet expounding his views on life was not Lamarck's purpose in defining 'nature': it was to differentiate nature from God on the one hand and from the physical universe on the other. Those who confuse God and nature, confuse the clockmaker with the clock, he asserted. To move from clock to clockmaker involved a deliberate and dramatic shift of conceptual level from natural effect to supernatural cause or agent. Lamarck objected to such moves. Natural theologians agreed that watch and watchmaker should be kept separate, but they wished to link them by treating nature as the signature of God. Thus when Lamarck separated watch and watchmaker, it was to banish the latter from the domain of science; when natural theologians did so, it was to distinguish between cause and effect, which, although distinct, are also of necessity linked. Therefore, Lamarck also rejected any suggestion that science should concern itself with the clockmaker. The very character of nature itself suggested to him that its workings were not the *direct* expression of divine will. Natural theologians agreed that maker and made were two different orders of being, but for them one was meaningless without the other. According to Lamarck, the difference in levels was crucial because one was accessible to human knowledge, whereas the other was not.[17] Lamarck thus implicitly rejected the idea that makers necessarily inscribe their marks on the end-product – the central tenet of natural theology.

Lamarck's distinction between nature and God was, of course, largely epistemological. But it signalled a larger style of thought. For him the different levels of creation and production were so profoundly distinguished – the one pure will acting out of time, the other visibly labouring – that to confuse them was to make a major category mistake. Nature does not contain evidence of another order of being; it merely reveals its own laws. For Lamarck this undercut the idea of God as author or nature as a book. His use of phrases like 'supreme author' was, I would suggest, mere convention.

Like other natural philosophers, Lamarck had to account for the origin and nature of activity. Matter was purely passive, all activity comes from nature. Yet in truth he did not account for the origin of activity beyond invoking a divine creative act in the beginning that made nature this way. For Lamarck, activity in nature derived from motion on the one hand, and laws governing motion on the other. Nature acts through space and time – significantly, Lamarck calls these the 'metaphysical entities which together constitute nature' – and the important thing about them was that they could be understood by the human mind in a way that God could not be.[18]

Having reiterated his definitions of nature, the universe, and so on, Lamarck shifted key to take up one of his favourite themes, the *moral* importance of the study of nature, including man. He did not himself use the term 'moral' here; instead, he claimed that knowledge of nature would help man know about his physical being, his preservation, his behaviour, and his relations with his fellow men.[19] Whereas Lamarck kept God and nature neatly distinct, man and nature are hopelessly intertwined. Man is part of nature because he is subject to its laws, yet he is the only part of nature capable of understanding it. He ought to study nature for his own well-being, yet he perpetually

misunderstands and projects his feelings on to nature. Lamarck used
the example of monstrosities to illustrate this last point:

> Certain irregularities in nature's acts, certain monstrosities which
> appear contrary to the ordinary workings of nature, reversals in
> the ordering of physical objects . . . are nonetheless the proper
> product of nature's laws and of the prevailing circumstances. . . .
> The word *chance* expresses only our ignorance of causes.[20]

Lamarck's repudiation of natural disorder does not, of course,
mean that he saw everything as being under the governing hand of
God. On the contrary, it expresses his commitment to coherent explanation based on universal natural laws. Chance has no place in
such laws. We attribute disorder to nature when something annoys
or distresses us. What follows in Lamarck's discussion of 'nature' is,
not surprisingly, an elaborate justification of the study of nature by
that creature whose place within it, and, as scientist, outside it, is so
complex. The programme set out here and in many other places in his
writings linked natural history with natural philosophy. The empirical
study of nature – Lamarck did much natural historical work during
his life – only made sense within a larger philosophical framework.
This broad programme engaged with a wide range of issues and was
a blend of epistemological, metaphysical and moral considerations.
He found social questions in general and ethical ones in particular to
be integral to his science.

Lamarck's 'psychology' was central to his philosophy of nature. It
is best understood in terms of contemporary interest in the science of
man. Indeed, he shared many of the central interests of that field,
especially as articulated by his contemporaries, the *idéologues*, who
placed themselves in the traditions of Locke and Condillac by insisting
that all knowledge derives from the senses.[21] Lamarck's commitment
to this position is clearly vital, as it spurred him to think through a
naturalistic account of the nervous system, and to reject any mental
faculties, such as will and imagination, not strictly compatible with
such an account. He downgraded those human activities that come
under what he called 'the field of fictions'. By contrast, knowledge of
nature acquired the status of the paradigm of knowledge, and all of it
was valuable to human beings. Lamarck took a dim view of those
who strayed beyond the 'the field of reality', declaring: 'it is forbidden
for man to leave it'. He moved effortlessly from what it was possible
for human beings to know, given their organization and that of nature,
to how they ought to conduct themselves generally. In so doing, his
natural philosophy spoke to a wide range of social, moral and ethical
questions. It has to be said that Lamarck was no moral philosopher;
he displayed a simplistic view of what the study and knowledge of

Lamarck had social concerns in mind but Darwin (just wrote it
as he saw it.

nature promised. 'Man', he said, 'would conform more easily to natural laws, and escape from evil of all kinds.'[22]

Through his dialogue with other perspectives on God, man and nature, Lamarck attempted to elaborate a world view in which his transformism played an integral part. Organic change was a paradigm case of nature's labours, of the successive production of bodies, to be explained naturalistically in terms of adaptation, habit, and so on. In this sense Lamarck gave nature a history even if his understanding of history lacked elements such as contingency. We can, I think, legitimately see Lamarck as both consciously constructing an alternative to a number of positions that he saw as determined by prior theological commitments, and as offering an entire philosophical package, a blueprint for society as well as for science – in other words, a cosmology.[23]

In Lamarck's account, the history of nature has no real beginning. The history of life *has*, even if no date can be assigned to the appearance of the first organized being. He consistently emphasized the antiquity of the earth, but he had no means of conceptualizing the initial act that brought the universe into being. We might say that this was because he was content to let the traditional biblical account stand, possibly to avoid giving unnecessary offence, or we might suggest that conceptualizing a unique event within a naturalistic perspective designed to explain routine occurrences is extraordinarily difficult in itself. Either way, Lamarck did not offer a new view of the very beginning of existence. On the other hand, he did suggest that new beginnings were *constantly* occurring. His commitment to spontaneous generation – the ceaseless re-enactment of the history of life – to organic change and to the perpetual decay and reconstruction of physical bodies rendered the notion of a first, unique beginning less significant than it was in the Judaeo-Christian tradition.

The little beginnings characteristic of Lamarck's natural philosophy had no meaning for the human condition; they simply arose from the normal operations of nature's laws. Ultimately, Lamarck was able to derive a moral code from his philosophy of nature, but spontaneous generation contained no human mythic implications in the way that, for example, the story of the Garden of Eden does. He did not offer a strictly comparable account, although it could be seen as a loose alternative to traditional creation myths. But it offered a different kind of explanation, by emphasizing work-a-day natural processes rather than cosmic meanings.

As part of his cosmology, Lamarck attempted to account comprehensively for a wide range of human phenomena, including suffering. I have already noted his fierce exclusion of chance and error from nature. From this it followed that, because man is a part of nature, things that go amiss for him cannot be explained in these terms. Nor, of course, can they be explained by some outside agency such as a devil; they have to be understood naturalistically. It must be admitted that Lamarck's attempts to produce a naturalistic account of social inequality, for example, were crude and simplistic. Yet this may be less important than the fact that he attempted to deal with such issues within his natural philosophical framework.

An interesting example is Lamarck's explanation of suicide. He claimed that it must result from a disturbance of the nervous system. Yet because aversion to death was, for him, a basic propensity (following well-established traditions in political philosophy in which the preservation of life was an important instinct), he was reduced to saying that 'suicide is the result of an unhealthy state in which the ordinary laws of nature are inverted'.[24] In Lamarck's *own* terms this was hardly a satisfactory response because he presented natural laws as fixed, universal and incapable of inversion.

Lamarck's attempts to account for inequality naturalistically were probably influenced by Rousseau. Like Rousseau, Lamarck criticized the social state for its inequality and artificiality, and for the tendencies of groups to exploit each other. Rousseau's *Discourse on the Origins of Inequality* (1754) had not only offered a model of how gradual change can be explained naturalistically, but had expressed powerful images of nature and society and the antagonism between them.[25] Lamarck and Rousseau also shared an interest in how man came to be as he is, and were both deeply ambivalent about the human race. Indeed, Lamarck's transformism is notable for the ambivalent status it gave to man. In the order of production, man is very much a newcomer. Yet, at the same time, he occupies a special position as the only part of nature capable of understanding it. This privilege came 'naturally'; it is a mere fact of organic superiority. However, man is also uniquely destructive. Only he can cause the extinction of other species. Man represents the summit of nature's powers and the only element of disruption. Lamarck thought of the negative side of human beings as the product of society rather than of nature. And because nature is still in the process of unfolding, the human race cannot be the goal towards which nature/God works. Man's ascendency would prove temporary if more perfect beings emerged.

Although, formally, God takes priority in Lamarck's philosophy, He is absent from the history of nature and of life. In so far as there is conflict in the world, it is within nature – among men and between human beings and the rest of nature. This conflict can be used pragmatically to help achieve a better understanding of where the human race went wrong, but it has no cosmic meaning. Lamarck's refusal to give unquestioned centrality to human beings was consistent with his resistance to any suggestion that nature is a book to be decoded in the service of revealing its author. Nature embodies no personality – it remains blind, limited, law-like and labouring. There is no book of nature and hence neither authors nor readers.

Natural theology

Although many of the central arguments found in natural theological writings go back to the ancient world, it was not until the second half of the seventeenth century that the genre became well-established. At a popular level it persisted throughout the nineteenth century, with the *Bridgewater Treatises*, published in the 1830s, boasting among their authors some of the most distinguished names of the period. Paley's *Natural Theology* represents a cogent summary of the central tenets of a natural theological approach that flowered in the eighteenth century.[26] By the time of its publication in 1802, there were not only a number of well-established and distinct traditions of natural theology, but also a body of opinion strongly critical of it. A full history of natural theology in the eighteenth century remains to be written, and until it is, placing Paley in his full intellectual context remains difficult. However, his extensive influence together with the systematic way he approached his subject qualifies him as an exemplar of one style of thinking about God as the author of nature. Another style is represented by Lamarck's compatriot, Chateaubriand, and it is also appropriate to draw from his work in order to reveal the range of themes implicit in natural theology.

Natural theology took as its central image God the designer or contriver, and it was to this that Lamarck intentionally created an alternative. Natural theological language implied an analogy between God's creative acts and man's productive ones, or, strictly speaking, between the relationship God has with nature and the ones human beings have with artefacts. Although it had powerful rhetorical potential, the analogy was by no means unproblematic, as can be seen, for example, in the Leibniz–Clarke controversy, where Leibniz objected to an argument that 'supposes God to perceive things ... by a kind of perception, such as that by which men fancy our soul perceives what passes in the body'. At the end of the century, however, Thomas Reid revived the direct analogy between man and God, which had been subject to sustained attack, when he likened human intentional movements to the will of God.[27] In many other natural theological writings the analogy was used with little self-consciousness. By inviting their readers to imagine God as artisan and master craftsman, natural theologians blurred the very boundaries between creation and production that Lamarck was so concerned to strengthen. At the same time they conveyed a sense of the profound gap between God and human beings. There were a number of different ways in which natural theologians achieved this double effect of similarity and difference, often generating considerable conceptual tensions in the process.

Chateaubriand, a romantic, reactionary, aristocratic novelist and diplomat, who was also deeply committed to Catholicism, came to natural theology with a set of assumptions quite distinct from those of the middle-class, Protestant, rationalist divine and moral philosopher Paley. Chateaubriand relied heavily on visual reactions to nature as the trigger of sublime thoughts of an emotional and spiritual kind. The wonders of nature draw human beings, nature and God together in a transcendent movement. Nature is dense with visual indicators of the divine presence. It is also conceptualized as a language:

> Among men, tombs are the pages of their history; nature, by contrast, only prints on living things; she/it needs neither granite nor marble to make eternal that which she/it writes.[28]

Behind the inscribing hand of nature lay the ultimate author, God. Here Chateaubriand could unite God, man and nature by showing how they all write, yet he could also show that they do so in different ways. God, like man, makes things, but unlike man, His productions are not limited.

Chateaubriand used natural signs to link human life to its providential Creator by showing how the seasons have their markers that were recognized by people long before formal calendars existed. He characterized such knowledge as a form of *divination*. Chateaubriand's natural world is the occasion for human sensual pleasure, a temple for spiritual contemplation and a source of knowledge of basic rhythms and processes. It carries meaning in a number of different ways; it signifies at many different levels. Chateaubriand also used the familiar arguments about adaptation, design and purpose, yet he deployed none of the artisanal or technological vocabulary that is so central to Paley's writings. Hence his description of fish as 'real hydrostatic machine[s]' stands out as exceptional.[29]

In sum, Chateaubriand presented the similarities and differences between the human and divine by using the powerful visual impact of nature to draw the mind from that which is clearly and intentionally made to the being capable of such creation. That for him God's role was associated with sensual and spiritual delight is revealed in his choice of the word *ordonnateur*, which includes, among other more sombre meanings, organizer of festivites and master of ceremonies. Yet Chateaubriand's was no domesticated God – He is in places of wild beauty and of solitude, a God of the desert not of the workshop. The contrast between Chateaubriand and Lamarck could not be more marked. Lamarck characterized nature principally as 'limited', because produced, in always being subject to definable laws, whereas Chateaubriand refused any such constraint on the freedom of God's creativity, which he *saw* everywhere. In an apt image, Lamarck, whose descriptions lack visual richness, presented nature as blind. The blindness of nature suggests that it is self-contained; unable, as it were, to look outside itself. Nature is also without intention and hence incapable of generating the providential signs that Chateaubriand discovered in abundance.

While Paley, too, conveyed wonder and pleasure in his arguments, his images are quite different from Chateaubriand's. He also stands in a somewhat different relationship to Lamarck in that many of his arguments constitute a direct repudiation of the central tenets of transformism. I am not suggesting that this was intentional, although it is possible that he knew, and sought to refute, Erasmus Darwin's theories.[30] Paley, being steeped in eighteenth-century natural history traditions and in turn-of-the century English reactions to events in France, knew exactly where the points of threat to his system lay. Where Chateaubriand appealed to the reader's emotional response, inviting them to think not in logical terms but in terms of emotionally charged visual correspondences, Paley took a more rationalistic route.

As is well known, Paley likened God's relationship to nature to that of human beings to their products, by presenting Him as a designer, contriver, carpenter, gardener, and so on. At the same time, he reinforced the differences between God and man by depicting 'art' (in the eighteenth-century sense) as inferior to nature with respect to the skill it displays. Man and God are both makers, a point Paley stressed in order to show that the maker is always *unlike* that which he makes. In both cases, the signs of craftsmanship have a material existence and hence act as signifiers of the author. Yet any similarity between God and man ends when the level of skill is considered. The degree of fit between a humanly produced mechanism/machine and the purpose it is intended to fulfil cannot compare with God's – a discrepancy exemplified for Paley by the difficulty a workman would have designing something to do all the things a spine can:

> The chain of a watch, (I mean the chain which passes between the spring-barrel and fusee,) which aims at the same properties, is but a bungling piece of workmanship in comparison with that of which we speak [i.e. the structure of the spine].[31]

Or again:

> In no apparatus put together by art, and for the purposes of art,
> do I know such multifarious uses so aptly combined, as in the
> natural organisation of the human mouth; or where the struc-
> ture, compared with the uses is so simple. . . . The mouth . . . is
> one machine.[32]

This last sentence makes clear why Paley insisted on using man-made
objects in his argument. We, as readers, can imagine the purposiveness of producing objects in which every detail serves some rational
end. At the same time, we experience ourselves as separate from the
tools we use, the process of manufacture and the end-product. Paley's
repeated stress on the mechanical character of organs reinforces the
side of his argument that lures the reader into identifying human and
divine capacities, while his stress on the fine, supremely beautiful fit
between form and function reinforces a sense of human inadequacy
in the face of God's superior skill.[33] Both God and man, therefore,
can be agents, causes, designers, authors and contrivers. Yet God
alone has a single, total, overarching plan – a fact that Paley thought
explained similarities between living things – and the aesthetic capacities to generate true beauty.

   Although Paley recognized that plants and animals were quite
distinct, his emphasis on the gap between maker and made put them
on a par with inanimate objects – both are 'passive, unconscious
substances'.[34] Now the profound differences between Paley and
Lamarck can be appreciated. For the former, a creative designer acts
upon matter in a manner modelled on intention and will. By contrast,
Lamarck moved the action away from a transcendental level and into
nature, and more particularly into organic nature. Paley, it is clear,
was perfectly aware of such arguments, for he explicitly countered
them by ridiculing the idea that the parts themselves could contribute
to their own form – a proposition that was absurd to him because he
saw the plan and materials used to execute it as separate categories.
Once activity was allowed to be present in natural things themselves,
the uniqueness of God was subtly undermined. Paley's language of
machines made the possibility that habit and the environment contributed to organic form, as Lamarck argued, seem preposterous.

   Lamarck's task was to generate a fresh language appropriate to a
naturalistic account of the historical changes organic forms had undergone. There were no suitable pre-existing models in natural philosophical discourse for him to draw upon, and there was certainly
none that had the empirical richness or sensual immediacy of Paley's
writings. Like Chateaubriand, Paley appeals to the eye. But he does
so not in a spirit of pleasure, but of rational instruction:

A plain observer of the animal oeconomy may spare himself the disgust of being present at human dissections, and yet learn enough for his information and satisfaction, by even examining the bones of the animals which come upon his table.[35]

Paley put the same point more succinctly when he said, 'in every part of anatomy, description is a poor substitute for inspection'.[36] Like watches, plants and animals, pipes, automata, engines, flood gates, pumps and pistons were easy to visualize. But when Lamarck tried to describe the actions of the nervous system without lapsing into mechanistic explanations he had no such fund of images to draw upon. Paley, in other words, could employ a language of machines because he treated physical mechanisms as signs of God. Lamarck rejected the language of contrivance because it could not adequately express the dynamism within nature and especially in organic beings, which he did not perceive as machine-like. In repudiating assumptions and forms of argument based on a system of decoding nature to reveal God, Lamarck cut himself off from a long-established tradition. His careful definitions of words such as creation and production were attempts to recast the terms of reference of natural philosophy so that a naturalistic, transformist world view could emerge. Comparing Lamarck with the natural theologians, we can see not only that they employed different images and metaphors, but that these were both rooted in and expressions of different modes of inference, methods and metaphysical systems.

The natural theologies of Paley and Chateaubriand, and the natural philosophy of Lamarck, expressed broad cultural concerns. The nature of God was clearly of prime interest to Paley, whose language was blatantly anthropomorphic. Lamarck, by contrast, considered the subject not amenable to rational discourse. The difference between them cannot be described solely in terms of content – *what* they believed – because it was also a matter of *how* they conceived the human mind to arrive at its ideas. It is significant that Paley must be seen as both a rationalist and an empiricist: that is, he wanted all beliefs to be arrived at through the use of reason, and, just as fervently, he wanted the direct observation of nature to form the basis for religious experience.

Both Chateaubriand and Paley used the idea of the world of nature as a temple.[37] Temples were places where, for Chateaubriand, the individual soul communes with God; for Paley, where rational contemplation of the deity occurs. Because Paley held that God's existence and character must be totally demonstrable by a blend of reason and observation, he could plausibly stress the mechanical qualities of natural objects. God is a maker whose contrivances contain no mystery. Chateaubriand, however, held that there is mystery in profusion; the wonders of nature suggest the divine creator more by virtue of emotional than of rational force. Whereas Paley's natural theology was closely tied to his moral and political philosophy, Chateaubriand's was linked with his fictional writings and his political commitment to the reinstatement of legitimate authority – both religious and secular – in France. A similar point can be made with respect to Lamarck, whose denial of God as part of natural philosophy and assertion of a secular, naturalistic approach to society, were expressions of well-established clusters of ideas with political dimensions.

These ideas were of course manifested as much in literary and political writings as in scientific and medical ones. The concepts of creation and production, in particular, were at once unstable in their meanings, hence inviting linguistic control, and immensely rich in their connotations, hence encouraging associations to be made between human, natural and divine domains. But these concepts were not simply espoused, rejected and fought over. They were dense with a variety of both epistemological and metaphysical qualities. They occupied a central place in European thought in the late eighteenth and early nineteenth centuries because they spoke to a wide range of deeply felt commitments.

In this period the ideological implications of a naturalistic approach to organic phenomena were widely debated. The peculiar quality of the discussions in the biomedical sciences during the 1790s is now becoming clear. Some of the more dramatic general features of the decade are well known, and their ideological impact has been widely discussed: the French Revolution, and particularly the execution of the king and the reign of terror, the crisis of the Poor Laws, the impact of European war, artisanal radicalism, and the consolidation of American autonomy. Broadly, the decade saw, first, successful challenges to authority, especially of a patriarchal nature; then, a reaction against these challenges, culminating in the first decade of the nineteenth century. Hierarchical forms of power were reasserted, social stability sought, and the interests of the propertied middle class defended. Integral to this process of reassertion were the imperialism of Napoleon, the revival of Catholic values in France, especially the Concordat with the Pope and the writings of Chateaubriand, the rediscovery of hierarchy and traditional values, and the flowering of 'Romanticism' in Britain and France with its celebration of an inner life and of individual creative genius. The impact of such large-scale social and cultural movements on science and medicine is beginning to be charted in work on early phrenology, the Lunar Society of Birmingham, the organization of French science and the 'birth of the clinic'. Historians of philosophy have confirmed the special quality of this period during which the political stakes invested in the world of ideas were particularly visible.[38]

It is because ideas of authority and power at this time were so closely bound up with prevailing ideas of hierarchy that we can trace shared themes in the biomedical sciences, natural theology, philosophy, literature and political thought. There were several, intertwining hierarchies built into the thought of the period, through which the respective powers of God, nature and the human race were conceptualized: God, King, fathers, women, children and servants; the chain of being; God, Pope, clergy, laity; mind and body; perfect animals ... least perfect animals; First, Second and Third estates. To evoke one of these was to evoke others. The overlap between the hierarchies derives not just from the fact that they evinced a shared preoccupation with power, order and authority, but also from their expression in a common language. There was a limited set of guiding metaphors that encouraged the transfer of meanings between the series – rulership, fatherhood, authorship, sovereignty, jurisdiction, ownership and degrees of spiritual or physical perfection. But to the sense of how deeply fraught overlapping vocabularies of power were in this period, we must add another dimension if we are to appreciate fully the significance of the creation/production distinction. It was not only part of a rich language; it was also a pivot around which stories were told, stories best understood as myths of creation.

Creation/production in a cultural context

Creation myths are stories about beginnings and about how things came to be.[39] They do not just account for the origin of the universe, but also for the place of human beings within it, their relations with their Gods and with worldly forms of authority. Thus they must provide acceptable accounts of origins and speak to predominant human preoccupations, especially the nature of suffering. By providing frameworks of meaning, they address political, social, psychological, theological and aesthetic questions in such a way that these elements are closely bound together. There are a number of reasons for viewing Lamarck's transformism in this mythological context. He wrote at a time when myths of creation were being reconceptualized as primary vehicles for thinking about the nature of creation and for interpreting human existence. Indeed, it is possible to see evolutionary theories in general as offering their own versions of creation myths.[40] There were already many rich texts that sought to articulate these matters, but one that appeared in the same year as Lamarck published his definition of nature is particularly striking. There can be no doubt that Mary Shelley's *Frankenstein* raises questions about the distinction between creation and production.[41] It does so most obviously by showing the limitations of human capacities when applied to ends normally attainable only through the power of God. When a person strives to be a creator, the prerogative of God, he becomes in fact a mere producer.

It has been argued that *Frankenstein* must be seen in the context of the Romantic preoccupation with creation myths, and more particularly with attempts to rewrite, subvert and invert them.[42] In the novel, the 'modern Prometheus' clearly and knowingly transgresses the boundary between creation and production. He makes, or produces, a being that should either be generated biologically through reproduction or be created by God. The point is not just that Frankenstein assumes God-like creative powers, but that in the process he undertakes the wrong sort of *production*. It is necessary, therefore, to differentiate between the two sorts of production at issue in *Frankenstein* – the one biological, the other artificial – and this requires two major distinctions. First, human biological (re)production results from the union of male and female. Two different elements produce a third; Frankenstein, by contrast, undertakes his production alone. Secondly, a foetus grows as a unity; it is always an integral whole, whereas Frankenstein's creation is pieced together from different bodies, stitched into a spurious integrity by its maker's hand. The monster was denied the experience of growth.[43] To these two kinds of production we can contrast a third, the one evoked by Lamarck, for whom nature's actions are a form of labour. Production is what nature does; it is a system of laws, acting uniformly in time and space. Production, like labour, takes time, is sequential and is a material process.

There are a number of ways in which the novel develops the production/labour theme. It emphasizes the arduous and unclean labours Frankenstein undertakes, the difficulties he has in finding the right bits and piecing them together properly, and the poor results he achieves. The monster is not beautiful but repellent because he bears the marks of his method of assembly. Everyone who sees him treats him as abhorrent. His appearance is a cruel parody of the aesthetic qualities of nature. When the monster is heard telling his own story, it becomes perfectly clear that his appearance has caused the problem. His voice indicates a sympathetic person, aware, reflective and, initially, warm. The poignancy of his story rests precisely on this radical disjunction between his inner life and his external appearance; it is on the basis of the latter that people judge him.[44] His appearance indicates that he is neither a true creature of God, nor a natural child of a woman, but a mere product of a man.

Here the novel has peculiar relevance for science and for natural theology. The assumption that human internal qualities could be inferred from external appearances remained widespread in the life sciences and medicine of the early nineteenth century, although it was not without critics. Indeed, the physiognomic mode of inference had been a paradigmatic form of thinking for centuries.[45] The monster suffers and ultimately becomes evil because he is a victim of the unthinking application of this approach. There are two issues here, both of them germane to my theme. First, reactions to the monster cause us to question the propriety of a pattern of inference that moves from external traits to internal ones. Secondly, these reactions of revulsion are based on the accurate perception that there was indeed something amiss with the monster's appearance, which carries the signs of its improper construction. Mary Shelley thus called into question some settled assumptions by demonstrating the dramatic disparity between inner life and outward appearance. The relationship between inside and outside had traditionally been used to infer the presence of God in nature and likewise the presence of the soul or emotions in human beings. And because the mental life of human beings was strongly identified with God, there was more than a formal analogy involved. At issue was a form of thinking that sanctioned inferences from visual signifiers to an unseen signified.

The relationship between outside and inside, nature and God, seen and unseen, becomes all the more interesting in the light of a recent interpretation of Rousseau's *Discourse on the Origins of Inequality* as a major source of Romantic creation myths. From a non-physiognomical perspective, Rousseau suggested that a crucial moment in the development of the human race comes when people see themselves through the eyes of others; when, in other words, there is potential tension between self-perceptions and the judgements of others looking only at the outsides of their fellows:

> The savage lives within himself; the sociable man, always outside of himself, knows how to live only in the opinion of others; and it is, so to speak, from their judgment alone that he draws the sentiment of his own existence.[46]

For Rousseau, this psychological turning point is indicative of the growing artificiality of society. More elaborate social arrangements encourage the development of 'masks', which promote self-interest and greed. In explaining the nature and genesis of human society, Rousseau made extensive use of a distinction between an inner and an outer man.

Rousseau's struggles to define 'nature' were deeply influential, specifically because he offered a naturalistic account of changes in human existence that were *historical*. These changes were not, of course, historical in the sense of chronicling events known to have taken place in particular societies, but rather in the sense that they were a sequence of hypothetical processes, each emerging out of the previous one. This constituted a model for a transformist view of nature's history, not unlike Lamarck's. Indeed, Lamarck seized the opportunity to include human interaction within his scheme, and he did so in terms recognizably derived from Rousseau. Rousseau's discourse spoke to the creation/production distinction by focusing on the production of human society, a process to be rationally reconstructed by examining its successive stages. It offered a fresh view of the present state of human existence in relation to a distant, almost mythical past – a state of nature. By tracing changes as much for their psychological as for their social and political implications, Rousseau was able to address many of the central issues embedded in creation myths.

Thus the significance of judging the monster from his appearance can now be specified more precisely. It is his ugliness, in fact, that first impels his creator to reject him. Later, the monster comes to recognize his gruesome body as the cause of his outcast status. There is another context in which the monster's ugliness can be placed. In both natural theology and aesthetic theory it had become customary to compare the beauty of God's work in nature with the inferiority of human achievements. Makers can never equal the Creator.[47] Frankenstein's monster is ugly because he was produced by man and constructed by illegitimate means. His ugliness denotes the violation of the distinction between creation and production that led to his existence. The beauty of nature, by contrast, is not just the sign of God's superior handiwork; it also serves as a constant reminder of the aptness of God's intelligence in fitting form so perfectly with function. Beauty and adaptation are two sides of the same coin – a coin called 'creation'. The discrepancy between God's work and human productions is visually registered both in terms of aesthetic quality *and* in terms of functional fit. Those who refuse to accept the limited role human beings have been assigned are transgressors.

Therefore, anxieties about the distinction between creation and production were hardly confined to Lamarck's natural philosophy. Mary Shelley explicitly links both terms to science and to myth. Like other Romantics, she was deeply indebted to Rousseau, who was also a significant influence on Lamarck. The naturalistic method of Rousseau and Lamarck was historical and implicitly anti-physiognomic. It opened the way for a transformist mode of thinking that did not just assert organic change as a fact, but sought to reorientate the way people looked at nature and at society. In order to do this, the pre-existing alternative – natural theology – had to be disposed of. It was a difficult task, because this way of viewing nature and society had many ramifications across a wide spectrum of issues. My purpose has been to suggest that much of Lamarck's work was dedicated to this end, hence its preoccupations with definitions and method, with epistemology and classification, and its lack of visual engagement with nature.

The Lamarckian enterprise reassessed

Lamarck can usefully be seen as the main thinker in the first decade of the nineteenth century who attempted to generate a language through which a new science of living things, sensitive to historical change, could be developed. The preconditions for this science included the ability to encompass all organic phenomena, including human thought and behaviour, the repudiation of agents external to nature, the explanation of each level of organic complexity in terms of the pre-ceding one, and the basing of all faculties in organ systems. Lamarck's project therefore had to achieve two closely related goals: eliminating God, soul and vital principles from scientific discourse, and demon-strating the sufficiency of natural processes as their own explanation. To attain these, Lamarck had to undercut a physiognomic mode of thought, which drew people to search for natural signs, and then to decode them as evidence of a quite distinct and higher level of existence. For him it was necessary to stay with the phenomena and find the systematic relationships among them, both at the present and over long periods of time.

I want to insist on the historical specificity of Lamarck's project. At just the moment when he was developing his full transformism, Paley published his *Natural Theology*. Although both writers emerged from eighteenth-century 'rationalist' traditions, yet, with Lamarck's theories, there came a radical split between a physiognomic and an anti-physiognomic mentality: Paley saw visual indicators of a divine author; Lamarck used abstract terms to reconstruct nature's processes. Natural theology was based on the idea that nature contained the palpable signs of God's creative hand. It was, formally speaking, the equivalent of history painting, where great stories could be read in the canvas.[48] In order for natural theology to work effectively, how-ever, the evidence of design had to be utterly unambiguous, so that the story of the creation could readily be appreciated. This led Paley to stress repeatedly the mechanical nature of organic structures and processes. Just as human beings build their intentions into the objects they make, so does God.

Yet there was a danger here for natural theology. Paley's consistent use of a language of artisanship drew the reader inexorably to liken God to an artisan – an artisan whose work necessarily lacks mystery, because his design is totally transparent in what he makes. This language works against his own argument for God's transcendent existence and attributes, because it so readily identifies God with man. Paley sought to reintroduce the mystery of God in two ways. First, he argued that God has a total guiding plan, unlike human beings who plan only individual projects. There is magnitude and magificence in God's plan that sets it apart from human designs, giving it an additional dimension. Secondly, Paley pointed to the beauty in God's work, which far surpasses anything people can achieve. In its scope and its aesthetic qualities, creation surpasses production. While it was also possible to evoke God's mystery, as Chateaubriand did, by stressing an emotional response to the pro-vidential aspects of nature, seen less as mechanisms than as marvels, for both writers, an intense visual engagement with nature was none the less fundamental to their project of finding God in nature.

Lamarck's distinction between creation and production, far from a fruitless play with words, signalled his reaction against natural theological traditions and his perception that there had been a dan-gerous blurring of these categories. It also constituted a pledge, both methodological and metaphysical, to develop a new science. The warm engagement with nature, which has been found so characteristic of Charles Darwin, is absent in Lamarck.[49] The reason is not that he lacked empirical expertise, but that his project was an abstract one: to clear the ground, to make certain things sayable and think-able, which previously were not so. According to Lamarck, one must dwell with nature rather than read there the hand and mind of a being of a different order of existence. Previous attempts to interpret nature in materialistic terms were unsatisfactory because they ignored nature's dynamism, particularly that of living matter. Lamarck had to find a way of infusing nature with activity without suggesting that the activity denoted a separate agent. This Lamarck achieved in his natural philosophy principally by positing levels of organization that were, simultaneously, historical stages of development and degrees of physiological and anatomical complexity. It was fundamentally a linguistic project for him because of the significance he attached to correct terminology. His classifications were dedicated to showing that taxonomic order, the order of temporal development – that is, the order of production – and 'organization' coincide in a proper classificatory system where species, orders, phyla, and so on, are labelled in such a way that these considerations are expressed in their names.[50]

It is much easier to understand Lamarck's enterprise by reading him in the light of Paley, and vice versa. Lamarck had to write God out of nature and then reconstruct nature's own order. He did this by designating anything to do with creation as non-science, for science deals only with nature's labours. By contrast, Paley wanted to write God into nature by inviting us to *view* His handiwork, and this led him to stress the similarities between creation and production. The difference between Paley and Lamarck is not simply a matter of different beliefs, but of different modes of inference and, equally significant, different attitudes to evidence. For Paley, evidence derives from inspection of the visual signs in nature, which signify God's presence; for Lamarck, it derives from information, acquired by the senses certainly, but then subjected to elaborate rational reconstruction in terms of major abstract ideas deployed as organizing principles. These include his definitions of matter, nature and life, his distinctions between living and inert, vertebrate and invertebrate, faculty and property, and his concept of levels of organization, all of which inform his methodology.[51] Thus Lamarck gave no special status to the sense of sight, or to beauty, which is a real term in natural theological discourse.

Therefore, Lamarck's transformism can now be placed in a larger context. It spoke to a very wide range of issues of the time, although these often remained implicit. By redefining terms such as creation, production, life and nature, Lamarck tried to generate a language purged of unwelcome theological associations, to set himself apart from natural philosophical traditions that could not sustain a science of life rooted in change over time, that is, in production. He wanted to set in place a vocabularly appropriate not just to his scientific concerns, but to his social and political interests, and for this the distinction between creation and production was essential. It embodied his metaphysics, his methodology and his social philosophy. It could work in many different ways and with considerable force because the two terms touched people so deeply. Creation and production not only stood at the very heart of the Christian tradition, which makes a fundamental distinction between begetting (re-production) and creating; they also expressed concerns so basic to human existence that it is hard to articulate them without banality. What acts are people capable of? What kinds of things can they make? Is making babies, books, knowledge and machines the same kind of process? Can we imagine a non-human kind of making? What responsibilities does the maker owe to the made? The questions are as endless as the cultural possibilities they evoke. Lamarck's transformism, formulated in the absence of models for the kind of scientific discourse he wished to generate, raised just such questions, anwered them in an original manner, and did so in a context that gave debates about creation and production both a wide application and a sense of urgency.

Notes

In addition to the editor of this volume, numerous people have been kind enough to comment upon an earlier draft of this eassy. They all have my thanks. I wish to express my special gratitude to John Brooke, Geoffrey Cantor, Martin Donougho, Tony James and Bernard Lightman for their constructive, thought-provoking comments. I only regret that I have not been able to incorporate them all.

Throughout the essay I have, for the sake of simplicity, used the term 'man' as it was employed by eighteenth- and nineteenth-century writers, because to put quotation marks around it each time would be clumsy. Such a convention implies no approval of the term itself.

The translations from the French of Lamarck and Chateaubriand are my own.

1. John C. Greene, *Science, Ideology, and World View: Essays in the History of Evolutionary Ideas* (Berkeley: University of California Press, 1981), p. 18.

2. On the Nicene and other creeds, see A.E. Burn, 'Creeds', *Encyclopaedia Britannica*, 11th edn, VII, 392–400. A related example is the second verse of the hymn 'O Come, all ye faithful', which reads, 'God of God, / Light of Light, / Lo! He abhors not the Virgin's womb; / Very God, / Begotten, not created ...' (*The New English Hymnal* [Norwich: Canterbury Press, 1986], p. 65). This hymn, originally in Latin, dates from the eighteenth century. It echoes the Nicene Creed, which in modern versions refers to, '... very God of very God, begotten, not made ...'. This form of words raises just the issues addressed in this essay. The complexity of the verb 'to beget' is also relevant, as it embraces both creating and producing, as the *Oxford English Dictionary* makes clear. For an unusual slant on the status of Christ as 'begotten, not created', see Leo Steinberg, *The Sexuality of Christ in Renaissance Art and Modern Oblivion* (New York: Pantheon Books, 1983).

3. P.O. Kristeller, '"Creativity" and "Tradition"', *Journal of the History of Ideas*, 44 (1983), 105–13 (106).

4. On Lamarck in general, see Richard W. Burkhardt, Jr, *The Spirit of System: Lamarck and Evolutionary Biology* (Cambridge, Mass.: Harvard University Press, 1977); Leslie Jean Burlingame, 'Lamarck's Theory of Transformism in the Context of His Views of Nature, 1776–1809' (Ph.D. dissertation, Cornell University, 1973); and *idem*, 'Lamarck', in Charles Coulston Gillispie, ed., *Dictionary of Scientific Biography*, 16 vols. (New York: Charles Scribner's Sons, 1970–80), VII, 584–94. The most important recent work on Lamarck is

Pietro Corsi, *The Age of Lamarck* (Berkeley: University of California Press, 1988). For a brief introduction to Lamarck's ideas, see L.J. Jordanova, *Lamarck* (Oxford: Oxford University Press, 1984). Lamarck's principal exposition of the distinction between creation and production may be found in M. Le Chevalier de Lamarck, *Système analytique des connaissances positives de l'homme, restreintes à celles qui proviennent directement ou indirectement de l'observation* (Paris: A. Belin, 1820).

5. Camille Limoges, 'Pluche', in Gillispie, ed., *Dictionary of Scientific Biography*, XI, 42–4; Toby A. Appel, *The Cuvier–Geoffroy Debate: French Biology in the Decades before Darwin* (New York: Oxford University Press, 1987), pp. 13, 56; Jacques-Henri Bernardin de Saint-Pierre, *Paul et Virginie*, 1788 (Paris: Garnier-Flammarion, 1966); *idem, Paul and Virginia*, translated by John Donovan (London: Peter Owen, 1982), esp. p. 11.

6. Quoted in Appel, *Cuvier–Geoffroy Debate*, p. 151; see also pp. 158, 173, 191, 213.

7. François René de Chateaubriand, *Génie du Christianisme*, 1802; 2 vols. (Paris: Garnier-Flammarion, 1966), I, 149–93.

8. *Catalogue des livres de la Bibliothèque de Feu M. Le Chevalier J.–B. de Lamarck* (Paris: Barrois et Benou, 1830).

9. Burlingame, 'Lamarck's Theory'; *idem*, 'Lamarck'.

10. Full details of these and Lamarck's other writings may be found in the bibliographies of the following works: Marcel Landrieu, *Lamarck, le fondateur du transformisme, sa vie, son oeuvre* (Paris: Société Zoologique de France, 1909), pp. 448–70; Alpheus S. Packard, *Lamarck, The Founder of Evolution: His Life and Work* (New York: Longmans, Green and Co., 1901), pp. 425–45; and Burkhardt, *Spirit of System*, pp. 261–8.

11. Lamarck, *Système*, p. 20.

12. Ibid., p. 22.

13. Ibid., p. 27.

14. Ibid., p. 32.

15. Ibid., p. 35.

16. Ibid., p. 36. On the recalcitrance of language, see Gillian Beer, '"The Face of Nature": Anthropomorphic Elements in the Language of "The Origin of Species"', in Ludmilla Jordanova, ed., *Languages of Nature: Critical Essays on Science and Literature* (London: Free Association Books; New Brunswick, N.J.: Rutgers University Press, 1986), pp. 207–43.

17. Lamarck was particularly attentive to distinctions between conceptual levels. See Ludmilla Jordanova, 'La Psychologie Naturaliste et le "Problème des Niveaux": La Notion de Sentiment Intérieur chez Lamarck', in *Lamarck et son temps, Lamarck et notre temps* (Paris: J. Vrin, 1981), pp. 69–80.

18. Lamarck, *Système*, pp. 52–3.

19. Ibid., p. 55.

20. Ibid., p. 59.

21. Jordanova, 'Psychologie', pp. 72–6; *idem*, 'The Natural Philosophy of Lamarck in Its Historical Context' (Ph.D. thesis, University of Cambridge, 1976), pp. 20–3 and ch. 3; S. Moravia, *Il Pensiero degli Idéologues: Scienza e Filosofia in Francia (1780–1815)* (Firenze: La Nuova Italia, 1974); Fr.

Picavet, *Les Idéologues: Essai sur l'histoire des idées et des théories scientifiques, philosophiques, religieuses, etc. en France depuis 1789*, 1891 (New York: Burt Franklin, 1971); George Boas, *French Philosophies of the Romantic Period*, 1925 (New York: Russell and Russell, 1964), ch. 2.

22. Lamarck, *Système*, p. 79.

23. Adrian Desmond, 'Artisan Resistance and Evolution in Britian, 1819–1848', *Osiris*, 2nd ser., 3 (1987), 77–110, examines one way in which Lamarck's ideas were deployed in the service of a politically inspired world view. See also Desmond's essay, this volume.

24. Lamarck, *Système*, p. 226.

25. For Lamarck's direct citations of Rousseau, see *Système*, p. 94 and his *Histoire naturelle des animaux sans vertèbres*, 7 vols. (Paris: Deterville, 1815–22), I, 330. See also Paul A. Cantor, *Creature and Creator: Myth-making and English Romanticism* (Cambridge: Cambridge University Press, 1984), p. xvi and Jean-Jacques Rousseau, 'A Discourse on the Origins of Inequality', in *idem, The Social Contract and Discourses*, ed. G.D.H. Cole, 1913 (London: Dent, 1973), pp. 27–113. On Rousseau's explanatory strategies in the second discourse, see Marc F. Plattner, *Rousseau's State of Nature: An Interpretation of the 'Discourse on Inequality'* (DeKalb: Northern Illinois University Press, 1979). Plattner stresses Rousseau's indebtedness to contemporary science. To be sure, the discourse considered only human development in naturalistic terms, but Plattner's book, especially ch. 3, shows how marked the similarities between Rousseau and Lamarck were. Another treatment of Rousseau in the context of contemporary science is Michèle Duchet, *Anthropologie et histoire au siècle des lumières: Buffon, Voltaire, Rousseau, Helvétius, Diderot* (Paris: François Maspero, 1971), pp. 322–76. To a certain extent the whole of the *Système analytique* is a study of man; part 2 is of particular relevance, however, to the Rousseau–Lamarck comparison.

26. William Paley, *The Works of William Paley*, 4 vols. (London: Longman and Co., 1836–8); Charles Coulston Gillispie, *Genesis and Geology: A Study in the Relations of Scientific Thought, Natural Theology, and Social Opinion in Great Britain, 1790–1850*, 1951 (New York: Harper and Row, 1959); C.A. Russell, ed., *Science and Religious Belief: A Selection of Recent Historical Studies* (London: University of London Press, 1973); D.C. Goodman, ed., *Science and Religious Belief, 1600–1900: A Selection of Primary Sources* (Milton Keynes: Open University Press, 1973); John Hedley Brooke, 'Natural Theology in Britain from Boyle to Paley', in *idem et al., New Interactions between Theology and Natural Science* (Milton Keynes: Open University Press, 1974); *idem*, 'The Relations between Darwin's Science and His Religion', in John Durant, ed., *Darwinism and Divinity: Essays on Evolution and Religious Belief* (Oxford: Basil Blackwell, 1985), pp. 40–75; Charles E. Raven, *Natural Religion and Christian Theology*, 2 vols. (Cambridge: Cambridge University Press, 1953); L.J. Jordanova and Roy S. Porter, eds., *Images of the Earth: Essays in the History of the Environmental Sciences* (Chalfont St. Giles, Bucks.: British Society for the History of Science, 1979), esp. chs. 1–3; Richard Yeo, 'William Whewell, Natural Theology and the Philosophy of Science in Mid-Nineteenth-Century Britain', *Annals of Science*, 36 (1979), 493–516; M.L. Clarke, *Paley: Evidences for the Man* (London:

SPCK, 1974); D.L. LeMahieu, *The Mind of William Paley: A Philosopher and His Age* (Lincoln: University of Nebraska Press, 1976); 'William Paley', *Dictionary of National Biography*, XV, 101–7; David Hume, *Dialogues Concerning Natural Religion*, 1779 (New York: Hafner, 1948).

27. Brooke, 'Natural Theology', pp. 26, 48.

28. Chateaubriand, *Génie*, I, 164. I have put she/it because it is hard to assess the degree to which French writers intended the personification implied by 'she'.

29. Ibid., p. 157.

30. This is suggested by Clarke, *Paley*, pp. 96–7.

31. Paley, *Works*, I, 52.

32. Ibid., p. 67.

33. This argument was not only used by natural theologians; others also appreciated its general aesthetic importance. For example, see William Hogarth, *The Analysis of Beauty*, ed. Joseph Burke (Oxford: Clarendon Press, 1955). On the interpretation of Hogarth in the context of contemporary natural theology, see Ludmilla Jordanova, 'Physiognomy in the Eighteenth Century: A Case Study in the Relationships Between Art and Medicine' (M.A. thesis, University of Essex, 1986), pp. 48–51.

34. Paley, *Works*, I, 26.

35. Ibid., p. 51.

36. Ibid., p. 73.

37. LeMahieu, *Mind of William Paley*, p. 90; Chateaubriand, *Génie*, I, 171.

38. Roger Cooter, *The Cultural Meaning of Popular Science: Phrenology and the Organization of Consent in Nineteenth-Century Britain* (Cambridge: Cambridge University Press, 1984); Maureen McNeil, *Under the Banner of Science: Erasmus Darwin and His Age* (Manchester: Manchester University Press, 1987); Paul Weindling, 'Science and Sedition: How Effective were the Acts Licensing Lectures and Meetings, 1795–1819?', *British Journal for the History of Science*, 13 (1980), 139–53; Ian Inkster, 'Seditious Science: A Reply to Paul Weindling', *British Journal for the History of Science*, 14 (1981), 181–7; Dorinda Outram, *Georges Cuvier: Vocation, Science, and Authority in Post-Revolutionary France* (Manchester: Manchester University Press, 1984); Robert E. Schofield, *The Lunar Society of Birmingham: A Social History of Provincial Science and Industry in Eighteenth-Century Britain* (Oxford: Clarendon Press, 1963); Michel Foucault, *The Order of Things* (London: Tavistock, 1970); idem, *The Birth of the Clinic: An Archaeology of Medical Perception* (London: Tavistock, 1973); Boas, *French Philosophies*, ch. 1.

39. Cantor, *Creature and Creator*; Judith N. Shklar, 'Subversive Genealogies', *Daedalus*, 101 (1972), 129–54; David Maclagen, *Creation Myths: Man's Introduction to the World* (London: Thames and Hudson, 1977).

40. This theme is explored in Gillian Beer, *Darwin's Plots: Evolutionary Narrative in Darwin, George Eliot, and Nineteenth-Century Fiction* (London: Routledge and Kegan Paul, 1983).

41. Mary Poovey, 'My Hideous Progeny: Mary Shelley and the Feminization of Romanticism', *Publications of the Modern Language Association of America*, 95 (1980), 332–47; George Levine and U.C. Knoepflmacher, eds., *The Endur-*

ance of Frankenstein: Essays on Mary Shelley's Novel (Berkeley: University of California Press, 1983); Mary Shelley, *Frankenstein,* ed. Maurice Hindle (Harmondsworth, Middlesex: Penguin Books, 1985). *Frankenstein* appeared in French in 1821. A stimulating recent interpretation is David E. Musselwhite, *Partings Welded Together: Politics and Desire in the Nineteenth-Century English Novel* (London: Methuen, 1987), ch. 3.

42. Cantor, *Creature and Creator,* esp. ch. 4.

43. Beer, *Darwin's Plots,* p. 110.

44. There are two crucial episodes in the novel where the plot turns on the revulsion inspired by the monster. The first is at the moment of his 'birth', when Frankenstein is repelled by the creature he had intended to be beautiful (described at the beginning of ch. 5); the second is when the monster reveals himself to the De Lacey family (end of ch. 15). In the latter context, it is only the *blind* old man who offers him any compassion.

45. Jordanova, 'Physiognomy'; Norman Bryson, *Word and Image: French Painting of the Ancien Régime* (Cambridge: Cambridge University Press, 1981), ch. 2; Graeme Tytler, *Physiognomy in the European Novel: Faces and Fortunes* (Princeton, N.J.: Princeton University Press, 1982).

46. Quoted in Cantor, *Creature and Creator,* p. 126.

47. For example, Hogarth, *Analysis,* pp. 85–7; Paley, *Works,* I, 67.

48. Bryson, *Word and Image,* ch. 2; Thomas Puttfarken, *Roger de Piles' Theory of Art* (New Haven, Conn.: Yale University Press, 1985).

49. Beer, '"Face of Nature"', pp. 225–7.

50. Henri Daudin, *Les classes zoologiques et l'idée de série animale en France à l'époque de Lamarck et de Cuvier (1790–1830),* 2 vols. (Paris: Félix Alcan, 1926), II, chs. 8, 10; Burkhardt, *Spirit of System;* Jordanova, *Lamarck,* ch. 2. Cf. James Lee Larson, *Reason and Experience: The Representation of Natural Order in the Work of Carl von Linné* (Berkeley: University of California Press, 1971).

51. On 'organization', see Karl Figlio, 'The Metaphor of Organisation: An Historiographical Perspective on the Bio-Medical Sciences of the Early Nineteenth Century', *History of Science,* 14 (1976), 17–53.

3

Lamarckism and democracy: Corporations, corruption and comparative anatomy in the 1830s

ADRIAN DESMOND

Where did the doctrines of scientific naturalism originate, and why did they prevail? Did scientific naturalism simply furnish intellectuals with an arsenal of "ideological weapons", or were naturalistic ideas related more organically to social and religious traditions? If so, to what extent was there already a basis of appeal in English culture for crisis-stricken individuals in their bid to gain ascendancy within the intelligentsia?[1]

The Scottish fruit farmer, Patrick Matthew, was printing his *Naval Timber and Arboriculture* when news of the July Revolution broke in 1830. He closed the appendix almost in mid-breath, wrapping up its rag-bag of radical injuctions on free trade, aristocratic entail, competition and animal transformation, and applauded the republicans' 'efforts towards the regeneration of man'.[2] Events in France – as in Britain where the new Whig Government was committed to electoral, church and corporation reform – meant that the time for theory was past. The vestiges of *ancien régime* privilege were finally, he believed, to be wiped away – and on that note he shut his book.

I talk of a 'rag-bag' of radical doctrines. But the point of this essay is to show that Matthew's appendix was not a chance juxtaposition of radicalism and transformism. Even the prima facie evidence should make us sceptical of accepting the link as fortuitous. After all, Edinburgh University in the Regency period and later produced a number of bourgeois radicals who embraced the *idéologues'* self-developing nature: not only Matthew, but free-thinkers like the dandy Robert Knox, reclusive Robert Grant and phrenologist Hewett Watson.[3] This essay probes the link between their democratic politics, free-trade demands and doctrines of self-development. I will examine the way the radicals' materialist sciences harmonized with calls for a capitalist free market and how, in London, at the new University and private medical schools (the main focus of the essay), these sciences were

99

able to do more bureaucratic Benthamite work. The point is to understand why naturalistic anatomies were turning up in confrontationist contexts, where trading monopolies and civil disabilities were denounced, and an unimpeded access to the upper echelons of the civic power structure was demanded by the sons of the merchant dissenters.

I can illustrate some of these ideas by turning again to Matthew's 'rag-bag'. As a transformist he argued that all life had a 'circumstance-suiting power', enabling it to respond automatically to environmental changes by producing 'superior' forms. In the same way, in the social world, he believed that recent industrial developments – the advent of manufacture, division of labour, and the uniting of knowledge, technology and power under the entrepreneur's aegis – had thrown up a superior breed, the industrial capitalist. The new industrialist was more 'circumstance-suited' than the idle aristocrat and Kirk minister, relics of the old order who had outlived their usefulness. As a merchant, Matthew was blatantly stretching the laws of competition and selection across society and nature; they dictated his tree selection and culturing techniques as much as they justified his demands for a deregulated market that could absorb his produce. He insisted that society had to obey nature's strict law – and 'natural' competition promised social progress once the impeding and 'unnatural' customs of the nobility, from primogeniture to inherited privilege, were removed. These were an affront to nature that would not 'pass unavenged'.[4]

Transmutation was only one of the 'natural' laws useful to the radicals for purging 'unnatural' customs, and not the main one at that. In a medical context, Lamarckism was inevitably embedded in Etienne Geoffroy St. Hilaire's wider morphological framework. This was based on the theory of unity of composition – an extreme unity that sought structural relations between insects, molluscs and vertebrates. Indeed, Geoffroy and his doctrinaire disciples accepted a common structure for *all* organisms, however 'remote from each other in the scale'.[5] Previous studies of philosophical anatomy have touched on the professional uses to which the teachers put the new science, in their attempt to raise their stock as low-status outsiders in the medical world. Here I take a different approach and concentrate on the radical demagogues and their more extreme doctrines put to more political uses. I shall in fact argue that Geoffroy's morphological package attracted the medical radicals more than any other group, but what will really concern me here is the *context* of their appreciation.

As for the old 'corruptionists' targeted by the radicals, I will focus on the Tory–Anglican dynasties in their corporation 'pest houses'.[6]

Democratization was essential for the dissenters, inside and outside medicine; without it, the repeal of the Test and Corporation Acts in 1828 would have been meaningless, for they would still have been unable to compete for office. The effects of democratization were to be dramatic, as can be judged from the fact that the Whigs' 1835 Municipal Corporations Act, enfranchising the towns, effectively resulted in power passing to a new Nonconformist elite. At the same time there was an aggressive campaign to reform the chartered medical corporations – the Royal College of Surgeons (RCS) in Lincoln's Inn Fields, and the Royal College of Physicians (RCP) in Pall Mall, both run by closed oligarchies, trading on family connections and court patronage. Without doubt Royal College reform was *the* issue in the medical press throughout the 1830s. Radicals loathed the corporate 'toad eaters' in their 'conservatories of bigotry and ignorance',[7] and sought consistently to undermine their scientific and moral credibility. The 'reptile press', as opponents dubbed it, indulged in smear campaigns to discredit the sergeant-surgeons. Thomas Wakley's *Lancet* (founded in 1823, and selling four thousand copies a week in the late 1820s) demanded an end to the RCS Council's self-election, nepotism and discriminatory legislation. In Wakley's projection, medicine was to be democratic and responsive to a wider bourgeois-dissenting constituency, nepotism was to give way to the ballot, and trained specialists were to replace corporate appointees in key posts.

The tendency in the past has been to see 'scientific naturalism' come in with the Huxleys and Tyndalls in the 1860s, when it was used to demarcate the scientists' area of professional concern from the clergy's. In fact, it was well established among the 'marginal' men by the 1830s. Also, my reading of the medical record suggests that the introduction of the new naturalistic sciences about the time of the Reform Bill (1832) had more to do with these radical dissenting campaigns against Tory–Anglican privilege than with professionalization *per se*. (At least as it is understood by historians studying the 1860s, although Benthamites were of course intent on creating a new bureaucratic class of 'experts'.) This means that a more political framework is needed to tackle the naturalistic anatomies of the 1830s. Indeed, historians are acknowledging the fact. The importance of the new bourgeois civic sciences is now duly recognized: the nebular hypothesis, statistics and phrenology are all being studied intensively. Arguably, Lamarckism was only a more extreme form of this contemporary 'crisis knowledge', deployed at this time of changing industrial relations, when both the working-class demagogues and bourgeois democrats were making new political demands.[8]

The new Wakleyan radicals and urban Benthamites made their presence widely felt in the London institutions of science. The Zoological and Botanical Societies, Royal Institution, Royal Society, Geological Survey and statistical offices were all affected by their presence. Even the British Museum's aristocratic and Church Trustees were ignominiously investigated by a Select Committee headed by the soap maker MP, Benjamin Hawes, intent on pushing reforming zoologists on to the board.[9] Thus the Benthamite 'revolution in government' went far beyond the establishment of state bureaucracies, and factory, health, Poor Law and Church commissions; it directly affected the management and content of science. And in this respect, it was the Benthamites' flagship enterprise, the London University (LU), opened in 1828, which was conceived as the main educational instrument for creating the new class of 'experts' – transforming the sons of the *arriviste* middle-class dissenters into a professional elite to run the new legal, medical and state administration.

The London University

The dissenters and Jews were largely responsible for raising the £150,000 capital to finance the joint-stock University, where medical education was to supersede the exclusive corporation approach. That was not all the school's projectors hoped to supersede. At Oxford and Cambridge, medical education was expensive, poor in quality, and inaccessible to the dissenters. It was geared to cultivating what critics called the 'drawing-room ornament', a gentleman physician who ministered to the gentry. The pure physician's place was considered 'next to that of a Knight'.[10] In truth, the court physicians, like the sergeant-surgeons, *were* often knights, and all enjoyed considerable privileges. The King's physician, William Macmichael, justified the practice of reserving RCP Council seats for Oxbridge graduates because this schooling guaranteed their moral standing and familiarity with the gentry. By the 1820s, however, reformers of all shades baulked at the claim that 'moral' qualifications transcended medical expertise, condemning any sacrifice of science on the altar of Latin learning.[11] While the courtiers ministered to few wealthy clients and ignored the growing public health movement, the University and private medical schools were now to turn out a different type of professional: the improving GP and civic health expert who was to minister to a larger trading clientele.

As one might have expected from its Council of lawyers, merchants, medical reformers and MPs, the University was to train medical and legal personnel for Benthamite service. The educators were not interested in 'polite knowledge for an hereditary elite', but in providing 'professional tools' for the reforming bureaucrats. They were also to create a constituency of specialists, who were to become the new educated electorate after the reform of the professions. Courses at the University were cheap, secular and dominated by the medical, legal and economic sciences – in other words, quite different from those in the Oxbridge 'seminaries'. More specifically, the Benthamites intended medicine to be more academic and of a much 'higher character' than that taught in the corporations. The University's anatomy teachers were to lay greater stress on theoretical developments and continental science; and they were to take a rigorous comparative approach where the surgeons concentrated exclusively on man. In short, the students were to be exposed to a naturalistic philosophical anatomy that transcended the surgeons' 'superannuated system', based on craft practice, Paleyite design and expensive apprenticeships.[12]

The University's Education Committee deliberately recruited academically trained teachers fluent in the continental languages. Many were Edinburgh–Paris educated and viewed French anatomy through Scottish spectacles. The medical professors discussed in this essay give an idea of the scientific mix the Committee was after: James Bennett (anatomy demonstrator, 1828; professor, 1830–1) had run his own English anatomy school in Paris; Robert Grant (professor of comparative anatomy and zoology, 1828–74) and William Sharpey (professor of physiology, 1836–74) were Edinburgh graduates and former extramural teachers; and Jones Quain (professor of anatomy, 1831–5) was recruited from London's Aldersgate Street School (founded in 1826 by William Lawrence among others). All were philosophical anatomists and all were hired in part for their Francophile learning. Grant was highly regarded by Geoffroy, and a sponge expert whose seminal papers had been translated into French – a fact that raised his stock among reformers complaining of the backwardness of British science. Sharpey had studied in Paris and under Friedrich Tiedemann in Heidelberg, and had taught in his own school next door to Robert Knox in Edinburgh. Bennett designed the London dissecting theatres along French lines and he specialized in 'L'Anatomie Générale', or the science of tissue structure and organ homologies.[13] He worked closely with Grant, both men applauding the bureaucratic structure and comparative approach of Parisian anatomy. Through their joint effort they were to promote Geoffroy's science, teach the laws of organization common to all life, and move beyond the surgeons' human-based approach.[14] Students were to be immersed in the newer continental developmental sciences, and taught to view man both in terms of the laws of embryology and of the progressive history of life. This was what was meant by giving them a more 'philosophical' outlook than their hospital counterparts.

Bennett and Grant immediately showed their antagonism to the London corporations. Not surprisingly so in Bennett's case. His original Paris school (in a rented theatre in the hospital of La Pitié), although well attended by English students, had been forced to close partly as a result of RCS action (or inaction). The College had refused to recognize his certificates and stopped the Foreign Secretary from intervening when the hospital withdrew Bennett's facilities during a period of conservative reaction in 1825. Grant deplored these 'exclusive and obnoxious' corporation powers as 'contrary to reason, justice, expediency, and public good'.[15] The College of Physicians he castigated as a 'kind of aristocratic high Church Establishment' (because it only admitted Oxbridge graduates as fellows). He refused to acknowledge its right to retest Edinburgh licensees and, as a result, even though he was a doctor, he never obtained a London licence to practice, admitting bitterly in 1841 that he had been unable to write a prescription for fourteen years even 'to save a brother's life'.

Ironically for Bennett, many students had been driven to France in the first place by the RCS's attempts to contain or close down the local London schools, as we will see. In Paris during the tense period following Charles X's accession in 1824, with the rise of church power and at the same time of republican militancy, these students found an increasing polarization of approaches to comparative anatomy. Cuvier was subordinating structure to function, and branding serial development, recapitulation, transformism and unity of composition as related concepts founded on the false premise of nature's autonomy. Geoffroy, by contrast, offered a self-contained deistic explanation of life, in which the laws of homology, embryogeny, teratology and development were a consequence of the fundamental unity of structure of animals. His anti-Cuvierian and anti-vitalist rhetoric was aimed at the younger medical reformers and republicans.[16] No one was left in any doubt about the contingent political ties of his science; nor of Lamarck's, with Geoffroy now praising his laws of transmutation. During this period of religious revival and growing republican activity, Geoffroy's and Cuvier's relations degenerated, until the famous eruption of hostilities at the Académie in 1830, shortly before the Revolution.

Geoffroy's ideology, manifesting in the sovereignty of material laws, appealed to the visiting British radicals. They too stood opposed to cloisters and corporations, and denied the static Paleyite design on which the old religious and medical leadership based its moral authority. The radicals themselves accepted what Knox – the most famous

The London University on its completion in 1828. The medical school was to offer a new-style academic education, which embraced both comparative and philosophical anatomy. (Reproduced by permission of University College London Library.)

Scottish extramural teacher in 1830 – would later call 'a self-created, self-creating world' (an autonomous, self-developing nature pregnant with moral and political implication).[17] Matthew, who shared many of Knox's political and anti-Kirk views, also believed that 'living organized matter' had a 'plastic quality', which gave it a 'self-regulating adaptive disposition'. Like so many bourgeois reformers, he invoked a sort of political economy of the body, which responded to conditions through a series of self-adjustments.[18] We have seen that he applied the 'natural' laws of competition to his self-progressing biology and to vitiate all *ancien régime* monopolies. Just as competition in nature removed the runts and permitted species self-improvement, so in society it legitimized 'natural' free-market operations. For Matthew and Knox it was wholly 'natural' that the manufacturers should outlaw trade tariffs and aristocratic privilege, clearing the path for capitalist 'self-government'.

Imported into London, these self-developmental sciences slipped easily into a Wakleyan context. In so far as they highlighted the laws of competition and progress, they fuelled democratic demands for *concours* (open competition for posts, as in France). But they also served in a somewhat distinct bureaucratic Benthamite setting. Here the emphasis was less on *laissez-faire*, and more on replacing the old surgeon with a public health specialist armed with a new set of professional tools. At the University, Grant's and Bennett's courses centred on the cluster of concepts despised by Cuvier: Geoffroy's unity of composition, the Lamarckian animal series and foetal recapitulation. Like all the radicals, Bennett and Grant backed Geoffroy in his Académie clash and replaced Cuvier's discrete *embranchements* with a 'uniform' serialist taxonomy. In other words, following their French heroes, they provided a naturalistic explanation of form and development based on a kingdom-wide structural unity. This secular self-sufficient anatomy was widely praised as the sort of 'philosophical' science needed to replace the old surgeons' practice,[19] and from the late 1820s it spread rapidly through the private medical schools, Benthamite bureaucracies and radical medical unions.

There was of course opposition to the new science, some of it even from the older Whigs in the University itself. For instance, the pedestrian anatomy professor Granville Pattison had actually sparked student riots for his supposed incompetence and his *'disgusting indifference'* to the new developmental morphology; or, as a pupil said, for 'failing to reveal to us the researches of Tiedemann, Meckel, Serres, Geoffrey [*sic*] St. Hilaire, and others, into the laws of organization'.[20] Pattison deplored Geoffroy's *'extravagant'* theories and championed the practical approach of the local surgeons. But the student leaders echoed Benthamite claims about the *'low'* state of London anatomy and the need to overcome this old corporation mentality. Bennett had originally been hired as the anatomy demonstrator, even if he did end up teaching pupils. As the students now trickled away to study the new science under him, a horrified Pattison accused his demonstrator of threatening his livelihood. And when Grant intervened on Bennett's behalf, Pattison mocked him too for wasting the students' time with the 'unprofitable' speculations of foreign morphologists.[21] The press split along political lines, with the conservative *Medical Gazette* championing Pattison and the *Lancet* demanding that he be sacked. Student pressure actually resulted in Bennett's instatement as Professor in 1830 and Pattison's dismissal the following year. It was the end of an episode that illustrated the fierce feelings generated by the new science. After Bennett's premature death in 1831, the University was careful to pick anatomists with morphological credentials. Jones Quain arrived confirming that the new comparative approach was absolutely essential; indeed, that human complexity could only be unravelled by treating man as the end-product of a single taxonomic and foetal series.[22]

Philosophical anatomy was welcomed by the reformers for its reduction of the arcane facts of morphology to a set of 'general laws'. Reviewers talked of it ridding English physiology of the sort of vitalism that still afflicted the surgeon-baronets, and told the Oxbridge and hospital professors to stop revelling in their 'ignorance' of true science.[23] Rubbing the point in, the reviewing quarterlies – the reformist *British and Foreign Medical Review* and *Medico-Chirurgical Review* – ran lengthy papers on Geoffroy's and Serres' theories to show that the 'elements' of animal structure were 'everywhere identical' and 'disposed according to invariable rules'. Reviewers emphasized that homologies were the 'great principle' of nature, testifying to an overall unity of structure, and that a recapitulationist embryology was essential to the science of teratology for revealing the 'precise rules' that govern foetal retardation.[24] As we will see, this accent on the 'rules' of morphological development was very important to the Benthamites. Emphasizing that human and animal forms obeyed the same strict laws, interpretable by the new improving professionals, they could undermine the authority of the Anglican-corporation teachers, who were promoting Paley's natural theology and Cuvier's functional anatomy. The reformers insisted that higher anatomy, because of its lawfulness and universality, gave the general practitioner an understanding that raised him above the purely craft-based surgeon. The reform journals applauded the new anatomists for bringing nature, man and mind under the control of the laws of progress and development; for pointing out, in effect, that nature was a legislated *'process* of change'.[25]

The private medical schools

These naturalistic anatomies also swept through the radical private anatomy schools in the metropolis. Again, to understand the attraction of these Parisian sciences we have to consider the anti-corporation feeling, civil disabilities and dissenting allegiance of many of the school staff. The nonconformists were still handicapped in the 1820s. By law they were forced to marry and solemnize births and deaths in an Anglican church. They were barred from taking degrees at Oxford and Cambridge, and thus from the Fellowship of the College of Physicians with all the privileges this entailed (including access to many hospital posts). By the early 1830s, noncomformists numbered over four million and were fiercely resisting Anglican demands. In 1833, for instance, the clergy took out fifty thousand summonses against dissenters refusing to pay the tithe money. Militancy among the medical dissenters increased as they wrestled with Tory privilege and encountered the 'vineyard of corruption' in the corporations and hospitals.[26] Many of the medical teachers were in fact active in both the movement for the removal of disabilities and in the more specific campaign to democratize the medical corporations. They were demanding a share of power in the unreformed parliaments of medicine – the Royal Colleges.

Seventeen private medical schools existed in London about the time of the Reform Bill. Most offered cut-price courses, partly to draw away the hospital students, but also, as the physical-force radical George Dermott put it, to knock 'down *the golden bar of exclusion*' and attract a humbler clientele.[27] The school owners were less wealthy (hospital surgeons could command anything up to £1000 for an apprenticeship), and their family backgrounds were often in trade, provincial medical practice, or the nonconformist ministry. Dermott, proprietor of the Gerrard Street School in Soho, was a Wesleyan minister's son; the parents of his colleague, the Quaker phrenologist John Epps, owned butchers shops; and Richard Grainger, running his school in an old church in Webb Street, was the son of a Birmingham surgeon. Many of these 'marginal' men were unrespectable in the consultant's eyes. The begrimed Joshua Brookes in Blenheim Street might have been a good teacher and have owned a huge museum to rival the RCS's own, but he was still 'not a gentleman, and very dirty'.[28] His protégé, Dermott, was by his own admission neither a 'pretended gentleman nor a pretended surgeon'; he dictated notes in a gin palace, threatened the monopolists with fisticuffs, and turned up in class with radical petitions for his students to sign. These teachers, specializing in turning out a new sort of medicaι tradesman, simply failed to meet the knightly surgeons' social ideal.

Antagonism between these marginal men and the hospital elite was fierce in the 1830s. It was due partly to business competition – for students and cadavers – but there was also class and often religious friction. The surgeon-baronets and Anglican physicians deplored this cheap down-market education, believing it vulgarized a gentleman's profession, and they accused Dermott and his like of letting tinkers into the profession. The RCS councillor was a wealthy hospital consultant with a triple role, that of teacher, licensing examiner and medical legislator; and it was the corrupt use of this power that caused such widespread condemnation. Most critically, the surgeons attempted to stop the growth of the schools through legislation. From 1822 the RCS refused to accept summer course certificates (many schools taught summer courses after Brookes' pioneered a technique to preserve cadavers), and in 1824 it decreed that only certificates countersigned by a hospital surgeon would be accepted from students wanting the College diploma. Largely as a result, a number of schools collapsed in the late 1820s and early 1830s, including Brookes', while others like Grainger's went into decline.[29] The proprietors condemned this misuse of power for party ends, claiming that the new by-laws effectively redirected students to the surgeons' own hospital winter courses. The issue was perennially aired at medical rallies, where Bennett and Brookes (who died penniless in 1833) were raised to the status of martyrs.[30] Faced by what it considered College aggression, the private teachers' paper, the radical dissenting *London Medical and Surgical Journal* (1828–37), moved sharply leftwards in the early 1830s. It agitated for better pay for practitioners, demanded council elections, and campaigned for Wakley's return to Parliament (he fought Finsbury for the radicals, and was finally elected in 1835).

The schools' anatomical approach differed from the clinical education available in hospitals. Study of the *Journal* confirms that a developmental morphology was being sold to the GPs. It published and praised John Fletcher's morphological lectures. These were anti-vitalist and supported an extreme unity of structure, Fletcher (like Grant) emphasizing that the organs of the lowest animals were 'fundamentally identical with those of the highest'.[31] A merchant's son, Fletcher had himself progressed from the 'counting house' to the Argyle Square school in Edinburgh, where he taught physiology from 1828 until his death from lung disease in 1836. His *Rudiments of Physiology* (1835–7) was a leading example of the Geoffroyan genre, and praised by the *Journal* as the best work of the day.

The medical academies were no Anglican seminaries celebrating a rigidly stratified, static creation. These radical dissenters accepted the precepts of a self-empowered nature. As reductionists they explained organic wholes by their physico-chemical constituents, and as democrats they argued that political constituents were empowered to sanction a higher authority – a philosophy consonant with their ideas of congregational polity and lay democracy. Many took issue with clergymen using direct creation and Paley's natural theology as divine sanctions for the social order. They opposed Anglicans like the Cambridge geologist Adam Sedgwick, who spoke as if the safety of the church and state depended on a teleological biology proving an 'overruling Providence'.[32] Radical dissenters refused to accept this supernatural sanction for the status quo any more than they obeyed the magistrate's biblical injunction to 'be subject unto higher powers'. They denied that nature and society were supernaturally manipulated through what the Oxford geologist William Buckland called 'Creative Interference'. For radical nonconformists God had established self-regulating moral and physical laws at the Creation, disclosing them directly to the individual through His word and works. Indeed, because of this they argued that the state was quite wrong to sanction an 'official' priesthood between man and his creator, or to pass laws favouring one sect over another. We must remember that the whole debate was inextricably linked to the campaign for disestablishment. There being no divine interference, dissenters denied all supernatural sanction for a state-endowed clergy. At the height of the campaign, Epps damned the 'fornicating' church as itself 'a subverter of the good order of society' for its adulterous connection with the state, demanding the break up of this *'illicit embrace'*.[33]

The medical radicals thus promoted the sciences despised by the Anglican divines, wielding the works of Geoffroy, Lamarck and the phrenologist George Combe as symbols of a defiant nonconformity. In an age of electoral reform when, as Wellington complained, power was tilting towards the merchant dissenters, the spread of these 'self-development' doctrines was seen to presage the passing of exclusive Anglican authority. It was no wonder that Anglicans worried about Geoffroy's 'dark school' gaining ground, or that they saw 'ruin and confusion' in the transmutation popularized in publisher Robert Chambers' *Vestiges of the Natural History of Creation* (1844).[34] At the same time, some reformers praised *Vestiges* precisely because it provoked the creationist 'bigots'; while for many it provided an 'ennobling' vision of lawful development more consistent with progressive dissenting thought.[35]

The radicals' attempt to shake the Church's hold on science and state is apparent in their denunciation of the claims of Oxford and Cambridge to professional hegemony. The medical dissenters demanded parity with the priesthood in pay and representation, and promoted Combe's *Constitution of Man* – with its acceptance of the mental faculties as part of man's organization – in order to extend their physiological control over the clergy's 'moral' domain. Where Sedgwick execrated attempts to conflate mind and matter, the radicals insisted that there was no reason to 'separate the study of physical and moral states'.[36] At times, even direct 'hands off' notices were issued to the parsons: Dermott, a fierce mental materialist, warned them in no uncertain terms to leave the physiology of the brain exclusively to the new medical expert.

The leading private teachers provided damning evidence of corporation malpractice before Warburton's Select Committee on Medical Education in 1834, condemning the council's self-election system as corrupt. They wanted 'equality of rank', total suffrage and a surgical free trade, with hospital posts open to competition, not treated as private property. Even well-heeled moderates by the early 1830s were advising the 'medical *magnates*' to reform their rotten-borough practices before they sparked a July Revolution in medicine.[37] But the called-for elections would have given the Colleges of Surgeons and Physicians quite different constituencies. Whereas the sergeant-surgeons or court physicians responded to the gentry and wealthy merchants, new councillors elected by the GPs would have been responsible to rank-and-file demands, and would have threatened the lucrative interests of the consulting elite.

Many of these 'outsider' schools were staffed by Wakleyans and Benthamites demanding the state control of medicine and establishment of Bentham's Ministry of Health. Take Richard Grainger's Webb Street school. This had its Benthamite cadre; indeed, it had more than a Benthamite corpus, it had the corpse as well. For it was here in 1832 that the utilitarian leaders gathered to hear Bentham's friend and physician Thomas Southwood Smith (Webb Street lecturer in forensic medicine) deliver an oration while dissecting Bentham's body.[38] Schools like Grainger's provided a pool of specialist Benthamite talent. Smith was a Unitarian minister and Edinburgh-trained physician at the hub of London's utilitarian community. He served in the upper echelons of the Benthamite bureaucracy, working with Edwin Chadwick to control child factory labour and improve town sanitation. Grainger was also part of this community and active on a number of hospital and health commissions.

Such schools provided something of a crucible, in which the materialism of the older Unitarians was alloyed with the morphological approach of the young reformers. Grainger adopted the French morphological package underpinned by Geoffroy's unity of composition. Many others did too. The former Hôtel Dieu house surgeon Thomas King, a fierce radical once thrown out of the RCS theatre by Bow Street policemen, reopened Brookes' school in 1833 as a forum for 'general anatomy', repeating in his inaugural address the Geoffroyan dictum that there was 'no absolute line of demarcation between one Organism and another; they seem all to be formed upon the same general principles'.[39] These teachers used the progressive series of animal organs to explain the complex analogous organs in man. None was actually a Lamarckian, but all approached transmutation calmly. Fletcher's was a common tack. He distinguished between organs and organisms. He accepted an extreme unity of structure among the analogous organs – say, the liver as it progessed through the chain from polyp to man – and that 'the history of the advancement of each organ towards perfection is merely the history of the progressive development of an imaginary unity'. But this was different from saying that each organ*ism* is built from 'the one below it' by a serial transmutation. For the different organ 'lineages' do not necessarily progress at the same rate. Thus, relative to the other organs in the body, the liver's development might be 'quite irregular', making the 'mutual relations' of the various organs possibly very different at 'every step in the ascending scale'. So there need be no synchronized ascent within the organ-assemblages, even if such synchronicity was fundamental to Lamarck's scheme. Fletcher criticized the foetal evidence for Lamarckism in the same way. The human embryo's recapitulation of the ascending animal series did not necessarily support transmutation, because the organs during ontogeny also developed differentially. So the foetus never entirely corresponded to any adult animal lower in the series.[40] Likewise, King, while accepting the contiguity and common composition of organs, denied that foetal recapitulation was proof of transmutation. None the less, this Geoffroyan, recapitulationist and serialist framework did make the implementation of Lamarckism much easier. And in fact Grant supplemented his meagre university income by teaching comparative anatomy in the private schools, giving the series a literal Lamarckian gloss.

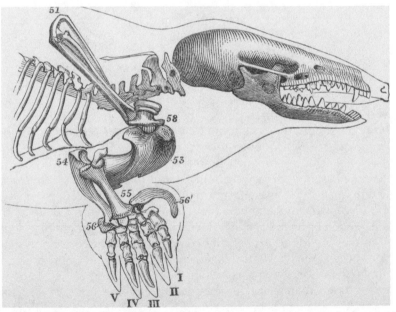

From the late 1820s all philosophical anatomists in Britain accepted a 'unity of plan' among vertebrates, according to which, for example, the mole's trowel (above) and the bat's wing (opposite) were composed of homologous bones. But some, notably the Lamarckians, went further with Geoffroy and postulated an extreme unity stretching from the invertebrates to man. (From Richard Owen, *On the Nature of Limbs* [London: Van Voorst, 1849], pp. 6–7.)

The Unitarian intelligentsia played a key role in establishing these reductionist physiologies and higher morphologies. Southwood Smith's meliorist *Divine Government* (1816; 5th edn 1866) expounded an extreme naturalistic Calvinism and claimed 'divine authority for a deterministic "Law of Progress"'.[41] He accepted the total genetic and environmental predetermination of life, in which even man's free will was an illusion. Like other medical Unitarians, he rejected vital laws and divine fiats. And he believed that, with active matter alive to the divine Presence – Unitarians accepted that God was '*actually* present in every part of the universe', inviting accusations of pantheism from Coleridge – life could be defined physiologically by the phenomena accompanying organization. Physiologists who acknowledged this could also accept a form of naturalistic development. It is well known, for example, that William Benjamin Carpenter, a Unitarian minister's son (like Smith, with a Bristol–Edinburgh background), went on to patch up *Vestiges*. The Benthamite social underpinnings of Smith's own developmental biology were vividly displayed in his *Divine Government*. Social reform and organic mobility were knitted together in Smith's theodicy. The Benthamite dissenter was duty bound to expose the evils of ill-health, poverty and factory exploitation and to reform society in line with predetermined divine intent. At the same time he was to awaken the poor 'to the need for self-development' by providing them with cheap education.[42] As commercial and medical entrepreneurs, the Unitarian elite believed that the working classes must be taught to strive in order to escape the Malthusian poverty trap. But for a medical Unitarian this social development – benefiting capitalist society by creating free-market mobility – was 'inseparably connected' with nature's own development, and subject to the same divine dispensation. All animals were driven to escape the evil of imperfection; 'striving' was therefore a natural act, enabling creatures to escape their lowly taxonomic station. In Smith's 'social Lamarckian' biology, all animals were 'continually advancing from one degree of knowledge, perfection, and happiness to another'. All beings were 'destined to rise higher and higher in endless progression' and 'to contribute to their own advancement'. This active and 'endless progression' illustrated in the popular *Divine Government* explains why some medical Unitarians accepted transmutation with equanimity; and indeed, why they preferred some forms of it to others.

For instance, not only were many Unitarian doctors *not* flummoxed by *Vestiges*, unlike their Anglican rivals, but some actually criticized Chambers' understanding of law-as-logos and took a more materialistic stance. The Unitarian intelligentsia could no more accept reified, active laws interposed between God and his creation than they could endorse Coleridge's 'half-crazed' idealism.[43] In Smith's words, organisms were not to be pulled from above, but they were 'to contribute to their own advancement'. Carpenter, who corrected many of *Vestiges'* early errors, typically eschewed vital agents and promoted a naturalistic Calvinism; for him, organisms were sovereign beings, internally fired by the divine Presence. His was a science of *'Law and Order'* rather than of outside 'Creative interference'. He enthusiastically accepted the new higher anatomy and even as a student promoted a structural and mental continuity that transcended Cuvier's discrete *embranchements*. He also exploited it in his educational strategy. He was a staunch supporter of the London University and campaigner for wide educational reform. He wrote for, and eventually edited, the *British and Foreign Medical Review*, a quarterly that fought to gain better opportunities for the dissenters' sons. The *Review* wanted secular institutions like the London University built right across the Empire. It also proposed that the metropolitan university itself be granted greater civic powers, for example by making it the sole examining and licensing body in England (thus stripping the corporations of both function and fees), something that would have given the Gower Street Benthamites a great advantage over the Oxbridge Anglicans. Science was integral to Carpenter's strategy to break the Oxbridge-corporation monopoly and claw more power for the dissenting *nouveau riche*. He wanted medical education in general standardized along London University lines, with study of the latest developmental anatomies substituted for the *'exclusive'* Oxbridge classics education.[44] The new-breed physiologist was to be taught a materialist science of 'general laws' that transcended the teleological science of small-scale adaptations so appealing to the Cambridge Paleyites. In the propaganda war, medical dissenters made it plain that the new practitioner was to be grounded in a science of greater scope and power than his Oxbridge counterpart.

Medical agitation peaked in the mid-1830s, when even moderate societies carried what Tories decried as 'wildgoose and wanton' motions, demanding the merger of the old estates of physic, surgery and general practice into one democratic faculty.[45] Medical unions too proliferated at this time. The most powerful in London was the aggressive British Medical Association (BMA), formed in 1836 by a group of South London GPs, but quickly embracing University and private school activists (Grainger, Grant and Marshall Hall), anti-Malthusian statisticians (William Farr), and journalists like Wakley. The BMA provided another crucible for the smelting of radical morphologies and democratic politics. Its leaders denounced Oxbridge-corporation monopoly (most were London or Edinburgh graduates

and ineligible for a Fellowship of the RCP) and the 1834 New Poor
Law (which was forcing down the workhouse GPs' pay), while demanding a one-faculty democracy. The Association's president declared that BMA activists were promoting science by removing 'the
trammels which the corporations had placed upon it'.[46] In fact, they
were promoting a *rival* science to gain the moral edge and commercial advantage over their corporation rivals. It included an academic
higher anatomy and Marshall Hall's mechanistic reflex arc (a form of
automatism attractive to the radicals). Typical, too, was the group's
environmentalism. This manifested itself as strongly in Grant's Lamarckism as in his pupil William Farr's environmental explanation of
poverty and disease, an explanation that was used to justify the
demand for government intervention in directing medicine to the
wider social ends ignored by the corporation consultants.[47]

To sum up, we have seen that much of this naturalized Calvinist
science ran counter to the natural theological ethos of Oxbridge-
corporation anatomy, in which function was seen to determine structure, adaptations were proof of providential design, and mind and
nature had a supernatural aspect. Many radical dissenters stripped
nature of this supernatural content. Epps praised the flamboyant
atheist John Elliotson for repudiating the superstitious churchmen
who would 'clothe palpable facts and sensible manifestations with
a spiritual halo; as if truth could be more sacred when thus surrounded'.[48] We have also seen that stripping away this spiritual garb
served a powerful religious and political purpose. It vitiated the
clergy's claim to a moral authority based on their mediating role in
natural theology, and was in line with the dissenters' belief in the
priesthood of all believers and the right to private interpretation of
the Bible. Dissenters berated the Oxbridge and hospital teachers who
promoted a Paleyite teleology for failing to 'penetrate beneath the
surface', for concentrating on the functional uniqueness of each form
and reducing anatomy 'to the sterile observation of facts, without
reciprocal connexion, rational analogies, or possible consequences'.[49]
An anatomy that cut the Paleyite ground from under established
religion and elite medicine was obviously useful to the Wakleyans
and dissenters, intent on storming the 'brazen walls of unjust mono-
polizing corporations'.[50] All the RCS elders, including Wakley's *bête
noire*, Sir Astley Cooper (RCS President 1827), head of a surgical
'family' founded on nepotism (seven of his relatives and apprentices
held posts in Guy's and St Thomas's Hospitals), employed Paley's
design arguments. Knox was not alone in castigating Cooper's anatomical theology as 'downright nonsense'.[51] Radicals insisted that it
was unable to compete with Geoffroy's science. The new morphology

could explain Paleyite anomalies such as rudimentary organs (male nipples) and arrested developments (hare lip); at the same time, embracing the whole of nature's plan and development, it provided the dissenting professionals with more sweeping powers of prognosis, prediction and control. In short, disgracing Paleyism with a piece of naturalized Calvinistic legislation was a potent means of undermining the courtiers' moral claims to medical authority. Nature itself was being shown voting for social reform.

Meeting the democratic threat

So the nonconformist and non-hospital teachers stood hammering on the corporation doors demanding an end to 'Old Corruption' and a medical emancipation bill. They directly threatened the interests of the elite surgeons, who stood to lose politically from council elections, financially from the withdrawal of their consulting privileges, and morally from an anti-Paleyite academic anatomy. The wealthy oligarchs now responded in kind. Just as the new 'class' of GPs were said to have 'completely deranged' the 'natural' estate system,[52] so their teachers were accused of destabilizing the profession with their petit-bourgeois cost-cutting and trade in base physiologies. The corporation elders and conservative press denounced the militants in the 'lecture-bazaars'. An un-Christian university was unthinkable to Anglicans; by 'converting science into a matter of traffic' it would cause a moral decline, with medicine passing into 'lower hands'. The University's foreign curriculum was depicted as unpractical and unpatriotic, and the surgeons – identifying their target precisely – demanded that the continental 'doctrines of *homologues*, and *heterologues*' be abolished in favour of sick-bed and clinical studies.[53]

The University's curriculum was far too mechanistic for the older Whig surgeons. RCS councillors were appalled at the relegation of providential design and substitution of a set of morphological laws for craft practice. The suave Charles Bell, the RCS professor of anatomy appointed to the LU chair of surgery by Henry Brougham (against radical advice), criticized this naturalistic morphology as the antithesis of the local surgeons' practical approach. He deplored Bennett's French affectations and accused him of propagating anatomical 'systems at variance with our own most approved opinions'.[54] Bell hated Geoffroy's 'cold' materialism. He had always hoped that the University would become a leading 'school of Design'. But it was not to be. Disappointed, he withdrew from Gower Street in 1830, criticizing the Council for taking the continental road to ruin. Nor did the other dissenting sciences escape his censure. He detested Southwood Smith's Unitarian pantheism. When Smith wrote his *Animal Physiology* for Brougham's Society for the Diffusion of Useful Knowledge, Bell, as a referee, branded its materialist tenets 'offensive' and pernicious.[55] The Society was, after all, designed to feed the working-class consumer with a wholesome knowledge, and Bell himself was promoting a more paternalist doctrine: that life was a divine 'endowment' – it *produced* organization, rather than resulting from it. These disputes suggest that the entire Broughamite educational empire suffered from a critical radical–Whig ideological split, with the radicals arguing for a more materialistic self-determining nature, and the Paleyites promoting a delegated divine power of arrangement.

But it was the reaction of the RCS Coleridgeans that provided the real scientific breakthrough in meeting the threat. They shifted the whole ground of the attack against Geoffroy and Lamarck. Coleridge himself disliked Paley's 'tricksy sophistry' and thought that belief in Christianity had to come from the heart.[56] He had no truck with the design arguments of the old-school surgeons – the Bells and Coopers – and attempted to move morphology on to a more sophisticated conservative plane. This shift was essential. RCS councillors had long been accused of being ignorant of foreign science, cut off on their 'Lilliputian' island in Lincoln's Inn Fields'.[57] Coleridgean patriots were now to turn the tables, fashioning an effective transcendental alternative to the democrats' materialist morphology, one that would return sovereignty to the godhead and authority to the traditional elite.

Foremost among the RCS idealists who challenged the democrats in the medical manufactories were Coleridge's close friend Joseph Henry Green and his protégé, the Anglican Richard Owen. Green was physician to St Thomas's Hospital, and the professor of anatomy (from 1824), councillor (1835–63) and President (1849) of the RCS, where Owen was to become Hunterian Professor. Like Coleridge, these College gentlemen condemned the '*plebification*' of science in the 'lecture-bazaars', seeing the radicals' 'Ouran-Outang theology' breed only further blasphemy and discontent.[58] They were to seize the intellectual initiative, identifying the social Lamarckians as the real enemies for proffering scientific support for popular sovereignty. The romantics embraced Coleridge's call for a 'national clerisy', comprising gentry, clergy and elite professionals, to be educated as a class at the ancient universities. They believed that this must strengthen the union between science and polity, and halt the erosion of traditional religious values. The Coleridgean ideal understandably appealed to the corporation gentlemen, already ministering to the gentry – indeed, in the case of the physicians, educated with them at

A typical lampoon in an age clamouring against 'Old Corruption'. Here, the radical University supporters proffering 'Sense and Science' are outweighed by the Tory bishops speaking for 'Money and Interest'. The bishops were about to found the Anglican King's College in London, a Coleridgean college whose rival medical curriculum was shaped largely by the elite consultants from the College of Physicians. (Reproduced by permission of University College London Library.)

Oxford and Cambridge. They were adamant that this training for the clerisy would ensure an acceptable future for the profession. By the same token they were horrified at the radical proposal that the Home Office should control teaching and licensing. The mere prospect caused the *Medical Gazette* to create alarmist images of a 'revolutionized' profession, as in Paris, where medicine was an arm of the state, gendarmes were used to quell student riots, and Cuvier had been forced to lecture under military protection.[59] Such propaganda was given a hard edge by the fact that London's radical leaders (including Wakley, King and Dermott) *had* been summonsed for occupying the RCS theatre and student militants had disrupted classes at the University. But the underlying reason for opposing any state takeover was that the corporations would lose their licensing revenues, the councillors their seats, and the consultants their ability to use hospital posts as party gifts.

The romantics were sensitive to the political threat from below. Green, a conservative reformer, wrote tract after tract on medical politics, justifying a paternalist system controlled by the pure surgeons. Owen's Hunterian job was actually created as a consequence of the council's anti-radical strategy. He had been hired to catalogue the College's anatomical specimens in 1827, following the public outcry about the state of the museum and library. (Twelve hundred members held a public meeting in 1826 at the Freemasons Tavern to complain about the situation and petition Parliament.) There was real bitterness about the state and accessibility of the museum and library, which housed John Hunter's preparations and manuscripts (left by the government in the College's care). Accusing fingers wagged at the councillors. As one critic put it baldly, 'these dog-in-a-manger rogues, who have stolen John Hunter's museum, take good care to keep the profession excluded from all effectual access to it'.[60]

Owen performed valuable services to a besieged council. By laboriously cataloguing Hunter's collection, he helped undo the damage done by Trustee Sir Everard Home's burning of many of Hunter's manuscripts (the extent of which only came to light during Warburton's Committee hearings in 1834). Owen's contribution, said President G.J. Guthrie in 1836, left the Hunterian Museum superior even to the Jardin des Plantes. The career consequences for Owen were enormous. The radicals had expected Warburton's Committee to recommend a total reorganization of the College. Many hoped to see the democratized institution placed under the Home Secretary's control, and a new regulative, naturalistic anatomy given prominence, preferably by the creation of a chair. But the councillors stole a march on the Benthamites. As part of their piecemeal reforms, they endowed their own chair in 1836 and installed Owen in it. And who better? A young Peelite Anglican 'raised' to abominate the 'savages' of the reptile press. He had been trained, as the Tory hawk Sir Anthony Carlisle said, to meet the 'hopes & wants of the College'.[61] Owen was a client of the Oxford divines and Tory gentry, and a staunch defender of the council before the Parliamentary Select Committees. He was also a man for whom Christian duty meant action against the working-class militants in more practical ways; hence his enlistment in the Honourable Artillery Company, which provided policing back-up during periods of Chartist violence. Above all, he was supremely competent in comparative anatomy, well able to tackle the radical Francophiles on their own morphological ground – and for the Coleridgeans this was paramount. Reforming conservatives now proclaimed that in Owen's hands their own school of 'the "science of life"' would be superior to any in the city.[62]

But how could the College's Coleridgean science reassert the authority of the traditional elite? How could it vitiate the self-developmental sciences of the Wakleyans and radical dissenters flying tricolour flags outside the corporation walls? The materialist mental physiologies, Lamarkian zoologies and Geoffroyan anatomies taught by the democrats all supposed an atomistic self-sufficient nature. They bestowed sovereignty on the individual and sanctioned progression from below. Coleridge himself had abominated a Jacobin philosophy of self-empowered atoms, believing that the sedition of the Regency period was nothing but the 'diseased' fruit of this Revolutionary French science. He had complained to the Prime Minister in 1817 that the language of 'Demiurgic' atomism was nothing but a 'fac-simile' of revolutionary rhetoric, with each atom having 'an equal right with all other Atoms to be constituent'. For *idéologues*, democratic authority stemmed from these self-existent atoms, or as Coleridge put it, the republican 'Ruffians' armed with this philosophy believed that they could make a 'convention' and demand an 'indefeasible sovereignty over their Governors'. But Coleridge denied that individuals carried a mandate or comprised the state; rather, the state was the superior authority that acted over them and commanded allegiance.[63] Green now articulated a new anti-radical philosophical anatomy that chimed with the Coleridgean view of religious and civil authority.

By the 1830s it was republican talk of Nature's autonomy and self-evolving powers that proved most worrying, signifying as it apparently did independence of God and traditional moral authority and a new onus on secular democratic development. Tory evangelicals, identifying Jacobin philosophy as the root of all profligacy, condemned Lamarckism for 'blasphemously' imputing 'a period of comparative imbecility to Omnipotence'.[64] Indeed, in the attacks on the radicals' philosophy the concept of a *self*-animating nature was singled out time and again. Tiedemann's imported transmutation was decried as a 'godless, self-existing, self-destroying, senseless, aimless crotchet'.[65] And when Owen denounced Lamarckism, it was for its 'self-developing energies'. There were different ways of combating the democratic 'disease'. The old Anglican clergyman William Kirby responded to Lamarck's designation of nature as a 'blind power without intelligence which acts necessarily' by proposing a hierarchy of *'inter-agents'* connecting God, society and nature, upholding all 'by the word of His power': a sanction of Anglican authority and clerical mediation that brought groans even from medical reviewers who held no Lamarckian brief.[66] But much more sophisticated was the romantic morphology developed by the RCS Coleridgeans.

In Coleridge's romantically unfolding nature, each 'living germ' had its potential realized by higher productive powers, acting through nature as an expression of divine will. The 'philosophic Naturalist' for Coleridge could thus never be a radical atheist; he was a 'minister of the *Logos*', a member of the National Church.[67] The anti-Lamarckian application of Coleridge's idealism was developed by Green. A study of his political pamphlets and zoology lectures reveals how his anti-democratic and anti-Lamarckian philosophies rested on this same romantic substructure. Execrating Lamarck's 'fanciful' theory, he denied any 'power in the lower to become, or to assume the rank and privilege of, the higher'. Were this not so, the 'cause' in nature would always be 'meaner and feebler than the effect', making 'blindness . . . the source of light' and 'ignorance . . . the parent of mind'.[68] For the same reason, he dismissed calls for College democracy, denying that the rank-and-file were empowered to mandate a 'higher' authority. The members' millennial demands for a uniform education, access to all posts and voting rights would 'vulgarize the profession'. For him, councillors were literally the illuminati – 'the suns of the system . . . and not mere mirrors, reflecting only the light they have been previously bestowed'. Their authority and office was sanctioned by a superior power; it was a case of 'appointment by a higher', as distinct 'from *election* by a supposed lower' source, in line with the traditional command structure of paternalistic society.[69] The GPs' intellectual elevation was to come from association with the gentlemen surgeons ministering to the upper ranks, those with leisure and education enough to pursue pure science. Green's philosophy thus illegitimated Lamarckism and with it the democratic demands. His own idea of nature as a vassal state, subject to what he called a 'descensive' spiral of power, fitted closely with the paternalist ideal of delegated supreme power. Democratic ascent was as meaningless as emergent evolution: one rather awaited the King' pleasure, the other depended on God's will.

Given the beleaguered College context, we can see how an ideal anti-Lamarckian anatomy chimed with Coleridgean social philosophy to provide a Tory bulwark against cultural 'Spoliation' and democratization. The Coleridgeans were using a '*moral copula*' to connect the 'is' of nature with the 'ought' of politics.[70] They were collapsing biological and social doctrines into a unifying philosophy of hierarchic powers, each dependent on the one above– and all deriving ultimate authority from the supreme Godhead. Nature, society and the professions were unfolding under the same 'predisposing power of a Divine providence'. In this patrician view, the 'interdependent' classes making up the 'organic whole' of society were kept in their rightful place by the same divine power that animated nature itself.[71] It was the same chain of authority through nature, Church and society, with the clerisy in its mediating role sanctioning the traditional social order.

The politics of evolutionary naturalism

It is well known that Richard Owen, destined to become England's greatest morphologist, was in the 1830s devising a variety of anti-Lamarckian strategies, while in the 1840s he went on to investigate what Green called 'the eternal Ideas, which are the regulating types and standards' of animal life – the ideal 'Archetypes'.[72] This essay is designed to give a fuller understanding of the reason why such corporation men rejected the democratic sciences of self-development. The major context in which these radical sciences flourished – in the Benthamite and dissenting medical manufactories, radical unions and Wakleyan press – suggests that they were clearly favoured by the dissenters and democrats. Given this fact, we can begin to answer the question raised in the epigraph. The doctrines of scientific naturalism, in comparative anatomy at least, originated in republican Paris, and were actively imported into London and incorporated into Benthamite and radical dissenting strategies at the time of the Reform and Municipal Corporations Acts. It was not solely that the marginal men were using philosophical anatomy for their professional advancement – aware of their lowly status as physiologists and intent on raising their intellectual standing by proffering new sweeping laws of organic form.[73] The fact that the new sciences were taken up so selectively in a politically charged context suggests that there was a deeper dimension. These scientific commodities appealed most strongly to younger reformers, many socially handicapped nonconformists and secularists, who were attempting to break the traditional power of the old corporation and Oxbridge oligarchs. Hence it was the Paleyite surgeons and privileged Anglicans, in the old universities and College of Physicians, who took the brunt of their attack, and who reacted by denigrating the new materialist anatomies. So again it is right to question whether the new naturalism was 'organically' related to religious traditions. In our study group it flourished among the Presbyterians, Quakers, Baptists and Unitarians. It became a form of 'crisis knowledge' for a radical dissenting intelligentsia locked out of the corridors of power. Democratization did come – in 1843 at the College of Surgeons, and with the opening up of the hospitals to new bourgeois blood – and it was followed by a gradual decline both in radical agitation (with a collapse of the medical unions and private schools) and interest in the more extreme forms of Geoffroy's philosophical anatomy. As late as 1846, the LU and Paris-trained Joseph Maclise might still doubt 'whether absolute differences of form have any real existence, except as forms under metamorphosis', or that the 'great interest' attaching to the subject had 'worn off with the sunset of the great mind of Geoffroy St. Hilaire'.[74] But by now the standard was passing to Owen in an age of Peelite compromise, and a more conservative form of archetypal anatomy was already in the making.

Notes

It is a pleasure to thank Frank Egerton, Pietro Corsi, Jim Moore, Evelleen Richards, Simon Schaffer and Jim Secord for showing me their various works-in-progress; and also the libraries of University College London and the British Museum (Natural History) for permission to study manuscript material.

1. James R. Moore, 'Crisis without Revolution: The Ideological Watershed in Victorian England', *Revue de synthèse*, 4th ser. (1986), 62.

2. W.J. Dempster, *Patrick Matthew and Natural Selection* (Edinburgh: Paul Harris Publishing, 1983), p. 110; Kentwood D. Wells, 'The Historical Context of Natural Selection: The Case of Patrick Matthew', *Journal of the History of Biology*, 6 (1973), 238.

3. Evelleen Richards, 'The "Moral Anatomy" of Robert Knox', *Journal of the History of Biology*, in press; Adrian Desmond, 'Robert E. Grant: The Social Predicament of a Pre-Darwinian Transmutationist', *Journal of the History of Biology*, 17 (1984), 189–223; Frank N. Egerton, 'Hewett C. Watson, Great Britain's First Phytogeographer', *Huntia*, 3 (1979), 91–2. Dr Egerton's forthcoming biography of Watson will throw more light on his free-thinking radicalism.

4. Dempster, *Patrick Matthew*, pp. 98–100, 108; Wells, 'Historical Context', 238–9. How much Darwin's own evolutionary views reflected the contem-

porary political struggle for 'open competition' is discussed by John C. Greene, 'Darwin as a Social Evolutionist' (1977), in *idem, Science, Ideology, and World View: Essays in the History of Evolutionary Ideas* (Berkeley: University of California Press, 1981); Silvan S. Schweber, 'Darwin and the Political Economists: Divergence of Character', *Journal of the History of Biology*, 13 (1980), 195–289; and Robert M. Young, *Darwin's Metaphor: Nature's Place in Victorian Culture* (Cambridge: Cambridge University Press, 1985).

5. Robert E. Grant, 'Lectures on Comparative Anatomy and Animal Physiology', *Lancet*, 1 (1833–4), 89. For sophisticated studies of philosophical anatomy and its professional uses, see L.S. Jacyna, 'Principles of General Physiology: The Comparative Dimension to British Neuroscience in the 1830s and 1840s', *Studies in the History of Biology*, 7 (1984), 47–92 and *idem*, 'The Romantic Programme and the Reception of Cell Theory in Britain', *Journal of the History of Biology*, 17 (1984), 13–48. Cf. Jacyna's more politically orientated analysis of physiology, which is closer in its approach to the present essay: 'Immanence or Transcendence: Theories of Life and Organization in Britain, 1790–1835', *Isis*, 74 (1983), 311–29.

6. 'Medical Reform', *Lancet*, 1 (1830), 564.

7. Robert E. Grant, *On the Present State of the Medical Profession in England* (London: Renshaw, 1841), p. 6. On the 1835 Act, see Derek Fraser, *Power and Authority in the Victorian City* (Oxford: Blackwell, 1979). On medical radicalism, see Ivan Waddington, *The Medical Profession in the Industrial Revolution* (Dublin: Gill and Macmillan, 1984); on its relationship to the new sciences, see Adrian Desmond, *The Politics of Evolution* (1989).

8. Roger Cooter, *The Cultural Meaning of Popular Science: Phrenology and the Organization of Consent in Nineteenth-Century Britain* (Cambridge: Cambridge University Press, 1984), pp. 85, 9–10; Simon Schaffer, this volume. On the more blatant political use of transmutation by working-class cooperators and atheists, see Adrian Desmond, 'Artisan Resistance and Evolution in Britain, 1819–1848', *Osiris*, 3 (1987), 77–110 and James R. Moore, '1859 and All That: Remaking the Story of Evolution and Religion', in R.G. Chapman and C.T. Duval, eds., *Charles Darwin 1809–1882: A Centennial Commemorative* (Wellington, NZ: Nova Pacifica, 1982), pp. 167–94.

9. *Report from the Select Committee on British Museum* (Parliamentary Papers, vol. 10, 14 July 1836); A.E. Gunther, *The Founders of Science at the British Museum 1753–1900* (Halesworth, Suffolk: Halesworth Press, 1980), ch. 8; James A. Secord, 'The Geological Survey of Great Britain as a Research School, 1839–1855', *History of Science*, 24 (1986), 223–75; Roy M. MacLeod, 'Whigs and Savants: Reflections on the Reform Movement in the Royal Society, 1830–48', in Ian Inkster and Jack Morrell, eds., *Metropolis and Province: Science in British Culture 1780–1850* (London: Hutchinson, 1983), pp. 55–90; Morris Berman, *Social Change and Scientific Organization: The Royal Institution, 1799–1844* (London: Heinemann, 1978), ch. 4; David Allen, *The Botanists: A History of the Botanical Society of the British Isles through 150 Years* (Winchester, Hants: St. Paul's Bibliographies, 1986), pp. 41–2; Adrian Desmond, 'The Making of Institutional Zoology in London, 1822–1836', *History of Science*, 23 (1985), 223–50.

10. George Clark, *A History of the Royal College of Physicians of London*, 3 vols. (Oxford: Clarendon Press, 1966), II, 656; 'The London University', *London Medical and Surgical Journal*, 3 (1833), 535–6.

11. *Lancet*, 1 (1838–9), 908. On Macmichael, see *Report from the Select Committee on Medical Education . . . Part 1. Royal College of Physicians* (Parliamentary Papers, vol. 13, August 1834), pp. 4–13, 32–43; S.W.F. Holloway, 'Medical Education in England, 1830–1858: A Sociological Analysis', *History*, 49 (1964), 301–2; M. Jeane Peterson, *The Medical Profession in Mid-Victorian London* (Berkeley: University of California Press, 1978), pp. 4, 8–9, 153.

12. 'Recent Improvements of Medical Education', *Quarterly Journal of Education*, 4 (1832), 4, 5, 10, 15; Berman, *Social Change*, p. 123; Pauline M.H. Mazumdar, 'Anatomical Physiology and the Reform of Medical Education: London, 1825–35', *Bulletin of the History of Medicine*, 57 (1983), 231ff.; H. Hale Bellot, *University College London, 1826–1926* (London: University of London Press, 1929), p. 144.

13. J.R. Bennett to L. Horner, 25 March 1828 and May 1828, University College London (UCL), College Correspondence 611, 614; D.W. Taylor, 'The Life and Teaching of William Sharpey (1802–1880), "Father of Modern Physiology in Britain"', *Medical History*, 15 (1971), 126–53, 241–59; Mazumdar, 'Anatomical Physiology', pp. 242–4. On the Scottish emphasis, see J.H. Burns, *Jeremy Bentham and University College* (London: Athlone Press, 1962), p. 7; Bellot, *University College*, p. 47; and C.W. New, *The Life of Henry Brougham to 1830* (Oxford: Clarendon Press, 1961), p. 375.

14. James R. Bennett, *Lecture Introductory to the Course of General Anatomy: Delivered in the University of London, on Wednesday, October, 6, 1830* (London: Taylor, 1830), p. 9; Grant, 'Lectures', pp. 95, 97; *idem, On the Study of Medicine: Being an Introductory Address delivered at the Opening of the Medical School of the University of London, October 1st, 1833* (London: Taylor, 1833), p. 10.

15. Grant, *Study of Medicine*, p. 17; *idem, Present State*, pp. 6, 10, 49, 54–61, 97–8. On Bennett's French school, see Russell C. Maulitz, 'Channel Crossing: The Lure of French Pathology for English Medical Students, 1816–36', *Bulletin of the History of Medicine*, 55 (1981), 491–5 and *Report from the Select Committee on Medical Education . . . Part 2. Royal College of Surgeons* (Parliamentary Papers, vol. 13, August 1834), pp. 36–7.

16. Dorinda Outram, *Georges Cuvier: Vocation, Science and Authority in Post-Revolutionary France* (Manchester: Manchester University Press, 1984), pp. 111–12; Toby Appel, *The Cuvier–Geoffroy Debate: French Biology in the Decades before Darwin* (New York: Oxford University Press, 1987), pp. 155–6, 173, 176; *idem*, 'Henri de Blainville and the Animal Series: A Nineteenth-Century Chain of Being', *Journal of the History of Biology*, 13 (1980), 304–5; Georges Cuvier, 'Nature', in *Dictionnaire des sciences naturelles*, vol. 34 (1825), pp. 261–8.

17. Robert Knox, 'Contributions to the Philosophy of Zoology', *Lancet*, 2 (1855), 218; Richards, '"Moral Anatomy" of Robert Knox'; Philip F. Rehbock, *The Philosophical Naturalists: Themes in Early Nineteenth-Century British Biology* (Madison: University of Wisconsin Press, 1983), pp. 46ff.

18. Dempster, *Patrick Matthew*, pp. 98–100, 106.

19. 'Outlines of Comparative Anatomy', *Medico-Chirurgical Review*, 23

(1835), 376; Robert E. Grant, 'Animal Kingdom', in R.B. Todd, ed., *Cyclo-paedia of Anatomy and Physiology*, 5 vols. (London: Sherwood, Gilbert and Piper, 1836–59), I, 107; Bennett, *Lecture*, pp. 8, 12ff.

20. 'London University', *Lancet*, 2 (1830–1), 763; Alex Thomson, 'A Sop for Cerberus!', *London Medical and Surgical Journal*, 5 (1830), 437–47; *Statements Respecting the University of London, Prepared, at the Desire of the Council, by Nine of the Professors* (London, 1830), pp. 20–1.

21. 'University of London', *Lancet*, 1 (1831–2), 86; Granville S. Pattison, *Professor Pattison's Statement of the Facts of his Connexion with the University of London* (London: Longman, 1831), pp. 4–8.

22. Jones Quain, *Lecture Introductory to the Course of Anatomy and Physiology, Delivered at the Opening of Session, 1831–32* (London: Taylor, 1831), pp. 10–11; *idem, Elements of Descriptive and Practical Anatomy* (London: Simpkins and Marshall, 1828), pp. 2–6, 22–7.

23. 'Outlines of Comparative Anatomy', *Medico-Chirurgical Review*, 23 (1835), 376–7; 'On Philosophical Anatomy', *Medico-Chirurgical Review*, 27 (1837), 84–7.

24. 'Saint Hilaire on Teratology', *British and Foreign Medical Review*, 8 (1839), 4–7.

25. 'German School of Physiology', *British and Foreign Medical Review*, 5 (1838), 86, 89, 100.

26. George Dermott, 'Mr. Dermott on Hospitals and Dispensaries', *London Medical and Surgical Journal*, 6 (1835), 20–1. On the summonses, see Elie Halévy, *The Triumph of Reform, 1830–41* (London: Benn, 1950), p. 150.

27. George Dermott, *A Lecture Introductory to a Course of Lectures on Anatomy, Physiology, and Surgery, delivered at the School of Medicine and Surgery, Gerrard Street, Soho* (London: Fellowes, 1833), pp. 20–1; M.J. Durey, 'Bodysnatchers and Benthamites: The Implications of the Dead Body Bill for the London Schools of Anatomy, 1820–42', *London Journal*, 2 (1975), 200; Z. Cope, 'The Private Medical Schools in London (1746–1914)', in F.N.L. Poynter, ed., *The Evolution of Medical Education in Britain* (London: Pitman Medical, 1966), pp. 89–109.

28. C.L. Feltoe, ed., *Memorials of John Flint South* (London: Murray, 1884), p. 106; 'Mr Dermott', *Medical Times*, 16 (1847), 619; J.F. Clarke, *Autobiographical Recollections of the Medical Profession* (London: Churchill, 1874), ch. 12.

29. *Report from the Select Committee on Medical Education ... Part 2*, pp. 191–4; Zachary Cope, *The Royal College of Surgeons of England: A History* (London: Blond, 1959), pp. 43–4; Dermott, *Lecture*, pp. 19–21.

30. 'Joshua Brookes', *Lancet*, 2 (1832–3), 722; *Report from the Select Committee on Medical Education ... Part 2*, p. 202.

31. John Fletcher, *Rudiments of Physiology*, 3 parts (Edinburgh: Carfrae, 1835–7), I, 36; *idem*, 'Substance of a Course of Lectures on Physiology', *London Medical and Surgical Journal*, 6 (1834–5)–8(1835–6), *passim*; 'Dr. Fletcher's Rudiments of Physiology', *London Medical and Surgical Journal*, 8 (1835–6), 756–8.

32. Charles Coulston Gillispie, *Genesis and Geology: A Study in the Relations of Scientific Thought, Natural Theology, and Social Opinion in Great Britain,*

1790–1850 (New York: Harper, 1959), pp. 169–70; Raymond G. Cowherd, *Politics of English Dissent* (New York: New York University Press, 1956), p. 66. William Buckland, *Geology and Mineralogy considered with Reference to Natural Theology*, 2nd edn, 2 vols. (London: Pickering, 1837), I, 586. On the opposition to liberal Anglican views on natural theology and university admissions from the extreme latitudinarian wing of the Church itself, see Pietro Corsi, *Science and Religion: Baden Powell and the Anglican Debate, 1800–1860* (Cambridge: Cambridge University Press, 1988).

33. John Epps, *The Church of England's Apostacy* (London: Dinnis, 1834), p. 3.

34. John Willis Clark and Thomas McKenny Hughes, eds., *The Life and Letters of Reverend Adam Sedgwick*, 2 vols. (Cambridge: Cambridge University Press, 1890), II, 86; Gillispie, *Genesis and Geology*, pp. 169–70.

35. 'Vestiges of the Natural History of Creation', *Medico-Chirurgical Review*, 1 (1845), 147, 157; [William Benjamin Carpenter], 'Natural History of Creation', *British and Foreign Medical Review*, 19 (1845), 155–81. On Epps' use of Lamarck, see John Epps, 'Essay on the Gradual Development of the Nervous System, from the Zoophyte to Man', *London Medical and Surgical Journal*, 2 (1829), 40, 43, 151, 155.

36. Michael Ryan, 'Introductory Lecture to a Course on the Principles and Practice of Medicine, delivered in the Hunterian School of Medicine, Great Windmill-Street. – Session 1836–37', *London Medical and Surgical Journal*, 10 (1836–7), 411; George Dermott, 'On the Organic Materiality of the Mind', *Lancet*, 1 (1828–9), 40.

37. 'Medical Reform', *Medico-Chirurgical Review*, 14 (1831), 573–4; *Report from the Select Committee on Medical Education ... Part 2*, pp. 98, 203.

38. Thomas Southwood Smith, *A Lecture Delivered over the Remains of Jeremy Bentham, Esq., in the Webb-Street School of Anatomy & Medicine, on the 9th of June, 1832* (London: Wilson, 1832).

39. Thomas King, *The Substance of a Lecture, designed as an Introduction to the Study of Anatomy considered as the Science of Organization* (London: Longman, Rees, Orme, Brown, Green, and Longman, 1834), p. 7; Richard Dugard Grainger, *Elements of General Anatomy* (London: Highley, 1829), pp. 77–81; *idem*, *Observations on the Structure and Functions of the Spinal Cord* (London: Highley, 1837), p. 106.

40. Fletcher, *Rudiments*, pp. 13–16, 36–8; King, *Substance*, 7–8.

41. F.N.L. Poynter, 'Thomas Southwood Smith – the Man (1788–1861)', *Proceedings of the Royal Society of Medicine*, 55 (1962), 384; Thomas Southwood Smith, *The Divine Government*, 5th edn (London: Trübner, 1866), p. 7; [*idem*], 'Life and Organization', *Westminster Review*, 7 (1827), 208, 212.

42. Smith, *Divine Government*, pp. 65–6, 78–9, 99–102, 104.

43. R.V. Holt, *The Unitarian Contribution to Social Progress in England* (London: Allen and Unwin, 1938), p. 343; [Carpenter], 'Vestiges', p. 160; *idem*, *Animal Physiology* (London: Orr, 1843), pt. 2, viii, 541–2; *idem*, 'On Unity of Function in Organized Beings', *Edinburgh New Philosophical Journal*, 23 (1837), 97–8.

44. [William Benjamin Carpenter], 'Dubois and Jones on Medical Study',

British and Foreign Medical Review, 9 (1840), 177, 183, 203; [*idem*], 'Physiology an Inductive Science', *British and Foreign Medical Review*, 5 (1838), 338–9; Dov Ospovat, *The Development of Darwin's Theory: Natural History, Natural Theology, and Natural Selection, 1838–1859* (Cambridge: Cambridge University Press, 1981), p. 12.

45. 'The One Faculty', *London Medical Gazette*, 13 (1833–4), 404.

46. 'The Second Anniversary Meeting', *Lancet*, 1 (1838–9), 83.

47. M. Eyler, *Victorian Social Medicine: The Ideas and Methods of William Farr* (Baltimore: Johns Hopkins University Press, 1979), pp. 23–4, 155–8, 198–200; M.J. Cullen, *The Statistical Movement in Early Victorian Britain* (Hassocks, Sussex: Harvester, 1975), pp. 36–8. On this support for Hall's reflex arc, see Grainger, *Observations*, ch. 3; *Lancet*, 1 (1846), 391–3, 418–20; Charlotte Hall, ed., *Memoirs of Marshall Hall* (London: Bentley, 1861), chs. 4–5; and Diana E. Manuel, 'Marshall Hall, F.R.S. (1790–1857): A Conspectus of his Life and Work', *Notes and Records of the Royal Society*, 35 (1980), 146–53.

48. [John Epps], 'Elements of Physiology', *Medico-Chirurgical Review*, 9 (1828), 100.

49. 'Saint Hilaire on Teratology', *British and Foreign Medical Review*, 8 (1839), 5; 'German School of Physiology', *British and Foreign Medical Review*, 5 (1838), 86, 89, 100.

50. *London Medical and Surgical Journal*, 1 (1832), 17–18.

51. Robert Knox, 'Contributions to Anatomy and Physiology', *London Medical Gazette*, 2 (1843), 501–2, 529–31.

52. 'Present State and Prospects of the Medical Profession', *London Medical Gazette*, 13 (1833–4), 212.

53. 'Necessity of Lectures being Practical', *London Medical Gazette*, 22 (1838), 778; 'The Conversation in Gower Street', *London Medical Gazette*, 13 (1833–4), 49; 'Another Joint-Stock School of Medicine', *London Medical Gazette*, 22 (1838), 474.

54. C. Bell to Lord Auckland, 12 Nov. 1830, UCL, College Correspondence, p. 46.

55. C. Bell to T. Coates, 2 Sept. 1829, UCL, SDUK Correspondence. Cf. [Charles Bell], *Animal Mechanics, or, Proofs of Design in the Animal Frame*, in *Library of Useful Knowledge, Natural Philosophy. IV* (London: Baldwin and Cradock, 1838), pp. 33, 44–50 with [Thomas Southwood Smith], *Animal Physiology* (London: Baldwin and Cradock, 1829–30), pp. 1–2.

56. S.T. Coleridge, *Aids to Reflection* (London: Bell, 1913), pp. 168, 271–2.

57. 'State of the Medical Schools of London', *London Medical and Surgical Journal*, 2 (1833), 311.

58. S.T. Coleridge, *On the Constitution of the Church and State according to the Idea of Each*, 1830 (London: Dent, 1972), pp. 51–3; Joseph Henry Green, *An Address delivered in King's College, London, at the Commencement of the Medical Session, October 1, 1832* (London: Fellowes, 1832).

59. 'Medical Reform', *London Medical Gazette*, 11 (1832–3), 89–92. On the summonses, see *Lancet*, 1 (1830–1), 785–97.

60. 'Professor Tiedemann in London', *London Medical and Surgical Journal*, 8 (1835–6), 379.

61. A. Carlisle to R. Owen, 26 Feb. 1835, in British Museum (Natural History), Owen Correspondence, vol. 6, fo. 298; R.S. Owen, ed., *The Life of Richard Owen*, 2 vols. (London: John Murray, 1894), I, 86; Guthrie's testimony in *Report from the Select Committee on Medical Education* ... *Part 2*, p. 47.

62. 'Sir Benjamin Brodie's Hunterian Oration', *London Medical Gazette*, 19 (1836–7), 972.

63. E.L. Griggs, ed., *Collected Letters of Samuel Taylor Coleridge*, 6 vols. (Oxford: Clarendon Press, 1956–71), IV, 758–62.

64. Quoted in Desmond, 'Making of Institutional Zoology', p. 167.

65. 'On the Physiology of Man', *Medico-Chirurgical Review*, 30 (1839), 452, criticizing F. Tiedemann, *A Systematic Treatise on Comparative Physiology*, translated by J.M. Gully and J.H. Lane (London: Churchill, 1834), pp. 13–15; Richard Owen, 'Report on British Fossil Reptiles. Part 2', *Report of the British Association for the Advancement of Science 1841* (1842), p. 202.

66. 'Kirby on Instinct', *Medico-Chirurgical Review*, 23 (1835), 400–13; and 24 (1836), 79–83; William Kirby, *On the Power Wisdom and Goodness of God as Manifested in the Creation of Animals and in Their History, Habits and Instincts*, 2 vols. (London: Pickering, 1835), I, xxxiii–iv; Jacyna, 'Immanence', pp. 325–6.

67. T.H. Levere, *Poetry Realized in Nature: Samuel Taylor Coleridge and Early Nineteenth Century Science* (Cambridge: Cambridge University Press, 1981), p. 106.

68. Joseph Henry Green, *Vital Dynamics: The Hunterian Oration before the Royal College of Surgeons in London, 14th February 1840* (London: Pickering, 1840), p. 108.

69. Joseph Henry Green, *Distinction without Separation. In a Letter to the President of the College of Surgeons on the Present State of the Profession* (London: Hurst, Chance, 1831), pp. 16ff.

70. David Bloor, 'Coleridge's Moral Copula', *Social Studies of Science*, 13 (1983), 614.

71. Coleridge, *Constitution*, p. 91; Green, *Address*, p. 3.

72. Green, *Vital Dynamics*, p. xxviii. On Owen, see Evelleen Richards, 'A Question of Property Rights: Richard Owen's Evolutionism Reassessed', *British Journal for the History of Science*, 20 (1987), 129–71; Nicholas Rupke, 'Richard Owen's Hunterian Lectures on Comparative Anatomy and Physiology, 1837–55', *Medical History*, 29 (1985), 237–58; Adrian Desmond, *Archetypes and Ancestors: Palaeontology in Victorian London, 1850–1875* (London: Blond and Briggs, 1982; Chicago: University of Chicago Press, 1984); *idem*, 'Richard Owen's Reaction to Transmutation in the 1830s', *British Journal for the History of Science*, 18 (1985), 25–50.

73. Jacyna, 'Principles', pp. 60–3.

74. Joseph Maclise, 'On the Nomenclature of Anatomy', *Lancet*, 1 (1846), 299; Cooter, *Cultural Meaning*, p. 85.

4

The nebular hypothesis
and the science of progress

SIMON SCHAFFER

Hermione:	I have a better theory.
Hermogenes:	Tell it me.
Hermione:	In the beginning was the nebulous matter. Its boundless and tumultuous waves heaved in chaotic wildness, and all was oxygen, and hydrogen, and electricity. Such a state of things could not possibly continue, and as it could not possibly be worse, alteration was here synonymous with improvement. Then came –
Hermogenes:	Now it is my turn to say Stop! Stop! Do let us be serious.

John Herschel (1865)[1]

In this parable the 73-year-old astronomer John Herschel poked gentle fun at the nebular hypothesis. He was writing in the first number of G.H. Lewes's *Fortnightly Review,* many of whose contributors accepted the hypothesis as an explanation of the origin of the Solar System. The nebular hypothesis gained currency in Britain during the 1830s, notably through the work of the Scottish political economist and astronomer John Pringle Nichol. In 1837 Nichol wrote of the implications of his cosmology: 'In the vast Heavens, as well as among phenomena around us, all things are in state of change and PROGRESS.' Nichol claimed to read the message of 'a life to come' in 'splendid hieroglyphics', written 'on the sky'. The nebular hypothesis carried many messages. In contrast to Herschel's reaction, for example, a member of Nichol's audience reflected more seriously on 'the dim, yet high suggestions' of the nebular hypothesis: 'the telescope brought us ever new, electric, telegraphic tidings of Him of whose goings forth were of *old* – from everlasting – and which were *new* to everlasting as well'.[2] The nebular cosmogony was a connected set of stories about the possible origin and likely history of the Solar System. It seemed blessed with the excitement and authority of telescopic astronomy. Thus the hypothesis was an important site at which the Victorians worked out their differing views of the progress of their world. In this essay I shall examine the relation between Herschel's satire and Nichol's enthusiasm in order to explain the interests served by this picture of cosmic progression in early Victorian Britain.

In its simpler versions, the nebular hypothesis was taken as giving an astronomically proper account of the origin of the Solar System through the action of natural law upon a condensing and rotating gaseous nebula of gravitating matter. It was claimed that as this cloud contracted, rings of matter would be precipitated into space at regular intervals. Each ring would then make a planet. The central condensed cloud produced the Sun. In this form, both Robert Chambers, in 1844, and Herbert Spencer, in 1857, gave the nebular cosmogony pride of place in their respective accounts of development in the world. Spencer said it exemplified 'the law of all progress'.[3]

The high status achieved by astronomy in Victorian Britain was an important part of the nebular model's appeal. The model could call on William Herschel's heroic observations of nebulae and on the appearances of comets' tails and the zodiacal light. Phenomena saved included the fact that all planets and most of their satellites moved round the Sun in the same plane and direction. These seemed invulnerable items in the astronomers' armoury. In his remarkable paper on the nebular hypothesis, published in the *Westminster Review* in July 1858, Spencer pointedly contrasted the mythic roots of the claim that 'the planets were originally launched into their orbits by the Creator's hand' with the eminent pedigree of the nebular hypothesis – Immanuel Kant, William Herschel and Pierre Simon Laplace. 'To have come of respectable ancestry is *prima facie* evidence of worth in a belief as in a person; while to be descended from discreditable stock is, in the one case as in the other, an unfavourable index.' The potency of the history of astronomy did persuasive work for other disciplines.[4]

An interdisciplinary historiography has been used to explore this persuasiveness of nineteenth-century astronomy. Many historians of evolution have appealed to cosmologies in the century before 1859 for evidence that cosmic and biological evolution were mutually supportive concepts. John Greene refers to 'the cosmic evolutionism of Kant and [William] Herschel' and the 'Kant–Laplace nebular hypothesis' as marks of a 'speculative temptation to derive the present structures

J.P. Nichol's sketch of the mechanism by which rings of nebulous fluid are deposited from a rotating nebula, the primary stage of planetary formation according to his nebular hypothesis. 'There are almost infinite chances against the condensation of any large or original Nebula, without the occurrence of circumstances which would throw off numbers of such rings.' (From Nichol, *Views of the Architecture of the Heavens*, 3rd edn [Edinburgh: William Tait, 1839], p. 177.)

of nature from previous states of the system of nature by the operation of natural laws'. Greene found a 'just analogy' between Charles Darwin's hostility to special creations and Kant's advocacy of the nebular hypothesis.[5] Metaphorical debts like this are catalogued to show the influence on evolutionists of astronomers, statisticians and political economists. Much ink has been spilt on the problem of whether the concession that Darwin read and used political economy damages the purity of his science. Similarly, historians have attended closely to the cosmological imagery in the rhetoric of the *Origin of Species*.[6] Greene argues that, 'like every other scientist, Darwin approached nature, human nature and society with ideas derived from his culture, however much his scientific researches may have changed those ideas in the long run'. But it is not clear how best to analyse these cultural resources. Simple homologies prove little. A more sophisticated approach is necessary to clarify the relations of different disciplinary communities and the interests served by transpositions of concepts among them.[7]

In the 1830s, when the nebular hypothesis first appeared in Britain, disciplinary boundaries were not completely porous: hard work was needed to justify transgressions of these frontiers. Neither were these boundaries self-evidently natural, for they were being made. Concepts were being made too. Their ownership was hotly disputed. The nebular hypothesis gives an excellent example of these features of social life. The cosmogony was an *artefact*: a consciously fashioned tool with distinct persuasive purposes. The makers and users of this artefact suggested, with Spencer, that it had been taken from secure celestial mechanics and so carried astronomical authority. Historians of evolution have usually accepted this claim. The result has been an assumption that, in this case, a simple lesson taught by the high science of astronomy entered the evolutionary debate. However, the 'nebular hypothesis' had to be shifted from its original context to acquire its polymorphous character. During the 1830s and 1840s its senses included astronomical accounts of the construction of the nebulae, as debated by John Herschel and his colleagues; cosmogonic stories about the origin of the Solar System, such as that told by William Whewell in 1833 to illustrate the compatibility of divine creation and supposed natural laws; and general models of the universal progressive development shown in the heavens, on Earth and in human society.[8]

I shall examine the last of these versions. Its most influential exposition was that of Nichol, its most public that of Robert Chambers in his anonymous *Vestiges of the Natural History of Creation* (1844). I argue, first, that the nebular hypothesis, in the sense of a general model of universal progress, was not imported from astronomy. Yet its proponents claimed astronomical sanction for their work. Secondly, I identify the interests of these proponents in making such an hypothesis, using Nichol's career to illustrate the activities of the journalists, political economists and educators who espoused the nebular cause. Last, I suggest the place of the nebular hypothesis in a 'science of progress' that was a central project of such groups in early Victorian Britain.

Nebular hypotheses: in search of an origin

The establishment of the pedigree of the nebular hypothesis demanded careful reinterpertation of astronomy's immediate past. The cosmogony was not a ripe fruit waiting to be picked from the astronomers' tree. Writers of the 1830s made a nebular hypothesis out of materials found at the end of Laplace's magisterial *Exposition du système du monde* and in William Herschel's papers on 'the construction of the heavens'. The status of the two men was a powerful resource. Auguste Comte lectured on Laplace's views in Paris in August 1831 and gave a mathematical defence of them in a paper for the Académie des Sciences in January 1835. Comte's argument reached Britain in the late 1830s through reports by David Brewster and J.S. Mill.[9] Other influential presentations appeared in Whewell's Bridgewater Treatise of 1833, where, significantly, the term 'nebular hypothesis' was coined, and in Nichol's very popular *Views of the Architecture of the Heavens*, a preview of which appeared in the *Westminster Review* in July 1836. Whewell baptized the nebular hypothesis by claiming that it still demanded 'an intelligent Author, an origin proceeding from free volition not from material necessity'; Nichol, by contrast, legalized it, arguing that the hypothesis suggested 'a moral I am extremely anxious to impress', which would 'bring us to right views of that stupendous ORDER within which we live, and of which our own beings constitute a part'.[10]

Nichol and his allies made their nebular hypothesis an object of a moral and a natural science. Stellar progress was pressed into the service of political reform. The allegedly astronomical credentials of the hypothesis mattered because astronomy was fit for public consumption – the 'pattern science', as Whewell put it in 1834. Struggles between astronomers and the Royal Society's leadership in the 1820s showed the sensitivies of such boundaries in the age of Reform. Popular lecturers often appealed to astronomy for evidence of their beliefs. New sciences, such as statistics and political economy, justified themselves in the same way. Much was made of the astronomical expertise of the *doyen* of the statisticians, Adolphe Quetelet. In 1835, he noted that, 'having just observed the progress made by astronomical science in regard to worlds, why should we not endeavour to follow the same course in respect to men?'[11] The Ricardian J.R. McCulloch told his London audiences that 'the political economist is to the statistician what the physical astronomer is to the mere observer'. Adam Sedgwick, preaching at Cambridge in 1832, countered that political economists should be like 'the early observers of the heavens', whereas the wise legistator was the right equivalent of the 'physical astronomer'. These various remarks did not mark a deep division between statisticians and political economists – many of the latter, such as McCulloch and Nichol, were advocates of both disciplines. But the contrast illustrates the ways in which an appeal to astronomical authority could do important persuasive work.[12]

This was why it was important for the advocates of the nebular hypothesis to ground their conviction in astronomy. Writers in the 1830s usually appealed to William Herschel's observations of the nebulae, published after 1791, and to Laplace's probabilistic calculations about planetary orbits, decisively announced in the fourth (1813) edition of his *Système*. They claimed that Herschel's work proved the existence of 'true nebulosity', clouds of luminous fluid that no telescope would ever resolve into stars but that would, in time, rotate and condense into stars. They held that Laplace had demonstrated that some similar simple physical cause must have produced the coplanar and codirectional motions of the planets in this Solar System. The two suggestions were welded into a powerful story about the astronomical status of the nebular hypothesis.[13] This was a deliberate accomplishment. Herschel's nebular research, the 'natural history of the heavens', was not a recognized part of late eighteenth-century astronomy, to which work on stellar and nebular structure was not added until the accomplishments of German observers after 1810. Similarly, Laplace's arguments were based on connections between celestial mechanics and philosophical probability. His initial evidence relied on observations of zodiacal light and Saturn's rings, not nebulae. By the 1810s, he was seeking arguments from probabilities, which could 'push back' evidence of design.[14] Astronomers such as Wilhelm Olbers, Friedrich Bessel and Karl Gauss remained sceptical of Laplace's claims and excluded them from their new stellar astronomy. Olbers cited Herschel's observations of the retrograde motion of the moons of Uranus to deny the inductive appeal of Laplacian cosmogony. Despite Laplace's notion that Herschel's testimony strengthened the nebular cosmogony, Herschel himself was well aware of the important differences between their work. These differences were useful to working astronomers when they sought to make their discipline secure. The 'nebular hypothesis', as an account of planetary origin, did not figure in their enterprise.[15]

The testimony of astronomers about the existence of true nebulosity was extraordinarily malleable. The stars gave no unambiguous lesson. Their message was always interpreted to fit the local interests of protagonists in the contests about progress in the Universe and in Society. The reaction of British astronomers and natural philosophers, including John Herschel and the Edinburgh professors John Robison and John Playfair, also illustrates this connection between hostility to some forms of the nebular hypothesis and the establishment of the bounded astronomical community. Robison told William Herschel that he saw striking contrasts between nebular astronomy and Laplace's dangerous conjectures. Robison's colleague, Playfair, was noted for his controversial support for Laplacian science, especially as an analogue for Huttonian geology, of which he was spokesman.[16] But he excluded the nebular cosmogony from his work on probability, geology and the science of heat. John Greene has drawn attention to the contrast between Playfair's enthusiasm for Laplacian physics and his critique of the nebular hypothesis. Playfair and his colleagues made their office the description of natural systems, not a reconstruction of their origins. In his influential *Dissertation on the Progress of Physical Science*, Playfair proscribed theorizing on origins. Charles Lyell adopted Playfair's normative account of astronomy as a model for conduct in geology.[17] William Herschel's son published the same message in his *Preliminary Discourse on the Study of Natural Philosophy* (1830), declaring that, just as geology could not reach 'to the creation of the earth', so astronomy was 'confessedly incompetent to carry us back to the origin of our system'. This ban segregated the nebular hypothesis as a cosmogony for 'our system' from the proper work that Herschel and his father conducted on the nebulae.[18]

John Herschel's role is particularly significant because of his filial regard for nebular astronomy, his efforts to introduce Laplacian science into Britain and his exemplification of what it was to be an astronomer in the 1820s and 1830s. In speeches at the Astronomical Society and the British Association, in his authoritative astronomy textbooks and in papers for the Geological Society and other learned bodies, Herschel emphasized the veracity and independence of nebular astronomy, and the seductive danger of the nebular hypothesis. First, there was always some reason to suppose the existence of true nebulosity. He recorded this in his observing books of the 1820s and, a decade later, at the Cape of Good Hope, he saw luminous fluid juxtaposed with real stars in the Magellanic Clouds. Even when, in June 1845, he faced reports that the Earl of Rosse's giant new telescope in Ireland had resolved some nebulae into stars, Herschel insisted that evidence for true nebulosity might remain.[19] Yet, second, Herschel never allowed that true nebulosity implied the truth of the nebular formation of our Solar System. Against the protests of writers such as Nichol, Herschel insisted that this cosmogony could only be accepted if individual objects were seen condensing, obviously an impossible demand. According to Herschel, no evidence for nebular rotation existed. So while cautioning his audience that Rosse's 'resolutions' did not destroy true nebulosity, he also warned them against accepting the rash conclusions of Comte, Mill and the author of *Vestiges*. The nebular hypothesis and nebular astronomy 'stand, in fact, quite independent from each other'. The former was necessarily 'mysterious'; the latter was based on 'perfectly legitimate principles'. Thus 'legitimate' astronomy was sundered from the dangers of specu- lative stories about the Solar System's origin and progress.[20]

'Legitimacy' was the key term in Herschel's programme. London geologists began to tell dynamical stories about the slow cooling of an originally incandescent Earth. In these stories the nebular hypothesis seemed a useful supplement to the suggestive theories of Elie de Beaumont on episodes of mountain building and J.B. Fourier's bril- liant analysis of heat flow. Herschel warned that geologists should learn the right lessons from the astronomers. Verifiable change in the eccentricity of the Earth's orbit was a legitimate geological explanans, whereas nebular condensation was not.[21] Herschelian nebular astron- omy was not to be blackened with the sins of Comte or *Vestiges*. Instead, it could be fitted into the project enshrined in the charter of the Astronomical Society, which was drafted by Herschel and Francis Baily in 1820, and then reaffirmed as part of physical astronomy in George Airy's report to the British Association in 1831.[22] This astron- omy found neither 'the evidence of a beginning, nor the prospect of an end', in the Huttonian catchphrase cited by Herschel in 1830. The fact that physical astronomy had nothing to say about the 'first and last things' of theology only served to secure the divine origin and blessed future of humankind.[23] The line between physical astronomy and the moral philosophy of creation and salvation traced a prudent if tenuous division of labour. However, while the astronomers were interested in preserving this division, those who made a nebular hypothesis in the 1830s and 1840s displayed interpretive ingenuity in using astronomers' work for their own ends.

The revelations of astronomy

Conflicts of interest at the astronomical observatories and among the early Victorian public explain much of the response to the lessons of the telescopes. The reports from the great reflectors completed in 1839 and 1845 at Rosse's observatory at Parsonstown show how this process worked. The chief astronomer working there was Thomas Romney Robinson, a Church of Ireland divine. He loathed the re- formers' uses of the nebular hypothesis, argued that its fate depended on evidence for true nebulosity, and did his best to destroy this evidence. Robinson reckoned that any report of a nebula being resolved into stars would undermine the evidence that a true nebular fluid existed in space. So while Rosse himself counselled caution in rashly interpreting 'resolutions' of nebulae as death-blows to the existence of true nebulosity, Robinson packaged reports from Parsonstown to do maximum damage to the nebular cosmogony. In November 1840, soon after inaugurating these nebular observations, Robinson told a cheering audience at the Royal Irish Academy that the great telescope would destroy the nebular hypothesis. His aims were not abandoned even when Rosse reported observations of spiral structures in several of the nebulae, suggesting apparent fluidity.[24] Nebulae that Robinson claimed were resolved, such as the one in Orion, are not now thought to be stellar. Robinson and observers such as W.C. Bond at Harvard may have been misled by a back-

A spiral nebula in the Dog's Ear in a sketch shown by the Earl of Rosse to the British Association in 1845. Rosse's 6-foot mirror displayed this spiral structure – as such, it was used as evidence for true nebulosity both by Whewell and Nichol, who declared: 'There is nothing to which we can liken it save a scroll gradually unwinding, or the evolutions of a gigantic shell!' (From Nichol, *Thoughts on Some Important Points relating to the System of the World*, 2nd edn [Edinburgh: John Johnstone, 1848], p. 73.)

ground of faint galactic stars. But Robinson's publicity campaign made his resolutions count as a standard for other instruments. Bond established the worth of Harvard's new Munich refractor by 'resolving' Orion in autumn 1847: 'I see no other way in which the public are to be made acquainted with [the telescope's] merits.'[25]

As a 'gentleman of science', Robinson was in a good position to use the British Association to make the Parsonstown telescopes symbols of triumphalist astronomy – the 'universal wonder of the world', as John Phillips told Whewell in 1843. Visits to Rosse's instruments were common, but the lessons of the resolutions were not unambiguous. Airy denied the certainty of Robinson's claim that no true nebulosity existed. In 1848 Whewell was reported as believing that the telescopes would not 'make important discoveries'. A contemporary commented acidly, 'I doubt whether even a Master of Trinity has a right to legislate on the subject.'[26] Regardless of his scepticism, Whewell was typical in using just the reports that suited his interests. In 1854 he needed evidence for the singularity of our inhabited world, so he reproduced Rosse's pictures of spiral nebulae and Herschel's observations of the Magellanic Clouds to suggest that true nebulosity was common and there were few, if any, distant stars capable of maintaining planets.[27] Whewell's great antagonist, David Brewster, was equally pragmatic. He attacked Whewell's use of the hypothesis in 1834, but then accepted Comte's arguments for it in 1838 and prophesied that Rosse's instruments would reveal planets forming from nebulae. When *Vestiges* used the hypothesis to prove the law of progress, Brewster again revised his view and cited Rosse's work to undermine true nebulosity and subversive philosophy. When Whewell used Rosse's spirals to evince nebular fluidity, Brewster countered with the Parsonstown 'resolutions'. There was never a moment when supporters or opponents of true nebulosity were at a loss for evidence.[28]

Nichol's relationship with the Parsonstown observatory is a splendid example of this ingenuity. In 1837 his *Architecture of the Heavens* made the Orion nebula the centrepiece of his picture of cosmic progress. He claimed that this picture had 'the respect of all and the silent acquiescence of many Astronomers'. Robinson was furious with this claim, telling Rosse in April 1841 that Nichol's book was nothing but a catchpenny presentation of nebular enthusiasm. Robinson held that the resolution of Orion ought to damage Nichol's faith. But in 1846, after the 'resolution', Nichol reconstructed his story to make the Orion observations irrelevant: 'The nebular hypothesis remains precisely where in my last volume I attempted to place it.' Nichol excitedly recalled the winter of 1844–5 when Rosse tried and failed to resolve Orion with his 3-foot mirror. At Christmas 1845 Nichol was the first to point the 6-foot reflector at the great nebula: 'It was still possible then, that the Nebula might be irresoluble by the loftiest efforts of human art.'[29]

Nichol's enthusiasm and envy for Rosse was second to none. His re-equipment of his Glasgow Observatory to match those of Herschel and Rosse drove him to bankruptcy. He even shipped some of Glasgow's best instruments to Parsonstown. On his return from Ireland in January 1846, Nichol asked Robert Peel to give Rosse a British earldom, comparing Parsonstown with Tycho's Uraniborg.[30] But on 19 March 1846 Rosse told Nichol that Orion was resolved. Nichol published Rosse's letter in the *Glasgow Herald* and spoke of his 'anxiety' about this report. But his book of the same year showed why the resolution did not matter. He claimed that Rosse had reached the limit of telescopic power. Laplacian cosmogony was independent of Herschelian nebular astronomy – Rosse only challenged the latter. Analogy suggested that nebular fluid might exist: 'We are distinctly prohibited from confounding the *unexplained* with the *inexplicable*.' Finally, as Nichol declared ever after, Rosse's spirals were better

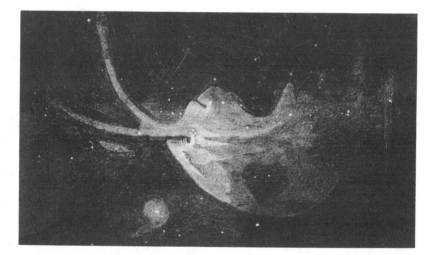

Nichol's reproduction of John Herschel's sketch of the Orion nebula. In 1837 Nichol viewed this object as a key case of true nebulosity. He quoted Herschel's view that the nebula 'suggests no idea of stars, but rather of something quite distinct from them'. Even when Rosse reported that he had resolved this nebula into stars in the spring of 1846, Nichol remained convinced that there was still decisive evidence in favour of the existence of truly nebular fluid. (From Nichol, *Thoughts on Some Important Points relating to the System of the World*, 2nd edn [Edinburgh: John Johnstone, 1848], p. 104.)

evidence for true nebulosity than was the Orion nebula against it. Evolution mattered more than resolution.[31]

Rich resources for Nichol's views continued to be available during the 1850s. German astronomers reported changes in the Orion nebula's shape, indicating that it was unlikely to be a distant star cluster. Spencer used this material in discussing the nebular hypothesis in 1857–8. He cited Herschel's reports of the Magellanic Clouds and disparaged Robinson's aggressive interpretation of the resolutions.[32] From the 1860s, convinced evolutionists such as William Huggins were infuriated that Rosse made out nebulae to be little better than 'cosmic sandheaps'. Huggins showed that many nebulae had gaseous not stellar spectra. Only after this conjuncture did the debates on the meaning of nebular development become a secure part of physical astronomy.[33] But public interest in the debate predated this change. For example, the journalist, political economist and opium eater Thomas de Quincey seized on the nebular hypothesis in the 1840s as an item of public concern. While at Nichol's observatory to write his *Logic of Political Economy* (1844), he joked that any of his daughters, and even the hearth-rug, would know what this 'nebular hypothesis' meant. In September 1846, after hearing of the Orion resolution, he published a remarkable paean to Rosse and Nichol in which, like his host, he distinguished between Rosse's accomplishments and the standing of the nebular cosmogony: 'We ought no longer to talk of astronomy but of *the astronomies.*' De Quincey and Nichol were confident that 'yet other and more fearful *nebulae* may loom in sight that may puzzle even Lord Rosse'. The 'revelations of astronomy', as Brewster called them, mattered to a wide public. They were resources not proofs. An examination of the public purposes of the astronomers and their audiences explains how these resources were used.[34]

A science of progress

Astronomy was 'progressive' in several senses. It 'gave the most violent shock to the prejudices and long-received opinions of men'. It also described cosmic progress. The former sentiment was Lyell's, copied in 1838 by Darwin when noting the sufferings of scientific martyrs.[35] The latter directionalist implication was made part of a 'science of progress' that would describe and manage social development. This science provided the setting in which Nichol and his allies made their nebular cosmogony. The science of progress appeared in government offices, lecture theatres, journals and popular texts of the reform movement in politics and education that developed during the

1820s and 1830s. Newly regimented knowledges emerged and were institutionalized at the nascent London University and within voluntary groups of political economists, statisticians, physicians and educators. Reformers found active, if sceptical, allies among the 'Philosophic Radicals' banded round J.S. Mill, John Roebuck and the *London Review*. In their self-appointed role as 'experts', reformers dominated such diverse organizations as George Porter's Board of Trade, Edwin Chadwick's Poor Law Commission, Thomas Wyse's Central Education Society and Lord Brougham's Society for the Diffusion of Useful Knowledge. 'The practical results of a tested system' were to replace 'the theories of an imaginary philanthropy'. Cities such as Edinburgh had their equivalents in radical journals and organizations of lecturers, phrenologists and economists.[36]

Such was the landscape of Nichol's career. He sought a form of knowledge that could bolster his political campaign and a platform from which to convey that knowledge. By 1836 he had found the right place and the right theory: the Glasgow astronomy chair and the nebular hypothesis. Nichol and his colleagues argued that social reform was both possible and necessary: possible, because society was ordered by malleable forces; necessary, lest the gains of Reform be lost to conservative reaction or plebeian anarchy. Education was a key tactic because popular instruction produced a stable society and a bulwark against reaction. The right knowledge was specified in detail: its centrepiece was political economy, whose truth was guaranteed by scrupulous surveys of social facts such as those commended by McCulloch and Quetelet. Demands for security accompanied those for change. So reformers had to show what kind and what rate of change was to be expected. Here nature was an indispensable authority. Political economy taught 'the operation of certain natural laws'. Its matters of fact instructed the 'dangerous classes' about the differences between these laws, against which protest was futile, and unnatural evils, which reformers might remove. The science of progress aimed to discriminate between these two features of the social order.[37]

It was hard to make such a science because the term 'progress' was contested. In the 1830s, any science had to show that it commanded an elite consensus. John Herschel and Robert Peel both spoke to patrons of local 'reading rooms' of the virtues of diffusion of sciences that inculcated a 'progressive' picture of the social and moral order. The Oxford Tractarian John Henry Newman answered that no one 'was made to do any secret act of self-denial or ... steeled against pain or peril by all the lore of the infidel La Place'.[38] 'Progress' had sectarian senses as well. Nichol reached Glasgow soon after the sanguinary end of a strike of the city's skilled trades, suppressed by the Tory sheriff Archibald Alison. This strike, and reaction to Chadwick's harsh New Poor Law, soon bred a Chartist response, represented in Glasgow by the weaver William Thomson's *Chartist Circular*. Thomson declared that 'by *politics* we mean the *science of human progression'*. Chartists like Thomson and Bronterre O'Brien, and their Tory enemies such as Alison, found the reformers' version of this 'science' silly and dangerous. O'Brien satirized Brougham's 'useful knowledge' in 1833; the next year Alison picked up O'Brien's remarks in a paper on 'the progress of social disintegration'. If knowledge were power, it was dangerous giving it to the plebs; if not, then it was stupid to suppose it could quell riot. Any reformist 'science of progress' had to picture nature and nature's laws in a way that did not look so threatening or so impotent.[39]

Given these burdens, the reformers ingeniously stressed natural law's inevitable effects and certainty. Close friends of Nichol, such as the Edinburgh phrenologist George Combe and the Philosophic Radical John Roebuck, agreed with him that 'both the moral and physical departments of the world' had been 'constituted by the Creator on the principle of a progressive system'. Roebuck said that knowledge of the laws of this system would make citizens 'docile and patient citizens under a good government and . . . irresistible enemies to a bad one' because they would know the difference between corrigible evil and incorrigible necessity. Combe, Nichol and Roebuck all saw 'the *few*' expert intellectuals as the wardens of this process of reconciliation with 'the *many* whose doom was physical toil and mental apathy'.[40] In April 1848 Nichol's allies among the Edinburgh journalists confronted Europe-wide revolution with the claim that 'the better progress of which the age makes its boast lags lazily', while 'there is a downward progress that never flags'. They argued that the cure demanded not 'the debris of professional men', but 'the ablest, where the most persuasive talents are required'. Nichol enlisted in this cause. His strategy demanded that the public be persuaded of the secure truths of a system of natural law that was necessarily progressive but nowhere in disorder. This was the model supplied by the nebular hypothesis. Many of his contemporaries shared Coleridgean scepticism of 'lecture bazaars' and the 'plebification' of science. Nichol did not. He sought instead to become an exponent of a science of progress, both in his journalism and political economy and in his astronomy and public lectures.[41]

Nichol's progress: from economic journalism to the nebulae

Nichol's impact on his audience is well attested. Darwin, Chambers, Combe and Mill were all persuaded by Nichol of the significance of the nebular hypothesis. In 1841 Edgar Allan Poe included Nichol and the Orion nebula in Dupin's celebrated chain of inference in 'Murders in the Rue Morgue' and then, after hearing Nichol's New York lectures of 1847–8, Poe wrote a full-blown nebular cosmogony in his *Eureka*.[42] Nichol's success was part of a deliberate search for a reformist vocation. In the year of Reform, 1832, Nichol was a 28-year-old Montrose schoolmaster. He had lectured on mathematical and natural sciences, and may have been in Paris in 1830 to witness the celebrated debates between Georges Cuvier and Etienne Geoffroy Saint Hilaire on the analogies and development of organic types. Nichol shared with his Scottish contemporaries Robert Knox and Robert E. Grant a commitment to the ideological lessons of new French sciences.[43] Between April 1832 and February 1836, Nichol worked hard as a radical journalist, writing on political economy for *Tait's Edinburgh Magazine*, for Cochrane's liberal *Foreign Quarterly Review* and, from the summer of 1834, for Mill's *London Review*. He planned essays on political economy, geology and the theories of Quetelet and Thomas Chalmers. He commented in detail on Mill's political economy and logic, and published on statistics, tithes, free trade and the Corn Laws, aligning himself with what he called 'the English Gironde', the Philosophic Radicals grouped round Mill in London. These were the interests that dominated Nichol's strategy even after 1836 when, with Radical support, he gained the Glasgow chair.[44]

In his work with the Radicals, Nichol defined his friends and his enemies and so made his career. His economic journalism of the 1830s helps explain why he made the Glasgow Observatory a centre for the diffusion of the science of progress in lectures on politics and education in the 1840s. The nebular cosmogony that Nichol began to present in 1836 stressed that all the observed phenomena of the heavens were the effects of a *single natural law* and that *this mundane system*, including organic life and human society, had originated through the self-same process. Nichol did not use the nebular hypothesis merely as a metaphor for the workings of society. He publicly connected the processes visible in the history of the universe with the immediate fate of humanity. Years before his first statements on cosmogony, his political economic theory already mapped the social system as a balance of dynamic change and law-like order. Reformist intervention had to be balanced by *laissez-faire*. 'Scientific and moral principles' could correct 'mere conventional laws', but any intervention in these laws had to respect 'the opposing principle of non-interference'. Nichol advocated repeal of the Corn Laws, steeling himself to congratulate Peel when this was effected. The Laws bred *unnatural* evils and their repeal would correct the cause of misery without the revolution that misery might prompt. 'An adjustment of actual interests will have *preceded* the demand for change', and so '*innovation* is altered into adaptation.' The only alternative – preservation of aberrant corruption – would merely lead to 'a banishment of stability . . . and a new period of vacillation'.[45]

Nichol's journalism taught that the task of reform was the production of order and progress at one and the same time. In 1835 Nichol used the message preached by Quetelet earlier that year: 'civilisation more and more contracts the limits within which the different elements relating to man oscillate'. Nichol went further. Reformist intervention should aim to contract these limits deliberately. To show that such a limitation might be socially progressive, it was necessary to redefine the meaning of the term 'progress'. Nichol said that Ricardian political economy properly defined social stages as not '*chronological* but *social* epochs'. Each epoch could be measured by economic resources. Nichol defined the '*momentum* of society' as the product of population size multiplied by the period during which citizens could be economically active. Moral conduct could increase '*momentum*' by lengthening life; intellectual power aided moral conduct by raising productivity. Nichol explained that industrial culture would always triumph over the expanding populations of barbarism because of this 'joint cultivation of the *intellectual* and *moral* powers'. He applied this insight to the industrial cities, where intellect outstripped morality. Reform should aim at 'energy with safety and health with rapid growth'. Political economy proclaimed the scientific message – law-like progress was to be balanced with non-intervention in truly natural affairs – and promised an economic and moral pay-off if the lesson were learned by the restive urban population and its representatives.[46]

Alliance with Mill and the Radicals gave Nichol the opportunity to propagate this message. It also provided him with opponents, churchmen and dons whom he viewed as partisans of corrupt establishments in education, politics and economics. The attack on issues that 'divide and disorganise British Society' was hammered home 'scientifically', using political economy as a weapon. When Mill launched the *London Review* he hired Nichol to give the journal 'the scientific character which Playfair gave the *Edinburgh*'. Nichol supported Mill's aggressive 1833 tract on church endowments, which demanded nationalization of the universities and turned Coleridge's argument for the clerisy into an appeal for state education. Churchmen were the least able to produce the facts that the science of progress needed. Nichol wrote that 'we anticipate little improvement until the civil authority shall relieve the ecclesiastical of the charge' of making records. Instead, Nichol drew on Porter's statistical surveys of 'the progress of the whole social system', sent to him from London by Mill.[47]

Priestcraft was also seen as the chief opponent of Ricardian truth. Mill and Nichol attacked the metaphysics and economics of this 'sinister interest'. In the autumn of 1834 Mill told Nichol of his notorious response to Sedgwick's *Discourse on the Studies of the University*. Sedgwick's bitter attack on utilitarianism was inspired by his colleague Whewell and had been delivered at Trinity College chapel two years earlier. In the spring of 1835, Nichol joined in this controversy with an assault on Whewell's close friend, the economics professor, the Rev. Richard Jones. Whewell had defended Jones's anti-Ricardianism by appealing to the close analogy between political economy and mechanics. He questioned Ricardian equilibrium models, displayed the fallacious physicalist reasonings in which political economists indulged and so showed that the Ricardian axioms of 'the moral and social principles of men's actions and relations' must 'fall to the ground'.[48] Using Mill's essay 'On the Definition of Political Economy', Nichol responded to Whewell and Jones by arguing for the consonance of Ricardian rent theory with 'the received canons in the science of experiment'. Abstraction was as legitimate in political economy as in the mathematical sciences. As a theorist, Ricardo was in the same class as other martyrs of reason, such as Galileo. Although Nichol recognized that 'by the infusion into abstract speculation of a degree of feeling amounting almost to vindictiveness, political economy does not seem to be benefited', yet he demanded to know whether it was 'necessary, that when an inquirer goes out of the common road, the Church should always be the first to send the blood-hounds after him?' The answer was yes. Mill counselled Nichol to moderate his attack, with some reason. Whewell told friends such as Jones and Archdeacon Hare that the journal in which Nichol published had the 'professed object ... to destroy the Church and democratize the nation'.[49]

What was at stake between Nichol's allies and his enemies in this period were rival models of proper science. Could the 'laws of human nature' be established as the basis of a science of progress? In January 1834, Mill and Nichol began planning 'a much more elaborate work on Logic' that would place such a science on a firm basis. Mill clarified the political role of the new logic in the economics tracts he sent Nichol: 'In whatever science there are systematic differences of opinion – which is as much as to say, in all the moral or mental sciences, and in Political Economy among the rest ... the cause will be found to be, a difference in their conceptions of the philosophic method of the Science.' Mill and Nichol worked together closely in the preparation of the first edition of the *System of Logic*.[50] The two Radicals discussed sciences that might provide resources for this logic, including Laplace's cosmogony, Combe's phrenology and Comte's positivism. In 1838 Mill drafted a long passage on the nebular hypothesis for his book. Nichol's enthusiasm and Comte's sums persuaded him that the cosmogony was an illustration of 'legitimate reasoning from a present effect to its past cause'. Nichol was named and his hypothesis treated with great respect. The reference was removed in publication, and the status of the hypothesis was damaged by John Herschel's attack on Comte and Mill.[51] For his part, Nichol remained sceptical of Comte, despite his fascination with the nebular hypothesis. Mill was equally hesitant about the evidence for phrenology in its Combean version. Yet these sciences afforded important models for the science of progress. Understanding their shared principles and the lessons they taught could help the reformers attain scientific authority. During 1835 and 1836 it became clear to Nichol that the nebular hypothesis was particularly appropriate as a bulwark of his campaign.[52]

Nichol's progress: the facts of humanity and the universe

Nichol's new job was the culmination of a lengthy search for a secure post with Radical backing. In February 1833 Mill and William Tait tried to get Nichol the Paris political economy chair in succession to J.B. Say. Despite the support of James Mill and Nassau Senior, they were unsuccessful. The following year, James Mill and Senior told Nichol that 'a Scotch University would suit you' and advised him to publish 'in some work more known and talked of among the people on whom such things depend'.[53] The tactics paid off. During the autumn of 1835 pressure from his London friends and from events in Scotland encouraged Nichol's move towards cosmology. Nichol and Combe launched an organization based in Edinburgh that would run lectures throughout the country and, as Nichol hoped, 'stir up a commotion in many districts'. Nichol was keen to make his name as a public lecturer on improvement and scientific reform, and publicly committed himself to 'an Institution of *practical* science' to train engineers and scientists. During October 1835 he gave a successful series of lectures on astronomy and on social progress, and in the same month he answered Mill's request for accessible material on science and reform with his celebrated paper on the nebulae, demanding its urgent publication in the *London Review*.[54]

While the career of the projected Edinburgh Society for Aiding the

General Diffusion of Science was brief, that of Nichol had received a decisive boost. Combe told the Lord Advocate that Nichol was politically safe, though 'bordering more on radicalism than Toryism'. Radicals such as the MP and fellow Scot Joseph Hume ('a good friend') were powerful allies in Nichol's application for the Glasgow astronomy chair when it became vacant in January 1836. The Liberal press campaigned for Nichol, Tory newspapers protested, and help was prudently sought from the great evangelical leader Thomas Chalmers. Nichol was relieved to learn that 'the clique of the Astronomical Society have no particular person in their eye', and his success marked the establishment of the Glasgow Observatory as a base from which the science of progress could now be propagated. He told Combe in December 1836, 'I want to become of some importance as a *professional man* because I believe that a professional station is necessary to give a man political and moral weight in society.' Nichol had 'certain grand objects' of politics and morality in view. While he believed he 'should have thrown into some office with closer relations to morals and politics', yet he saw the Observatory and Glaswegian support as crucial: 'My heart is set on having one of the most magnificent establishments in the Empire.' In 1848 he proposed to Combe the foundation of 'Professorships of Pedagogy' that would teach the abstract principles of learning just as schools of mechanics taught abstract principles to engineers. The 'darkness of dogmatic religion' would vanish. Everything was 'favourable' to making this new observatory his centre for moral advance and reformist science.[55]

In Glasgow, Nichol still maintained close links with Mill's *Review*, sending papers on astronomy and geology. He also sought jobs for his London and Edinburgh friends. He aided Combe's unsuccessful effort to win the Edinburgh logic chair; at Glasgow, he offered to depute for James Mylne, the ailing professor of moral philosophy, then tried to obtain the post for the Benthamite jurist John Austin. At the same time, Nichol worked up his nebular astronomy paper into the blockbusting *Views of the Architecture of the Heavens*. It ran through seven editions in as many years. Mill said that the book would make Nichol's career, while Combe wrote that it would be 'invaluable in a high degree as a means of destroying superstition'. He read Nichol as arguing from the majesty of the universe to the implausibility of Christ's incarnation 'in an obscure corner of an obscure world'.[56] By the end of the decade Nichol's empire-building was in full swing. He successfully lobbied for a Glasgow meeting of the British Association in 1840 and sponsored the triumphant geological work of his young protégé Andrew Ramsay. He introduced the novel physics of Fourier and Fresnel into the natural philosophy lectures he took over from his colleague William Meikleham. His student William Thomson recalled the powerful influence of Nichol's lectures, and when Meikleham died in 1846 Nichol first ensured a major curriculum reform at the College, then helped engineer the portentous appointment of Thomson to the chair.[57]

Nichol was obviously an impressive performer. For reasons connected with his reforms, this served him well in the 1840s. His spending on the observatory sent him into irretrievable debt: 'at every point it has almost crushed me', he said. Brewster hoped that Nichol's powerful new instruments would enable him to see the structures built by the inhabitants of the Moon. But when Nichol lobbied the Chancellor of the Exchequer, Thomas Spring Rice, for financial aid, he was told that the price of this subvention would be the submission of the Glasgow Observatory to Airy's programme of systematic magnetic and geodesic observation. Nichol's aims were different – he sought new advances in knowledge of the nebular cosmogony. Airy's programme was only effected in Glasgow when Robert Grant took over after Nichol's death. Within eighteen months of these fund-raising efforts, Nichol was broke and his finances under attack from the conservative Faculty. William Thomson learned from his father that Nichol would be forced to publish money-spinners in order to recoup his losses.[58]

So if Nichol was 'the prose laureate of the stars', as his friend George Gilfillan suggested, he was a laureate with commitment to public success. Large crowds heard him at Glasgow City Hall in 1844–5 and in New York in 1848. Several critics found Nichol's 'gorgeous style, gigantic diagrams and enthusiasm' rather suspect, both politically and morally. His 'very liberal and very unconsidered use of such words as inconceivable, infinite, eternal' simply 'deprives them of all such power, and at last for the infinite we mentally substitute greater than earth, and for eternal, longer than the earthly lifetime of man'. This was a well-aimed barb. Nichol's showmanship was aimed at demonstrating that his nebular hypothesis drew its force from 'infinite and eternal' astronomical truths, yet was immediately relevant to 'earthly man'.[59] The Glasgow Observatory, the Edinburgh presses and the northern lecture circuit were good pulpits from which to capture an audience for the science of progress. Nichol told his audience at Stirling Arts School in January 1849 that 'as society is at present constituted, the *teaching of the world* – the practical contact and dealing with the FACTS of Humanity and the Universe – is incomparably more effective in the evolution of mental culture than the *teaching of the school*'.[60]

The contrast between the 'teaching of the world' and the 'teaching of the school' was the key to Nichol's propagation of the nebular hypothesis. First, Nichol held that the cosmogony was appealing because its truth was manifest in the world, even if the blinkered pedagogy of the establishment denied this patent fact. Natural law governed the world, the nebular hypothesis was obviously an application of natural law, and so it was compelling and believable. Secondly, the nebular hypothesis was part of the science of progress, which showed how advance in the natural world would gradually help change the social order. Faith in moral advance needed the security of the evidence that the nebular hypothesis provided. Last, the science of progress was dependent on new sciences of political economy and education. Nichol claimed that teaching truths such as the nebular hypothesis would aid and illustrate educational and economic reform.

Each of these features was made manifest in his long series of astronomy books and lectures in the 1840s. The human mind was ineluctably disposed to accept evidence of natural law. Nowhere was this disposition more manifest than in the companion sciences of political economy and cosmogony. The world demonstrated that 'evolution, ceaseless and irresistible, advancing from the imperfect to the perfect, is the law of the Universe'. The nebular cosmogony was fortified against possible astronomical challenge. Nichol's 'measure of certainty when penetrating so far' lay 'simply in the assured permanence of law – in the degree of our conviction that the laws we know, and which uphold the harmonies lying around us, rule there also'.[61] Just as Ricardians used 'social epochs' to explain contemporary economies, so Nichol argued that the nebular cosmogony seemed to 'enable us more than almost any other possible conception, to extend our ideas of TIME'. History showed that apparently inexplicable structures in heaven and earth were really produced by the temporal action of a natural law. 'If . . . we fail to discern correlation between that fact contemplated and some other system', then 'we are forced by the strength of our convictions, that a place and import among Nature's harmonies must be found for it, to conclude unhesitatingly that its character must be determined through its relations, not with *Space* but with *Time.*' Nichol emphasized that this conclusion was irresistible. A phenomenon such as that of nebular formation seemed anomalous, but 'its very isolation among surrounding things, or its apparent incompleteness, constrains us to regard it as only one part or term of a Series, complete in itself'.[62]

Nichol's claim about the psychological power of the law of progress was based on Mill's 'metaphysical creed' of 1834. Both men agreed that succession and coexistence were 'ultimate laws of our mind or (what is the same thing in other words) of the phenomena of nature'. Mill used his principle in the version of the *Logic* that Nichol helped to edit. A 'state of society', like an astronomical structure, could only be understood as the result of previously acting law. 'The mutual correlation between the different elements of each state of society is therefore a derivative law, resulting from the laws which regulate the succession between one state of society and another.' Mill argued that:

> the fundamental problem of sociology is to find the laws according to which any state of society produces the state which succeeds it and takes its place. This opens the great and vexed question of the progressiveness of man and society; an idea involved in every just conception of social phenomena as the subject of a science.[63]

The science of progress demanded secure facts that linked the laws of the human mind and those of the universe. Nichol reckoned that the nebular hypothesis accomplished this task. It was a prized demonstration of 'the relation of this vast mechanism to the moral world'. The 'moral world' meant the world of intellectual advance. Such advance was to be treated as part of the new science. In his commentary on Willm's reformist *Education of the People* in July 1847, Nichol argued that, just as 'the common life of the masses of society' showed the workings of the law of progress, so the reformers, 'in our efforts to upraise these masses', were merely 'working along with the natural course of the world'. Cosmology taught that the 'elevation of man' was 'the world's most determinate *First Cause*; in seeking to advance it by education, we therefore act in harmony with manifold resistless agencies, nor, if the task be understood aright, is it possible but that we must prevail'.[64]

The force of Nichol's 'resistless agencies' derived from his claim that the *same* agencies governed the changes in the world and in society. Disciplines such as geology and comparative anatomy could therefore be used to add force to his nebular cosmogony and his reformist social theory. 'Any development considered as *progress*', he believed, was always marked by 'the intensifying and expansion of the powers of LIFE'. In 1837 he made much of William Herschel's natural historical imagery in describing nebular growth; he used Charles Daubeny's unfashionable chemical theory of change in volcanic action to extend the nebular cosmogony. As a directionalist, he accepted Lyell's limitation on explanation of 'existing and known causes' but he held that Lyell '*dogmatizes*' when rejecting 'the possibility of superior energies of these forces in former epochs'. The nebular cosmogony explained, and also depended on, such changes of force.[65] By the 1840s Nichol was using Geoffroy Saint Hilaire's transformism to deny a break in organic sequences between geological and anatomical series: 'Restricting our thoughts to the *vertebrated* races, it is easy to discern that creatures of rising functions, and a growing concentration of brain, somewhat regularly, and in due succession, come upon the scene. . . . The long course has terminated in the meantime with MAN – the latest product of this toil of ages.' The nebular cosmogony evinced the working of a natural law that generated society.[66]

Nebular hypotheses and their uses

Nichol's version of the nebular hypothesis was not an isolated statement of an astronomical truth. It appeared alongside reflections on the origin of life, the progress of humanity and the future of society. His cosmogony was part of a sectarian view of history and it had stiff competition. Audiences were used to a mixture of astronomy and moral philosophy, especially in Scotland, where both Chalmers and Thomas Dick were egregious performers in the field of astrotheology. Dick preached widely on the conventional messages that universal order was an exemplary contrast with social chaos, and that the heavens taught men the fitness of their social station. Chalmers was a more powerful and threatening contemporary. This was why he was such a useful recruit to Nichol's campaign for the Glasgow chair. He made his reputation in Glasgow in 1815 with a series of lectures on astronomy and revelation. Many of his subsequent prescriptions seemed akin to those of the Radicals: tithe abolition, free trade, diffusion of knowledge through new mechanics' institutes. But his political economy and natural philosophy were very different. Peelite social theory married the gloomiest Malthusianism with a moral condemnation of the indigent. Chalmers drew a sharp line between the search for natural law, which led to atheism, and natural histories, which at best taught lessons of divinely planned but transitory 'collocations'.[67]

Nichol planned an answer to Chalmers in the summer of 1834 and added hostile comments to his paper on the nebulae in 1836. Chalmers and some of his evangelical allies made the nebular hypothesis an enemy of proper morality. Nichol answered that 'collocations' did nothing but excite admiration. The nebular hypothesis was, for Nichol, the key example of the need to search for natural law-like causes. If Chalmers were right then he would aid the enemies of reform, because 'weak minds' would think that such contemporary arrangements were unchangeable. 'We have at once the reluctance of

the intolerant mob of Greece to permit the extension of physical law.'
In the 1840s Nichol preached that the mental need for law-like accounts of progress must be met, or else any citizen might rebel. If a
view of the unchangeable virtue of existing structures 'is employed to
prostrate his will, to show him that the relations of things are wholly
beyond his understanding, and therefore, that on the occurrence of
misery and misfortune he has no recourse but in self-abasement',
then 'the world would be a legitimate cause of Discontent'. Nichol,
however, denied that natural knowledge should breed discontent,
arguing that 'a sphere of intelligent action' must always be displayed
to the audience. Progress, not submission, was the right lesson to
teach.[68]

In challenging Chalmers, Nichol was also challenging some of the
Cambridge professors, such as Whewell and Sedgwick. The Cambridge men shared the evangelical scepticism of arguments from
natural law: Whewell declared this in his 1833 Bridgewater Treatise,
Sedgwick in his 1832 *Discourse*. When Whewell coined the term 'nebular hypothesis' he denied that it gave 'any resting place or satisfaction
for the mind', compelling the intellect to a First Cause divorced from
law-like nature. In the disciplinary order that Whewell outlined in his
magisterial *History of the Inductive Sciences* (1837) and re-emphasized
from the presidential chair of the Geological Society, the hypothesis
was assigned to 'cosmical palaetiology', one of those 'palaetiological
sciences' which showed that past causes were different in kind and
scale, and insufficient to explain origins. At the Geological Society,
Whewell influentially labelled the causes indicated by the nebular
hypothesis as 'ulterior', 'obscure' and so hard to integrate in 'geological dynamics'. Sedgwick agreed. His fellow geologists had 'a good old
pedigree without any need of being helped out by an illegitimate link
to a more god-like stock; and we gently hint to them, for their own
good, that they have enough to do on earth without attempting the
sky'. These were powerful strictures against any role for the nebular
hypothesis in *linking* disciplines. They rejected the adequacy of the
hypothesis as an account of law-like *origin*. They saw the search for
such laws as an invitation to atheist materialism, which would reduce
life to matter and humanity to beasts.[69]

The boldest challenge to this emphasis on disciplinary propriety
and on the difference in status between astronomy and the nebular
hypothesis was the appearance of *Vestiges* in 1844. The work came
from sources close to Nichol, was sometimes attributed to him, and
used his name and stature as an important resource. Nichol himself
believed that Chambers had plagiarized his ideas, and commented
in his *System of the World* (1846) that *Vestiges* had made the science of

progress and its lessons the ubiquitous concern of 'almost every popular periodical as having passed within the domain of common knowledge'.[70] But Nichol did not see the book as dangerous or as a distortion of his own teachings. On the contrary, he used the furore as an opportunity to encourage further work on his science. He was newly interested in the probabilities on which his cosmogony relied and in the physics used to describe condensation and rotation of the original nebulous cloud. Both issues were raised by Chambers and amplified in Nichol's *System of the World*. Preparing for this opportunistic book in February 1846, Nichol told William Thomson that 'I have been induced to return to this matter because of a foolish book called Vestiges of Creation and still foolisher reviews where attempts are made to speak of Atheism.' For Nichol, Chambers offered a 'foolish' version of what he still took to be the disciplinary principles of the nebular cosmogony and its secular implications.[71]

No doubt this was too sanguine. The *Vestiges* controversy made Nichol's version of the nebular hypothesis look even more like 'low' knowledge. The status of the scientist and 'popularization' of knowledge were at issue; the knowledge that drew disproportionate attention in reviews of *Vestiges* was the nebular hypothesis. During 1845, Herschel, Whewell, Sedgwick and Brewster all challenged its standing. Nichol's responsibility for this version and his use of it as a polemical weapon helps to explain their emphasis.[72] In treating the nebular hypothesis as silly or populist, elite savants such as Herschel were answering such usages. They could not allow the link between a cosmogony and a science of progress on Earth to survive in this carefully designed form. Nichol's was an unacceptable transposition of concepts. This transposition had its characteristic temples, the mechanics' institutes, where, as Nichol put it in 1849, 'they have undertaken to unfold the progress of external nature, and especially the relations of its Laws with human happiness and duty'. The creed preached there was based on political economy and cosmology, so ecclesiastical control had to be rejected: 'Alas, if our Nation must be subservient to its Churches', Nichol exclaimed. The creed of natural law had its saints – Nichol named Andrew Combe's popular medical texts as good examples of the work experts could do in aiding progress – and it had its martyrs, too. Citing Chadwick's sanitary reports, Nichol fulminated against the ignorance of natural law and social action displayed by 'a set of men in London ... with their private feast on Sunday, and their carriage to ride to Church in', who 'could, in the face of humanity, plead their vested interests in the practice of destroying ten or twenty thousands annually of their fellow countrymen'. Last, the science of progress had a vision of a

'Commonwealth' of the future where, as Nichol put it, 'no state worthy of the name would regret to see itself in time merging into a sounder and more fitting organisation'.[73]

The nebular hypothesis was a precedent to which Victorian naturalists might indeed appeal when making a story of law-like development of past into future states, but the very forms in which it was then visible were those where 'development' carried a troubling political message. If the nebular hypothesis were to serve as a resource, naturalists had to specify what they were *not* transposing as well as what they were. The implication is that nebular astronomy could be made to teach many different lessons.

Compare two ways in which the giant Parsonstown telescopes were used by lecturers on astronomy and morals. Nichol's enemy, Robinson, told Irish audiences about the divine role that the telescopes fulfilled: 'the *very passage of light* tells us that God is there'. He claimed that the function of nebular astronomy was to teach that 'confusions about nature and law are at the root of much of that infidelity which is at present endeavouring to flood the land'. Nichol, by contrast, gave his Scottish audience a vision of an economic and political future where experts would explain the prospects of the working classes as self-confidently as they now explained the revelations of Rosse's great instrument. Nichol argued that 'there is nothing whatever in its separate portions' to indicate the purpose of the giant telescope, yet he was able to explain clearly that this telescope taught nebular progress. Similarly, 'if one were looking simply at the separate portions, let me say, of British society', there would be no difficulty in seeing 'for what high end this singular organisation can be destined, or what share in it can well be apportioned to the workman, confined by the action of that singular principle, the division of labour, to the unceasing repetition of some simple act'. The point of Nichol's elaborate comparison of astronomical and economic expertise was that they were part of the same science and had the same progressive end.[74]

Notes

I am grateful for permission to cite manuscripts in the possession of the British Library (Peel Papers), Cambridge University Library (Kelvin Papers), the National Library of Scotland (Combe Papers), the Royal Astronomical Society (Herschel Papers), the Royal Society (Herschel Papers) and the Earl of Rosse (Birr Castle Papers). I have received generous help from Paul Baxter, Jim Bennett, Michael Crowe, Adrian Desmond, Michael Hoskin, Bernard Lightman, Jim Moore, Jack Morrell, Jim Secord, Steve Shapin, Crosbie Smith and Norton Wise.

1. J.F.W. Herschel, 'On Atoms', *Fortnightly Review*, 1 (May 1865), 81–4.

2. J.P. Nichol, *Views of the Architecture of the Heavens* . . . (Edinburgh: William Tait, 1837), p. 206; [George Gilfillan], 'Popular Lecturers No. 1 – Professor Nichol', *Tait's Edinburgh Magazine*, 15 (Mar. 1848), 145–53 (146).

3. [Robert Chambers], *Vestiges of the Natural History of Creation* (London: John Churchill, 1844), pp. 1–26; [Herbert Spencer], 'Progress: Its Law and Cause', *Westminster Review*, 68 (1857), 445–85, reprinted in *idem, Essays: Scientific, Political, and Speculative*, 3 vols. (London: Williams and Norgate, 1891), I, 8–62 (10).

4. [Spencer], 'Recent Astronomy and the Nebular Hypothesis', *Westminster Review*, 70 (1858), 185–225, reprinted in *idem, Essays*, I, 108–81 (108–11).

5. John C. Greene, *Science, Ideology, and World View: Essays in the History of Evolutionary Ideas* (Berkeley: University of California Press, 1981), p. 131; *idem, The Death of Adam: Evolution and Its Impact on Western Thought* (New York: Mentor, 1961), p. 268.

6. For example, S.S. Schweber, 'The Wider British Context in Darwin's Theorizing', in David Kohn, ed., *The Darwinian Heritage* (Princeton, N.J.: Princeton University Press, 1985), pp. 35–69; Robert M. Young, *Darwin's Metaphor: Nature's Place in Victorian Culture* (Cambridge: Cambridge University Press, 1985).

7. Greene, *Science, Ideology, and World View*, p. 124. On transpositions, see M.J.S. Rudwick, 'Transposed Concepts from the Human Sciences in the Early Work of Charles Lyell', in L.J. Jordanova and Roy Porter, eds., *Images of the Earth: Essays in the History of the Environmental Sciences* (Chalfont St. Giles, Bucks.: British Society for the History of Science, 1979), pp. 67–83.

8. Ronald L. Numbers, *Creation by Natural Law: Laplace's Nebular Hypothesis in American Thought* (Seattle: University of Washington Press, 1977); Stephen G. Brush, 'The Nebular Hypothesis and the Evolutionary Worldview', *History of Science*, 25 (1987), 245–78. My account follows J.H. Brooke, 'Natural Theology and the Plurality of Worlds: Observations on the Brewster–Whewell Debate', *Annals of Science*, 34 (1977), 221–86 (269–70) and *idem*, 'Nebular Contraction and the Expansion of Naturalism', *British Journal for the History of Science*, 12 (1979), 200–11.

9. P.S. de Laplace, *Exposition du système du monde*, 2 vols. (Paris: Cercle Social, 1796), II, 293–312; William Herschel, 'Astronomical Observations relating to the Construction of the Heavens', *Philosophical Transactions of the Royal Society*, 101 (1811), 269–336; August Comte, 'Premier mémoire sur la cosmogonie positive' (1835), in Comte, *Ecrits de jeunesse, 1816–1828*, ed. P. Carneiro and P. Arnaud (Paris: Mouton, 1970), pp. 585–608. See S.L. Jaki, *Planets and Planetarians: A History of Theories of the Origin of Planetary Systems* (Edinburgh: Scottish Academic Press, 1978), ch. 5 and J. Merleau-Ponty, *La science de l'univers à l'âge du positivisme* (Paris: J. Vrin, 1983).

10. William Whewell, *Astronomy and General Physics considered with reference to Natural Theology* (London: William Pickering, 1833), pp. 188–91 (189); [J.P. Nichol], 'State of Discovery and Speculation concerning the Nebulae', *Westminster Review*, 25 (July 1836), 390–409; *idem, Architecture of the Heavens*, pp. 166–7. Whewell's book appeared in March 1833. Nichol's first statement on the nebular hypothesis followed six mouths later, 'Recent Discoveries in

Astronomy', *Tait's Edinburgh Magazine*, 4 (Oct. 1833), 57–64.

11. W. Whewell to R. Jones, 19 Dec. 1834, in I. Todhunter, ed., *William Whewell, D.D., an Account of His Writings, with Selections from His Literary and Scientific Correspondence*, 2 vols. (London: Macmillan, 1876), II, 199; J.L.E. Dreyer and H.H. Turner, eds., *History of the Royal Astronomical Society, 1820–1920* (London: Royal Astronomical Society, 1923), pp. 7–10; Adolphe Quetelet, *A Treatise on Man and the Development of His Faculties* (Edinburgh: R. and W. Chambers, 1845), p. 9.

12. J.R. McCulloch, *Principles of Political Economy* (Edinburgh: William Tait, 1825), p. 39; Adam Sedgwick, *A Discourse on the Studies of the University* (Cambridge: Pitt Press, 1833), pp. 74–5.

13. This is the version given by G.B. Airy when presenting the Royal Astronomical Society's medal to John Herschel, printed in *Memoirs of the Royal Astronomical Society*, 9 (1836), 303–12 (305–6) and by Nichol, 'Discovery and Speculation concerning the Nebulae', pp. 399–401. It is rejected in 'Recent Astronomy', *British Quarterly Review*, 6 (Aug. 1847), 1–40 (10–13).

14. S. Schaffer, 'Herschel in Bedlam: Natural History and Stellar Astronomy', *British Journal for the History of Science*, 13 (1980), 211–39; M.E. Williams, 'Was there such a thing as Stellar Astronomy in the Eighteenth Century?', *History of Science*, 21 (1983), 369–85; P.S. Laplace, *Essai philosophique sur les probabilités* (Paris: Courcier, 1814), pp. 55–8; J. Merleau-Ponty, 'Situation et rôle de l'hypothèse cosmogonique chez Laplace', *Revue de l'histoire des sciences*, 29 (1976), 21–49; and 30 (1977), 71–2.

15. H.W. Olbers to F. Bessel, 10 July 1812, in A. Erman, ed., *Briefwechsel zwischen Olbers und Bessel*, 2 vols. (Leipzig: Avenarius and Mendelssohn, 1852), I, 336–7; Jaki, *Planets and Planetarians*, pp. 128–30; Schaffer, 'Herschel in Bedlam', pp. 230–3.

16. J. Robison to W. Herschel, 31 Jan. 1802, Royal Astronomical Society, Herschel MSS, W.1/13, W.131. See J.B. Morrell, 'Professors Robison and Playfair and the "Theophobia Gallica"', *Notes and Records of the Royal Society of London*, 26 (1971), 43–63.

17. J. Playfair, *Illustrations of the Huttonian Theory of the Earth* (Edinburgh: W. Creech, 1802), pp. 119–20, 437–40; Greene, *Death of Adam*, p. 90; C. Lyell to M. Horner, 27 Oct. 1831, in K.M. Lyell, ed., *Life, Letters, and Journals of Sir Charles Lyell*, 2 vols. (London: John Murray, 1881), I, 346–7.

18. John Herschel, *Preliminary Discourse on the Study of Natural Philosophy* (London: Longman, Rees, Orme, Brown, Green and Longman, 1830), pp. 282, 304–5. Lyell rejected the nebular hypothesis because 'we have never seen one step in the progression', so there was just as much evidence that a 'solid nucleus' could rarefy into a nebula (Lyell to C. Babbage, [Jan.] 1837, British Library, MSS Add. 37190, fos. 185–88, partly printed in K.M. Lyell, ed., *Life*, II, 9–10).

19. John Herschel, 'Observations of Nebulae and Clusters made at Slough', *Philosophical Transactions*, 123 (1823), 359–505 (500); D.S. Evans et al., eds., *Herschel at the Cape: Diaries and Correspondence of Sir John Herschel, 1834–1838* (Austin: University of Texas Press, 1969), p. 50; John Herschel, *Outlines of Astronomy*, 4th edn (London: Longman, Brown, Green, Longmans and

Roberts, 1851), pp. 598, 615. See M.A. Hoskin, 'John Herschel's Cosmology', *Journal for the History of Astronomy*, 18 (1987), 1–34.

20. John Herschel, 'Address of the President', *Report of the Fifteenth Meeting of the British Association for the Advancement of Science* (London: John Murray, 1846), pp. xxvii–xliv (xxxvi–xxxix).

21. John Herschel, 'On the Astronomical Causes which may influence Geological Phenomena', *Transactions of the Geological Society of London*, 3 (1830; pub. 1835), 293–9; P. Lawrence, 'Heaven and Earth: The Relation of the Nebular Hypothesis to Geology', in W. Yourgrau and A.D. Breck, eds., *Cosmology, History, Theology* (New York: Plenum, 1977), pp. 253–82.

22. Dreyer and Turner, *History of the Royal Astronomical Society*, pp. 3–7; G.B. Airy, 'Report on the Progress of Astronomy during the Present Century', *Report of the First and Second Meetings of the British Association for the Advancement of Science* (London: John Murray, 1833), pp. 125–89 (186–9). For criticism of Airy, see Whewell to W. Vernon Harcourt, 4 Nov. 1831, in J.B. Morrell and Arnold Thackray, eds., *Gentlemen of Science: Early Correspondence of the British Association for the Advancement of Science* (London: Royal Historical Society, 1984), p. 95.

23. Herschel, *Preliminary Discourse*, pp. 7, 281; *idem, Outlines* (1851), p. 599.

24. M.A. Hoskin, *Stellar Astronomy: Historical Studies* (Chalfont St. Giles, Bucks.: Science History, 1982), pp. 143–5; William Parsons, 3rd Earl of Rosse, *Scientific Papers* (London: Lund, Humphries, 1926), p. 108, 129, 132; J. South to Rosse, 11 Nov. 1840, Birr Castle, MSS K.1.2.

25. B.Z. Jones, 'Diary of the Two Bonds', *Harvard Library Bulletin*, 16 (1968), 49–78 (56, 61); B.Z. Jones and L.G. Boyd, *Harvard College Observatory, 1839–1919* (Cambridge, Mass.: Belknap Press, Harvard University Press, 1971), pp. 67–8; Numbers, *Creation by Natural Law*, p. 35.

26. J. Phillips to Whewell, 5 Sept. 1843, in J.B. Morrell and Arnold Thackray, *Gentlemen of Science: Early Years of the British Association for the Advancement of Science* (Oxford: Oxford University Press, 1981), p. 114; G.B. Airy, 'Address', *Report of the Twenty-First Meeting of the British Association for the Advancement of Science* (London: John Murray, 1852), p. xxxix–liii (xli); Wilfrid Airy, ed., *Autobiography of Sir George Biddell Airy* (Cambridge: Cambridge University Press, 1896), pp. 198, 221–2. For Whewell, see R.P. Graves, *Life of Sir William Rowan Hamilton*, 3 vols. (Dublin: Dublin University Press, 1882–9), II, 118–22, 629–30; III, 399.

27. [William Whewell], *Of the Plurality of Worlds: An Essay* (John W. Parker, 1854), pp. 211–14, 224; Michael J. Crowe, *The Extra-Terrestrial Life Debate, 1750–1900: The Idea of a Plurality of Worlds from Kant to Lowell* (Cambridge: Cambridge University Press, 1986), pp. 312, 320.

28. P. Baxter, 'Brewster, Evangelism and the Disruption of the Church of Scotland', in A.D. Morrison-Low and J.R.R. Christie, eds., *'Martyr of Science': Sir David Brewster, 1781–1868* (Edinburgh: Royal Scottish Museum, 1984), pp. 45–50 (47–9).

29. Nichol, *Architecture of the Heavens* (1837), pp. 127, 187; *idem, Thoughts on Some Important Points relating to the System of the World*, 2nd edn (Edinburgh:

John Johnstone, 1848), pp. iv, 107–10; T.R. Robinson to Rosse, 7 April 1841, Birr Castle, MSS K.5.4.

30. James Coutts, *History of the University of Glasgow* (Glasgow: Maclehose, 1909), pp. 388–9; J.P. Nichol to R. Peel, 29 Jan. 1846, British Library, MSS Add. 40583, fos. 270–2.

31. Nichol, *System of the World* (1848), pp. 31–4, 73; *idem, The Stellar Universe* (Edinburgh: John Johnstone, 1848), p. 102; *idem, Architecture of the Heavens* (9th edn, 1851), pp. 126, 135, 140.

32. Spencer, *Essays*, I, 116–17; *idem, Autobiography*, 2 vols. (London: Williams and Norgate, 1904), II, 21–2.

33. William Huggins, *Scientific Papers* (London: W. Wesley, 1909), 'Historical Statement', pp. 105–8.

34. A.H. Japp, *Thomas de Quincey: His Life and Writings* (London: John Hogg, 1899), p. 251; Thomas de Quincey, 'The System of the Heavens as revealed by Lord Rosse's Telescopes', in *Works*, ed. D. Masson, 14 vols. (Edinburgh: A. and C. Black, 1890), VIII, 7–34; David Brewster, 'Revelations of Astronomy', *North British Review*, 6 (Feb. 1847), 206–55 (206–9).

35. [Charles Lyell], review of G.P. Scrope, 'Memoir on the Geology of Central France', *Quarterly Review*, 36 (1827), 437–83 (475), noted by Charles Darwin in P.H. Barrett, ed., *Metaphysics, Materialism and the Evolution of Mind: Early Writings of Charles Darwin* (Chicago: University of Chicago Press, 1980), p. 74 (*N* 19e).

36. R. Johnson, 'Educating the Educators: Experts and the State, 1833–1839', in A.P. Donajgrodzki, ed., *Social Control in Nineteenth-Century Britain* (London: Croom Helm, 1977), pp. 77–101; M. Berman, *Social Change and Scientific Organization: The Royal Institution, 1799–1844* (London: Heinemann, 1978), pp. 100–23. For the Philosophic Radicals, see W. Thomas, *The Philosophic Radicals* (Oxford: Oxford University Press, 1979) and P. Richards, 'State Formation and Class Struggle, 1832–1848', in P. Corrigan, ed., *Capitalism, State Formation and Marxist Theory* (London: Quartet, 1980), pp. 49–78. For the 'tested system', see Andrew Scull, *Museums of Madness: The Social Organization of Insanity in Nineteenth-Century England* (Harmondsworth, Middlesex: Penguin, 1982), p. 108.

37. [J.R. McCulloch], 'Census of the Population', *Edinburgh Review*, 49 (1829), 1–34; Herschel, 'Quetelet on Probabilities' (July 1850), in *idem, Essays from the Edinburgh and Quarterly Reviews* (London: Longman, Brown, Green, Longmans and Roberts, 1857), pp. 464–5; William and Robert Chambers, *Political Economy for Use in Schools and for Private Instruction* (Edinburgh: W. and R. Chambers, 1852), p. 54.

38. John Herschel, *An Address to the Subscribers of the Windsor and Eton Public Library*, 2nd edn (London: Smith, Elder, 1834), pp. 5–7, 31–2; [John Henry Newman], *Letters on an Address delivered by Sir Robert Peel on the Establishment of a Reading Room at Tamworth* (London: John Mortimer, 1841), pp. 13, 39. I owe this reference to Jim Moore.

39. Dorothy Thompson, *The Chartists: Popular Politics in the Industrial Revolution* (London: Maurice Temple Smith, 1984), pp. 21–3, 125 (Thomson), 32 (O'Brien); Iorwerth Prothero, *Artisans and Politics in Early Nineteenth-Century*

London: John Gast and His Times (London: Methuen, 1981), p. 298 (O'Brien). See S. Shapin and B. Barnes, 'Head and Hand: Rhetorical Resources in British Pedagogical Writing, 1770–1850', *Oxford Review of Education*, 2 (1976), 231–54.

40. [John Roebuck], 'National Education', *Tait's Edinburgh Magazine*, 2 (Mar. 1833), 755–65; George Combe, *The Constitution of Man*, 6th edn (Edinburgh: John Anderson, 1836), pp. 12–13, 103–4, 225; Roger Cooter, *The Cultural Meaning of Popular Science: Phrenology and the Organization of Consent in Nineteenth-Century Britain* (Cambridge: Cambridge University Press, 1984), pp. 120–33.

41. [J.P. Nichol], *The Importance of Literature to Men of Business* (Glasgow: Griffin, 1852), p. v; 'The Wants of the Times', *Tait's Edinburgh Magazine*, 15 (April 1848), 215–18 (217); S.T. Coleridge, *On the Constitution of the Church and State according to the Idea of Each* (1830), in *Complete Works*, ed. John Colmer, vol. 10 (London: Routledge and Kegan Paul, 1976), pp. 68–70.

42. S.S. Schweber, 'The Origin of the *Origin* Revisited', *Journal for the History of Biology*, 10 (1977), 229–316 (252); M. Ogilvie, 'Robert Chambers and the Nebular Hypothesis', *British Journal for the History of Science*, 8 (1975), 214–32 (218); F.W. Connor, 'Poe and John Nichol', in R.A. Bryan *et al.*, eds., *All These to Teach* (Gainesville: University of Florida Press, 1965), pp. 190–208.

43. 'Obituary of John Pringle Nichol', *Monthly Notices of the Royal Astronomical Society*, 20 (1859–60), 131; [Agnes Clerke], 'John Pringle Nichol', *Dictionary of National Biography*, XIV, 412–13. For the Paris trip, see Nichol, *System of the World* (2nd edn), p. 264. For Paris and Edinburgh radicals, see Adrian Desmond, 'Robert E. Grant: The Social Predicament of a Pre-Darwinian Transmutationist', *Journal of the History of Biology*, 17 (1984), 189–223 (195–202).

44. Nichol's papers include: 'Incidence of Tithes', *Tait's Edinburgh Magazine*, 1 (May 1832), 224–8; 'Political Economy for Farmers', *Tait's Edinburgh Magazine*, 3 (May 1833), 191–8; 'Comparative Mortality of Different Populations', *Foreign Quarterly Review*, 13 (May 1834), 272–83; 'Tithes and Their Commutation', *London Review*, 1 (April 1835), 164–73; and 'Rae's New Principles of Political Economy', *Foreign Quarterly Review*, 15 (July 1885), 241–67. The 'English Gironde' is mentioned in J.S. Mill to Nichol, 10 July 1833, in F.E. Mineka, ed., *Earlier Letters of John Stuart Mill*, 2 vols. (London: Routledge and Kegan Paul, 1963), p. 164.

45. [Nichol], 'Rae's Principles', p. 245; *idem*, 'Tithes and Their Commutation', p. 171; Nichol to Peel, 29 Jan. 1846, British Library, MSS Add. 40583, fos. 270–2.

46. [Nichol], 'Tithes and Their Commutation', p. 172; *idem*, 'Comparative Mortality', pp. 273, 280–1; Quetelet, *Treatise on Man*, p. 108.

47. Mill to W. Tait, 7 Nov. 1832; Mill to Nichol, 16 Jan. 1833; Mill to Tait, 28 Feb. 1833; Mill to Nichol, 10 July 1833, in *Earlier Letters*, pp. 131, 136–7, 142, 164–7; [Nichol], 'Tithes and Their Commutation', p. 170; Mill, 'Corporation and Church Property' (Feb. 1833), in J.W.M. Gibbs, ed., *Early Essays of John Stuart Mill* (London: George Bell, 1897), pp. 161–98 (195). For statistics, see Mill to Nichol, 15 April 1834, *Earlier Letters*, pp. 220–3; [Nichol], 'Comparative Mortality of Different Populations', pp. 281–2; and G.R. Porter, *The Progress*

of the Nation in Its Various Social and Economic Relations (London: Charles Knight, 1836), secs. 1–2, pp. 2, 24–6.

48. For 'sinister interest', a Benthamite catchphrase, see Mill, 'Corporation and Church Property', p. 191. For Sedgwick, see Mill to Nichol, 25 Nov. 1834, Earlier Letters, pp. 237–9; [Mill], 'Professor Sedwick's Discourse', London Review, 1 (April 1835), 94–135. For political economy, see Whewell, Mathematical Exposition of Some Doctrines of Political Economy (Cambridge: J. Smith, 1829), pp. 3, 5; idem, Mathematical Exposition of Some of the Leading Doctrines of Mr Ricardo's Principles of Political Economy and Taxation (Cambridge: J. Smith, 1831), pp. 13, 15; and S. Hollander, 'William Whewell and John Stuart Mill on the Methodology of Political Economy', Studies in History and Philosophy of Science, 14 (1983), 127–68 (160).

49. [Nichol], 'Tithes and Their Commutation', pp. 166–7; [Mill], 'On the Definition of Political Economy', London and Westminster Review, 26 (Oct. 1836), 1–29; [Nichol], 'Fallacies concerning Tithes', Tait's Edinburgh Magazine, 2 (Dec. 1832), 316–22; Mill to Nichol, 15 April and 14 Oct. 1834, Earlier Letters, pp. 220–3, 234–7; Whewell to R. Jones, 24 April 1831, and Whewell to A. Hare, 15 Oct. 1838, in Todhunter, William Whewell, II, 118, 270–2.

50. Mill, 'Definition of Political Economy', pp. 128, 136; Mill to Nichol, 17 Jan., 14 Oct., and 26 Nov. 1834, Earlier Letters, pp. 209–12, 234–7, 237–9.

51. Mill, System of Logic (1843), in Collected Works, vols. 7–8, ed. J.M. Robson (London: Routledge and Kegan Paul, 1973–4), VII, 507–8.

52. Mill to Nichol, 7 Oct. 1835, Earlier Letters, pp. 274–6; Nichol's letter, 20 April 1836, in Testimonials on Behalf of George Combe ... (Edinburgh: John Anderson, 1836), p. 24; I.W. Mueller, John Stuart Mill and French Thought (Urbana: University of Illinois Press, 1956), pp. 107–12.

53. Mill to Tait, 28 Feb. and 30 Mar. 1833; Mill to Nichol, 10 July 1833 and 14 Oct. 1834, Earlier Letters, pp. 142, 147, 164–5, 234. See n. 10 above.

54. Nichol to G. Combe, National Library of Scotland (NLS), Combe MSS 7236 fo. 3; Steven Shapin, '"Nibbling at the Teats of Science": Edinburgh and the Diffusion of Science in the 1830s', in Ian Inkster and J.B. Morrell, eds., Metropolis and Province: Science in British Culture, 1780–1850 (London: Hutchinson, 1983), pp. 151–78.

55. Shapin, '"Nibbling at the Teats of Science"', p. 160, 163; Nichol to Combe, 16 Jan. 1836, NLS, Combe MSS 7240, fo. 88; idem, 18 Jan. 1836, fo. 192; idem, 19 Jan. 1836, fo. 190; idem, 13 Dec. 1836, fo. 201; idem, 21 Nov. 1836, fo. 203; idem, 25 Feb. 1848, Combe MSS 7296, fo. 78.

56. Mill to J. Robertson, 28 July and 6 Aug. 1837; Mill to Nichol, 21 Dec. 1837; Mill to S. Austin, 28 April 1837, Earlier Letters, pp. 344, 345, 363–4, 335; Coutts, History of the University of Glasgow, p. 383; Combe to Nichol, 9 Nov. 1837, NLS, Combe MSS 7387, fos. 468–9.

57. Morrell and Thackray, Gentlemen of Science, pp. 205, 211–12, 215, 219. For Nichol and Thomson, see S.P. Thompson, Life of William Thomson, Baron Kelvin, 2 vols. (London: Macmillan, 1910), I, 14; Coutts, History of the University of Glasgow, p. 384; W. Thomson to Nichol, 18 Mar. 1857, Cambridge University Library, MSS Add. 7342 N30.

58. Nichol to Combe, 8 Feb. 1847, NLS, Combe MSS 7287, fo. 48; Nichol to

J. Herschel, 4 Nov. 1838, 29 Jan. and 24 Feb. 1839, Royal Society, MSS HS.13, 131–3; J. Thomson to W. Thomson, 12 Jan. 1842, Cambridge University Library, MSS Add. 7342 T191. I owe this reference to Crosbie Smith. For funding, see Coutts, *History of the University of Glasgow*, pp. 388–90. For Brewster on lunar life, see Nichol, *Architecture of the Heavens*, pp. 39–40.

59. [Gilfillan], 'Popular Lecturers – Professor Nichol', pp. 148, 152; 'Recent Astronomy', *British Quarterly Review*, 6 (Aug. 1847), 1–40 (26). For another attack, see William Martin, *Exposure of Dr. Nichol, the Impostor and Mock-Astronomer of Glasgow College* (Newcastle: Pattison and Ross, 1839).

60. Nichol, 'Address delivered at the Soirée of the Stirling School of Arts', in *idem, The Importance of Literature to Men of Business*, pp. 212–52 (226).

61. Nichol, *Views of Astronomy: Seven Lectures delivered before the Mercantile Library Association of New York* (New York: Greeley and McElrath, 1848), p. 32; *idem, System of the World* (2nd edn), p. 270.

62. Nichol, *System of the World* (2nd edn), pp. 132, 198–9.

63. Mill to Nichol, 14 Oct. 1834, *Earlier Letters*, pp. 236–7; Mill, *System of Logic*, bk. 6, ch.10, para.2, in *Collected Works*, VIII, 912. In the second edition 'sociology' was replaced by 'social science'.

64. Nichol, *Architecture of the Heavens*, p. 206; *idem*, 'Preliminary Discussion on Some Points connected with the Present Position of Education in this Country', in J. Willm, *The Education of the People* (Glasgow: William Lang, 1847), pp. xi–lxxx (xix).

65. Nichol, *System of the World* (2nd edn), p. 247; *idem, Architecture of the Heavens*, pp. 219–23.

66. Nichol, *System of the World* (2nd edn), pp. 261–5; *idem*, 'Discovery and Speculation concerning the Nebulae', p. 405n; *idem, Views of Astronomy*, pp. 36–7.

67. J.V. Smith, 'Reason, Revelation, and Reform: Thomas Dick of Methven and the Improvement of Society by the Diffusion of Knowledge', *History of Education*, 12 (1983), 255–70; David Cairns, 'Thomas Chalmers's *Astronomical Discourses*: A Study in Natural Theology', *Scottish Journal of Theology*, 9 (1956), 410–21; S.J. Brown, *Thomas Chalmers and the Godly Commonwealth* (Oxford: Oxford University Press, 1982), ch.4.

68. 'Dr Chalmers', *Tait's Edinburgh Magazine*, 2 (Nov. 1832), 189–95; Thomas Chalmers, *On the Power, Wisdom and Goodness of God manifested in the Adaptation of External Nature to the Moral and Intellectual Constitution of Man*, 2 vols. (London: William Pickering, 1833), I, 34n, cited in Nichol, 'Discovery and Speculation concerning the Nebulae', pp. 406–9; *idem, Views of Astronomy*, p. 28; *idem*, 'The Present State of Education', pp. lxxvi–vii.

69. Whewell, *Astronomy and General Physics* (2nd edn, 1834), pp. vi, 185–6; *idem, History of the Inductive Sciences* (3rd edn), 3 vols. (London: John W. Parker, 1857), II, 229; III, 400, 516; *idem*, 'Presidential Address', *Proceedings of the Geological Society of London*, 2 (1838), 624–49 (645); [Adam Sedgwick], 'Natural History of Creation', *Edinburgh Review*, 82 (July 1845), 1–85 (18).

70. Nichol to Combe, 25 Feb. 1848, NLS, Combe MSS 7296, fos. 80–1; Combe to Nichol, 18 Mar. 1848, Combe MSS 7391, fos. 349–50; Nichol, *System of the World* (2nd edn), p. viii. See [Robert Chambers], *Vestiges*, pp. 12–19;

[*idem*], *Explanations* (London: John Churchill, 1845), pp. 7–13.

71. Nichol to W. Thomson, 1 Feb. 1846, Cambridge University Library, MSS Add. 7342 N28; Nichol, *System of the World* (2nd edn), pp. 130–2.

72. Richard Yeo, 'Science and Intellectual Authority in Mid-Nineteenth Century Britain: Robert Chambers and "Vestiges of the Natural History of Creation"', *Victorian Studies*, 28 (1984), 5–31 (17). See Herschel, 'Address' (June 1845), in *idem*, *Essays*, p. 677; [Sedgwick], 'Natural History of Creation', pp. 17–25; [David Brewster], 'Vestiges of the Natural History of Creation', *North British Review*, 3 (1845), 470–515 (480); and William Whewell, *Indications of the Creator*, 2nd edn (London: John W. Parker, 1846), 'Preface', pp. 26–8.

73. Nichol, 'Address at the Stirling School of Arts', pp. 233–4, 237, 252; *idem*, 'The Present Position of Education', pp. lix–x, lxxix–xxx.

74. Thomas Romney Robinson, *Light: Lecture delivered at the Dublin Young Men's Christian Association, 26 November 1862* (Dublin: Hodges, Smith, 1863), pp. 11, 13, 22; Nichol, 'Address at the Stirling School of Arts', p. 241.

5

Behind the veil:
Robert Chambers and *Vestiges*

JAMES A. SECORD

'Knowledge is power', the popular author and publisher Robert Chambers believed, 'because knowledge is property'.[1] Chambers' own most famous intellectual property was *Vestiges of the Natural History of Creation* of 1844, the best-selling evolutionary work in the

period before the *Origin of Species*. In *Vestiges*, Chambers made the
entire cosmos his territory. The book ranged from the nebular hypothesis of the formation of the Solar System, through the record
of life on Earth and the formation of new species, to the constitution
of the human mind.[2] This essay presents a new view of *Vestiges* and
how it came to be written.

Vestiges is sometimes simply dismissed as 'bad science', but its
significance cannot be doubted. As is well known, it brought the
concept of natural law before the middle-class reading public to an
unprecedented degree, not only in Great Britain, but in America,
Germany and Holland. In America the book sold over twenty editions; in Britain, at least thirteen. It called forth a bitter reaction,
especially from conservative Cambridge clerics like Adam Sedgwick
and William Whewell. In so doing it heightened divisions within
the scientific and religious community, dissolving old alliances and
cementing new ones.[3] And finally, *Vestiges* represents a key moment
in the colonization of Victorian science by the middle classes. George
Combe's *Constitution of Man* (1828), John Pringle Nichol's *Architecture
of the Heavens* (1837), Chambers' *Vestiges*, and the works of Baden
Powell, George Henry Lewes and Herbert Spencer, form a series that
began to unite an alternative scientific vision of progress, advocated
in newspapers and popular periodicals, with an elite science under
the control of gentlemen.[4] One might say that when Darwin finally
published, he engineered not a popular revolt but a palace coup within
the scientific elite. The wider revolution had been won, because the
'people', as Chambers always called them, were already on his side.

Yet *Vestiges* was published without an author's name on the title
page; Chambers did not claim his property. As in the case of Sir
Walter Scott and the Waverley novels, Chambers (a Scott protégé
and Edinburgh resident) was fascinated by anonymity. His first
book, *Illustrations of the Author of Waverley* (1822), was a collection
describing the originals of various characters in Scott's novels. Both
the title and the frontispiece (shown above) played with the idea of
revealing, or 'illustrating', the mysterious author. The veil on the
framed portrait was teasingly only half-removed, so that the reader
was left guessing who was hidden behind. During the process of
publishing his own anonymous work, Chambers enjoyed acting
under a similiar veil. 'What a treat all this is to secretiveness!', he
wrote soon after *Vestiges* appeared.[5]

Anonymity of this kind was commonplace in fiction, popular
journalism, and religious and political writing, but in science it was
highly unusual. Only the established facts of nature were anonymous in science; new ideas and theories were always attached to

individual names. Just how pervasive this aspect of science could be was demonstrated by public eagerness to pin down the *Vestiges* authorship. (Many, more appropriately than they knew, called it the greatest sensation since Waverley.) The authorship was divulged in the posthumous twelfth edition of 1884, although many people had guessed correctly before then, and accusing fingers had even been pointed at Chambers in print. But there is a fundamental difference between an acknowledged and an attributed text. Chambers tried to make sure that no one could ever be certain that he was the author.

The separation of text and author for forty years has had important consequences for our understanding of the reception of *Vestiges*. There is an immense difficulty in recapturing the sense of dislocation that the lack of an acknowledged author posed for contemporary readers. The usual keys to interpretation were missing; reviewers could never be sure who they were attacking, supporting or ignoring. The sex, class, religion and morals of the author had to be judged solely by clues in the text. Many, for example, thought the author must be a woman.

There have also been problems for our understanding of the genesis of *Vestiges*. Here the dislocation of contemporary readers becomes our own, for the book – to a surprising degree – still lacks a context. Almost no attention has been paid to the early history of the book or the development of its basic message in the work of its author.[6] The situation contrasts with the vast quantity of books and articles on the development of Darwin's theories, a difference that is only partially explicable through the longer-lasting impact of the *Origin*. Robert Merton's 'Matthew effect' operates as much among historians as among scientists: 'For unto every one that hath shall be given, and he shall have abundance: but from him that hath not shall be taken away even that which he hath.'[7]

The most important reason for the neglect of Chambers is simple. Popular science remains an unpopular scholarly subject. But there are other reasons, almost all closely related to the anonymity of the author. With the exception of a short preface to the tenth edition of *Vestiges* in 1853,[8] there are no autobiographical reminiscences and retrospective accounts. Even more discouraging is the poor quality of the biography written by William Chambers, Robert's brother. William had no sympathy with *Vestiges*, and his account barely mentions it.[8]

In the face of this apparent lack of printed information, one might hope that notebooks, memoranda and similar documents might be preserved among the Chambers family papers in Edinburgh. But materials of this kind do not exist. Concerned that he would be

discovered as the Vestigian, Chambers systematically destroyed all
notes and letters relating to the genesis of the book. Even at this level
of privacy, Chambers was careful to remove himself from the authorship. Some evidence does remain in the archives, including nearly
four hundred letters from Chambers about *Vestiges*, but this needs
to be pieced together with care.[10]

In approaching a figure like Chambers, it is essential to take into
account his own attitude towards how a theory like that of *Vestiges*
might best be constructed. This suggests that a detailed investigation
of the process of innovation would be misguided. Chambers had no
interest in what Frederic L. Holmes has called 'the fine structure of
scientific creativity'.[11] For it was precisely Chambers' principal criticism of specialist Victorian science that it focused too narrowly on
a restricted set of problems. Chambers approached the problem of
natural law, not like Darwin, as a leisured naturalist interested in a
complex body of technical literature, but as a popular journalist. The
key to understanding *Vestiges* is not therefore to be found in detailed
Darwin-style private notebooks, but in the lengthy row of Chambers'
published works from the 1820s onwards.

In short, the gestation of the *Vestiges* and its place in nineteenth-
century science can be understood only by using techniques developed for studying the Victorian press and the rise of popular
journalism. The development of Chambers' ideas is best evidenced in
what was perhaps the most public forum in early Victorian England:
Chambers's Edinburgh Journal, a weekly periodical founded by William
and Robert Chambers in 1832. The tone of the *Journal* is unmistakable: self-improvement, the progress of society, and rational, non-
sectarian entertainment. By 1847 Chambers had written nearly four
hundred leading articles. 'My design from the first', he said, 'was to
be the essayist of the middle class.'[12] How did *Vestiges* mesh with this
self-image, and how did the book come to be written? This essay will
locate *Vestiges* within Chambers' career, so that the publication of
a book on natural law can be seen as a consequence of the 'progressive
development' of the author himself.

The saints and the science of progress

Robert Chambers was born in 1802 into a prosperous middle-class
family. His father had been a successful manager in the Scottish
handloom industry, with over a hundred looms under his control.
But business decline and drink meant that young Robert and his
brother William had to begin earning their way at an early age. The
result was one of the classic Smilesian self-help stories of the nine-

teenth century, what Robert detested as 'the twice-told – nay, of the two-hundred-times-told story' of early hardship and ultimate success.[13] Robert commenced as an Edinburgh bookseller, setting up shop with a small shelf of books belonging to the family. He began independent authorship in 1822, with the pamphlet-sized anonymous *Illustrations of the Author of Waverley*. Other books on similar themes followed.

These works soon brought Chambers to the notice of Walter Scott himself, who provided the struggling young man with opportunities of various kinds. At an early stage, for example, Chambers earned an essential addition to his income as a copyist of fine documents, and Scott introduced him to important customers. 'My gratitude to you, respected Sir Walter, rises almost to a paroxysm, when I reflect upon the very great happiness you have been the means of conferring upon me, in relieving me from such a painful perplexity.' It was a 'mania of gratitude'.[14] In the literary culture of Edinburgh in the 1820s Chambers had gained a powerful patron.

In political terms the connection with Scott was not fortuitous, for Chambers (or 'Young Waverley', as he sometimes signed himself) was at this time a staunch Tory. From 1830 to 1832 he edited the Edinburgh *Advertiser*, which violently opposed the first Reform Bill. His publications, he assured the conservative publisher of *Blackwood's Magazine*, took 'a vigorous stand ... against the prevailing tide of radicalism'.[15] Chambers feared that the riots that marked the Reform Bill agitation were hurrying society towards catastrophe, as he confessed to Scott:

> This fervour is as fatal to literature as the irruption of the goths. Nor do I think it near an end: it is rather at a beginning. People formerly had a maxim, which history in all its ages showed to be good, that the great object of informed and civilised society was to keep the mob in check; but now the maxim is, that the government must reside in the mob. . . . The fiend, say I, take all the fools who are now hurrying us on to revolution and vandalism. But really this way madness lies; I must to business.[16]

In this charged atmosphere, Chambers' antiquarian writings possessed an inherent political dimension. In his 1828 history of mid-seventeenth-century Scotland, Chambers confessed to being unable to offer an even-handed treatment; 'good sense and good feeling' demanded that he 'give his countenance almost exclusively to the royal cause' rather than 'the rude cause of the populace'.[17] Like Scott's own histories and novels, works of this kind celebrated a world of deferential paternalism – a world that seemed to be slipping away.

But Chambers' attitudes were changing. In the preface to his selected writings, published in 1847, Chambers surveyed his early output and noted a sharp shift in his concerns. The shift had led him
towards science – and, implicitly, towards *Vestiges*:

> I loved the old tales and legends of my native country with the
> most passionate ardor, and delighted to gather up every little
> trait of bygone times. [My early] works were an effluence of
> mental youth, analogous to a green phase of the studious mind
> of England at the present day, which shows itself in a love of
> patristic reading and of Gothic architecture. The mind, in pro-
> gressive men, passes out of such affections at thirty, and the
> national mind will pass out of them when the time comes for its
> exercising its higher faculties.[18]

Although one would not want to date the change too closely, it is
not irrelevant that Chambers was thirty on 2 July 1832. This was only
a few weeks after the passage of the Reform Act, just after Robert
joined his brother in publishing the *Journal*, and a few months before
the death of Scott. Although Robert published historical and literary
works after that date, his interests were shifting towards the themes
of progress, utility, improvement and modernity that dominate his
later output. In politics he became a liberal Whig, emphasizing the
need for educational reform and harmony between the classes.

A closely related change, crucial to understanding *Vestiges*, involved Chambers' attitude towards religion. Throughout the early
years of marriage, Chambers and his wife Anne attended the Scottish
Kirk in the parish of St Cuthbert's in Edinburgh, where they held
fashionable pews in the first row of the gallery. Chambers' daughter
Eliza records the incident in the early 1830s that precipitated their
departure into the less evangelical fold of the Scottish Episcopalians.
The minister at St Cuthbert's directly denounced *Chambers's Edinburgh
Journal*, which was selling tens of thousands of copies every week.[19]
The preacher involved was almost certainly the Reverend David Dickson, the senior parish minister. Like many presbyterians, Dickson
detested the *Journal*'s neutrality on religious matters and its strongly
secular tone. His concern for this kind of issue extended back at least
two decades, as evidenced by a sermon he delivered in 1813 before
the Society for Propagating Christian Knowledge. As Dickson had
said on that occasion, 'Great as are the advantages of knowledge and
learning, they are wholly unavailing ... if separated from the spirit
and influence of vital, personal godliness.' 'Unsanctified knowledge',
he continued, 'instead of profiting us, will only increase our final
doom; for however useful we may be as members of society, we shall
be found guilty of having diverted our talents from promoting the

glory of God. . . .'[20] Precisely what Chambers, his wife and their friends heard at St Cuthbert's is not certain, but it must have been similar in tone, for they never returned to the church.

The sermon damning the *Journal* precipitated a shift that was probably inevitable anyway. Within the Scottish context, the social basis of membership of the presbyterian church was in the lower and lower-middle classes on the one hand, and the aristocracy on the other. Chambers had moved into the prosperous middle classes, who tended to be Episcopalians. His departure from St Cuthbert's would thus have seemed a natural move. By the time of writing *Vestiges*, Chambers had little respect for the presbyterians, especially those who joined the Free Kirk, which he saw as an obstructing group of ignorant fanatics. Religion was a drag to reform. As he noted to his friend Alexander Ireland, 'I believe this liberal view is advancing, but we are still far from being able to fight those dogs of clergy.'[21]

From the early 1830s onwards, Chambers maintained a belief in a deity, but only in the most distant and abstract sense. He had only perfunctory contacts with the formal church; as his wife noted, he could not be accused of being 'very guilty of church going', although she and the children seem to have attended with some regularity. When asked in later years why he kept two pews, each in a different church, Chambers jokingly replied that 'when I am not in the one, it will always be concluded by the charitable that I am in the other'.[22] As a moderate deist, Chambers disliked evangelical enthusiasm and doctrinal controversy. Religion occupied much the same place in his personal life that it did in *Chambers's Edinburgh Journal*: a minor one.

Because of this, it is misleading to speak of Chambers' inquiries into natural law as an attempt to construct a 'theodicy', as a number of scholars have done.[23] Justifying God's ways to man was simply not central to Chambers' personal project. But if Chambers did not need a true theodicy himself, he recognized that his middle-class readership did. In writing to his friend Ireland in June 1844 about possible publishers for the 'opus', he emphasized that the public's need for a providential gloss had been kept in mind. 'There is nothing in it of a worse character than geology is when you consider its inconsistency with the Mosaic record', he explained. 'And every effort is made that reason and common sense would at all admit of to keep smooth with the sticklers – though I daresay I shall not succeed with the extreme ones.'[24] Other letters reinforce the point that the explicitly religious aspects of *Vestiges* were tacked on to placate those evangelicals he contemptuously referred to as 'the saints'. 'I am happy to say', Chambers wrote in sending a final batch of manuscript, 'that I have been able at the end to introduce some views about religion which will help greatly, I think, to keep the book on tolerable terms with the public, without compromising any important doctrine.' References to divine providence made a secular pill easier to swallow.[25]

The older historical literature made the mistake of seeing conflict as inherent in the very nature of religion and science. Scholars have been rightly concerned to avoid this worn-out battleground, and have emphasized references to divine action in *Vestiges*. But in adopting the idea that conflict was never fundamental or logically necessary, it is important not to lose sight of the very real disputes that did take place. In early Victorian Edinburgh, there *was* a warfare between those who favoured a science of progress and those who looked to evangelical faith for salvation; and *Vestiges* was not on the side of the godly.[26]

The phrenological wedge

The two shifts so far discussed, towards political liberalism and anti-clericalism, form the broad background to *Vestiges*. The setting for these changes lies in the world of Edinburgh phrenology. Phrenology, which argued that the brain was composed of specific organs with specific mental functions, is now infamous as a kind of pseudo-scientific parlour game for judging character on the basis of bumps on the head. But in its early years, and certainly in the context that Chambers came into contact with it, phrenology was a serious study and the major agency for the introduction of naturalism into Victorian Britain. Its leading Scottish proponent was George Combe, whose widely selling *Constitution of Man* (1828) was in many ways a model for Chambers when he sat down to write *Vestiges*.[27] If histories of evolution are to include Chambers, as everyone who has written on the subject agrees, then they must devote at least as much attention to the phrenological movement from which his work sprang.[28]

From the early 1830s, Chambers' interest in natural law developed through close contacts with Combe and his circle. Important associates included Robert Cox, Combe's nephew and editor of the *Phrenological Journal* from 1830 to 1837; Cox's wife Ann; Combe's brother Andrew; and Combe's wife Cecilia, daughter of the celebrated actress Mrs Sarah Siddons. Other friends with liberal leanings included the novelist Catherine Crowe; the London physician Neil Arnott; the Glasgow professor of astronomy John Pringle Nichol; and the Manchester-based journalist Alexander Ireland, who was well known as an early champion of Emerson.[29]

The importance of the world of Edinburgh phrenology might be surmised from *Vestiges* itself, for the first edition adopted Franz Joseph Gall as its authority in discussing man's mental endowments, while later ones recommended the *Constitution of Man* for further reading. But the author's debt to the movement goes well beyond a single chapter. Phrenology was as much a way of life as a specific doctrine of mind, and the wider applications of the doctrine appealed most to Chambers.

It is significant in this context that Chambers' interest in natural science emerged from a phrenologically inspired educational programme in publishing. In 1835 the brothers Robert and William launched a series of textbooks for young children, *Chambers's Educational Course*, implicitly based on Combe's doctrines. Robert's most successful contribution was also his very first writing on science, a general *Introduction to the Sciences*, published in 1836. The book began with chapters on the 'extent of the material world' and 'the stars', and ended with 'man – his mental nature'. This book, anonymous like others in the series, was very widely used in schools. It had sold over one hundred and twenty thousand copies by 1849, even more than Combe's *Constitution of Man*.[30]

The *Introduction*, like other pre-Vestigian works by Chambers, drew on phrenology's lessons for living in accordance with natural laws – what he termed 'the philosophy of phrenology'. It did not, however, refer to what he called its 'organology' – the physiological and anatomical basis of the doctrine. In fact, the word 'phrenology' is not mentioned in the book, which adopted a vague compromise on the central question of the relation between mind and matter.[31] This was to be expected in an introductory schoolbook, for many parents would instantly equate phrenology with materialism and a denial of man's immortal soul. For similar reasons, almost no references to phrenology are to be found in *Chambers's Edinburgh Journal* before 1844.[32]

In this strategy Chambers differed fundamentally from Combe. From Combe's perspective, to write of the 'philosophy' of phrenology without the physiological views underpinning it was neither candid nor fair. As Combe complained privately to Chambers in 1835:

> I am not disposed to dispute that the philosophy of mind developed by Phrenology is, as you say, "the point of the wedge which is yet to rend asunder the mass of philosophical heathenism"; and I am willing to see it driven by all hands who are disposed to give it a blow. But agreeing with you in this estimate of it, I am anxious that it should be known, when used, to be the *Phrenological wedge*. What I objected to was, kind friends [in other words, Chambers himself] driving *a* wedge, & not giving

it a name. If the name is with-held, the heathenism appears to be rent by common wedges, & not by *the* wedge which really does the duty.[33]

Chambers, as a commercial publisher, was afraid to reveal his debt to phrenology lest the circulation of the *Edinburgh Journal* and the *Educational Course* be harmed. The money-losing *Phrenological Journal* was not a model to be followed. In his view, the basic message of the doctrine could be communicated, and received by readers 'as if it was a new gospel', without alienating them by the word 'phrenology':

> The process is analogous to that recommended by Duff for converting the heathen. First convince their understandings that a correct system of mind has been discovered, and there will then be no obstacle to the reception of its fundamental truths except what may still be presented by the habit of venerating former systems. When we reflect that some of the forms of heathenism survived in Scotland till the close of the last century, perhaps eight hundred years after Christianity had been acknowledged to all intents and purposes, we must not fret at the slow progress phrenology makes....[34]

Similarly, Chambers thought that Combe was unwilling to make enough concessions to contemporary religious feeling. Combe, he wrote in a letter, 'is too materialistic'.[35]

Despite these disagreements, the difference was basically one of strategy, not substance. Chambers accepted the essential tenets of phrenology and their significance for his growing interest in natural law. As he remarked in recommending Combe for the chair of logic at the University of Edinburgh, 'Phrenology appears to bear the same relation to the doctrines of even the most recent metaphysicians, which the Copernican astronomy bears to the system of Ptolemy.'[36]

The most telling indication of phrenology's importance for Chambers is provided by early plans for his own philosophical '*chef-d'oeuvre*'. He told friends in 1837 that his spare moments for the preceding two years had been spent on a manuscript treatise on 'the philosophy of phrenology' – 'a system of mind on the phrenological basis'. Towards this end he borrowed many books and periodicals, including a run of the *Phrenological Journal*.[37] Thus *Vestiges*, usually discussed as the leading pre-Darwinian evolutionary text, grew directly from Combeian phrenological soil.

Stars, strata and civilization

By the mid-1830s Chambers had abandoned his purely antiquarian interests as products of his intellectual infancy. From the staunch

Toryism of Walter Scott, he had turned to the advocacy of liberal (but always gradual) reform. He had left the Scottish Kirk and become a force to be reckoned with in the Edinburgh circle of George Combe. And, as evidenced by the children's book of 1836, he was beginning to evince an interest in science. It is not difficult to see how, having got this far, Chambers went on to write *Vestiges*. In 1824 the topic would have been abhorrent to him; by the mid-1830s, through a contingent series of circumstances, it was not. The rest of this essay will suggest how important aspects of the work were elaborated.

One point is clear. The idea of turning from the phrenological philosophy of mind to the entire natural world came from an exposure to the nebular hypothesis of the formation of the universe.[38] Here the direct influence of the author and lecturer John Pringle Nichol was critically important. The pages of the *Journal* indicate Chambers' intense enthusiasm for Nichol's *Views of the Architecture of the Heavens*. Published in 1837, this work described in vivid and accessible language the evolution of the universe and the formation of galaxies and stars. As the *Journal* said in a review, 'high and rational wonder has never been so delightfully associated with moral feeling'.[39] The very summer that the *Architecture of the Heavens* appeared, Chambers, Nichol and a number of their phrenological friends toured Ireland together for sixteen days; Anne Chambers jokingly warned her husband not to be carried away by his astronomical enthusiasm. What Chambers saw as Nichol's goal – to connect 'the mystical evolution of firmamental matter with the destinies of man' – became his own; he would carry the story further, into the living world of plants, animals and human origins.[40]

Once the nebular hypothesis had led Chambers to contemplate the application of a law of progress to the whole realm of nature, he rapidly began to explore other relevant sciences. Thus the day after Chambers returned from Ireland in the summer of 1837, he told a friend that he was 'in the commencement of a geology fever, and extremely anxious to make up a little collection of the appropriate objects'.[41] He began to explore not only the earth sciences but zoology and botany as well. Most pieces of the puzzle are evident in *Chambers's Edinburgh Journal*. There were essays on monstrosities, the progressive nature of the fossil record, the habits and instincts of animals, the learning abilities of dogs and pigs, the spontaneous generation of insects through electricity, and the effects of diet and exercise upon health.

As a result, a careful reader of the *Journal* would have been familiar with much of *Vestiges* well before it appeared. Especially striking was the large number of articles on the origins of races, nations, languages and civilizations. These elaborated a developmental model almost identical to that found in the chapter of *Vestiges* on the 'Early History of Mankind'. In the essay 'Gossip about Golf', Chambers even applied the model to his favourite sport, which he argued was an inevitable result of 'the existence of a certain peculiar waste ground called links'. Similarly, cricket was said to be a natural outcome of village greens in England.[42]

Comments in these essays indicate just how far Chambers had come from his High Tory antiquarianism. One article proclaimed that 'physiology alone could throw more light on the origin and progress of nations, within the bounds of one small volume, than could be done by a whole library of political history, or the united labours of a score of archaeological societies'.[43] This was a remarkable statement from an author whose best-known titles were *The Traditions of Edinburgh* and *A History of the Rebellions in Scotland*. Behind the wall of anonymity in the *Journal*, his volte-face was now complete: the new interest in science was now to be applied to history itself. Although he believed that scientific study of the past was just beginning, progress and physiology would ultimately subsume anecdote and antiquarianism.

As publication of *Vestiges* came closer, the *Journal* also featured articles on the aims and methods of natural science. There were comments on the fate of Galileo, the necessity of judging science by more than mere utility, and the benefits of speculation. It is hard not to see pieces like 'The Easily Convinced' of 1842 as attempts to soften the reception of the forthcoming bombshell:

> The constant cry is, give us facts and leave hypotheses alone. But it is not possible for any human being to go on constantly collecting dry unconnected facts. We require to be allowed a little generalisation by way of *bon-bons*, to encourage us in our tasks. And is not imagination often a means of leading on to fact? . . . And thus there is a utility and a final cause for even that mocked thing, credulity. The credulous are the nurses appointed for ideas in their nonage – which, if left to the tender feelings of the cautious alone, would for certain perish of cold and hunger, before ever they had shown their first teeth. The credulous catch them and foster them, and look out for their parishes, and get them comfortably brought on to their apprenticeships. By and by, they begin to kick about for themselves, and settle into respectable and useful members of society – but no thanks to the awful doctors who never have any thing to do with the intellectual bantlings that don't come into the world properly stamped and labelled.[44]

Well before he published, Chambers was acutely aware of the potential reception of his ideas, which were not at all 'properly stamped and labelled' by the standards of specialist Victorian science.[45] By this point the outline of a book had taken shape. It would, like his children's book, cover all aspects of the natural world from the formation of the Solar System to the origin of civilization, and it would do so from the standpoint of a law of development and progress.

The origin of animated tribes

How did Chambers come to incorporate organic evolution within his over-arching scheme? At one level, the answer seems obvious. For anyone anxious to replace divine intervention with law-like regularities, the origin of new organic beings needed an explanation. In 1837 Nichol had spoken of 'the germs, the producing powers of that LIFE, which in coming ages will bud and blossom', but had left the mechanism for an extension of natural law into the organic world to be inferred.[46] Chambers, in the excitement of reading the *Architecture of the Heavens*, believed that a gap had opened up between the planetary and the human realms, between the works of Nichol and the works of Combe. As the preface to the 1853 edition of *Vestiges* explained, he 'first had his attention attracted to the early history of animated nature, on becoming acquainted with an outline of the Laplacian hypothesis of the solar system'.[47] Once Chambers accepted the need for a naturalistic explanation, generation from pre-existing species was by far the most obvious alternative.

The adoption of this view, however, demanded a radical shift in Chambers' attitude towards organic evolution – 'transmutation', as it was usually known. The transformation can be followed in the changing emphases of articles in the *Journal*. These are unsigned and at least some of them may not be by Chambers, but they had to pass his close editorial scrutiny. They are, in consequence, unlikely to deviate from a position he would have found acceptable.

In its early years the *Journal* firmly opposed organic evolution. The subject was first mentioned in an article of November 1832 entitled 'Natural History: Animals with a Backbone'. This attacked transmutation and those who upheld it in no uncertain terms. Ignorance of the facts and knowledge of the unity of organic types, it explained, had 'tempted many philosophers to hazard the absurd opinion that man had his beginning in a minute animalcule, and has attained his present perfect condition from progressive improvement by reproduction'. The article expressed 'astonishment' that learned men like Erasmus Darwin and Lord Monboddo had advocated such heresies.

However, transmutation would never become popular: 'views like these can never be entertained by healthy minds, and it requires but little reflection to dispel such absurd theories'.[48]

A second essay appeared in September 1835, based on the discussion of Lamarck in the second volume of Charles Lyell's *Principles of Geology* of 1832. The article noted that some 'very eminent philosophers ... have boldly asserted that all the varieties of plants and animals which abound in nature originally sprang from one individual specimen of organized life; in short, that man himself, Socrates, Shakespeare, and Newton, were merely zoophytes in a state of high improvement and cultivation!' As before, this idea was roundly condemned. The limits of variability, the sterility of mules, and a host of other evidences precluded any possibility of evolution. Lyell's demolition of Lamarck was 'so satisfactory as to require us to say nothing in addition'.[49] This continued to be the *Journal*'s editorial policy through 1838, as indicated by some brief comments on these 'most absurd notions' prefacing an extract from Dr Clarke Abel's description of the orang-utan.[50] Even if Chambers did not write all these articles (the second at least bears hallmarks of his style), his initial opposition to transmutation is obvious.

Conversely, after 1838, Chambers never came out publicly as an evolutionist; this would have been dangerous. With its audience among educated artisans and the middle class, the *Journal* could scarcely sanction a doctrine associated with revolutionaries, atheists and Frenchmen.[51] The opposition to evolution expressed in its early volumes was of a piece with its advocacy of political economy, self-help and Malthusian restraint among the working class. 'Moral affairs of this kind', the *Journal* noted in 1838, 'proceed with all the irresistible force of the great physical laws: as well try to give a check to the law of gravitation itself, as to counteract the principles which regulate the rise and fall of wages.'[52] Although Chambers came to believe that utilitarian economic and political doctrines were compatible with transmutation, its radical associations were simply too damaging to be embraced in public. After 1838, the *Journal* next spoke out on the subject in 1860, when Chambers reviewed the *Origin of Species*.[53]

Even leaving aside the political and religious dangers of transmutation, the topic could be discussed in detail only by revealing other more basic matters, which were usually kept veiled in a family paper like the *Journal*. In this forum, subjects unsuitable for women, children and workers could be mentioned only with the utmost discretion. Chambers explained the problem to Combe, who was revising some lectures for publication in the *Journal*:

With regard to the duty of studying anatomy, I must really say
that I fear considerably for the reception of that advice, and
hope you will make every effort, in your contemplated
alterations, to introduce and handle the subject in a manner so
little startling as possible. . . . [F]or if they are told that one part
of the sex has stood the supposed horror of the thing, and not
been consequently looked upon as not comme il faut, they
would see the less difficulty in their also doing it. At all events,
pray try to make the passage as sweet as possible.[54]

When it came to sex and generation, the taboo was absolute. The
Journal became a byword for chastity; as the London journalist
Dudley Costello asked a later editor of the paper, Leitch Ritchie,
'How is the population of Scotland kept up? By immigration?'[55]

But during the late 1830s and 1840s, Chambers came as close as he
could to discussing evolution in public. Articles on related topics such
as progressionist geology could use language readily transferred to
the terms of transmutation. Within a few years Chambers became
more explicit. As he wrote in an 1842 piece on 'the educability of
animals', 'We become more confident in the improvability of our
own species, when we find that even the lower animals are capable
of being improved, through a succession of generations, by the
constant presence of a meliorating agency.'[56] The point was under-
lined by contemporary studies of the races of man. Instances of 'white
negroes' demonstrated 'that the rise of the white races of men out of
the black is within the range of possibility'.[57] Given the constraints
on publishing on subjects relating to sex, it is not surprising that the
Journal never spoke more directly. Of all the issues 'behind the veil',
this was the most darkly hidden.

Because the *Journal* could not publish on sex, generation or embryology, we are thrown back upon *Vestiges* itself for indications of
Chambers' inquiries into possible mechanisms for transmutation. In
addition, the autobiographical preface to the 1853 edition makes it
possible to link the public evidence into a more coherent account.
Chambers noted how he commenced inquiry with the basic assumption that the '"fiats", "special miracles", "interferences", and other
suggestions and figures of speech in vogue amongst geologists'
would have to be rejected. With this starting point, he examined
'such treatises on physiology as had fallen in his way' for evidence
of how species might have originated.[58]

In the late 1830s naturalistic physiological and anatomical doctrines were common currency among nonconformist medical men.[59]
Chambers thus had a wide choice of works to consult. He relied especially on three works in transcendental anatomy by authors who

had taught or been trained in Edinburgh. (His awareness of continental authors, such as E. Geoffroy St Hilaire, Lamarck, E. Serres,
Friedrich Tiedemann and Karl Ernst von Baer, seems to have been
entirely at second hand.) The first source, and probably the first to be
consulted, was Perceval Lord's *Popular Physiology* (1834).[60] This was
an elementary manual published under the auspices of the Society
for the Promotion of Christian Knowledge. Lord's anti-phrenological
book is just the kind of test that Chambers would have encountered
while researching his treatise on the philosophy of phrenology. Lord
explained how the human embryo passed successively through
stages resembling a fish, a reptile, a bird and a mammal; it then
discussed how the human brain 'not only goes through the animal
transmigrations we have mentioned, but successively represents the
characters with which it is found in the Negro, Malay, American, and
Mongolian nations'. This was the recapitulation doctrine that Chambers took as the heart of his hypothesis, expressed without qualification. *Vestiges* quoted these striking passages at length, turning
Lord's argument to an evolutionary end.[61]

A second source for Chambers' theory was a much more reputable
textbook, the Scottish medical lecturer John Fletcher's posthumous
Rudiments of Physiology (1835–7). This work also contained sections
critical of phrenology and discussed recapitulation at length. But
Fletcher went further and cautioned against using it to support
transmutation. The doctrine had 'nothing but the most vague and
rambling presumptions in its favour', especially in the face of the
functional integrity of each species. Individual organs, such as the
brain, could form a perfect developmental series, but whole organisms could not. Differences in the timing of embryonic changes ruled
this out.[62] Despite the denial of transmutation, Chambers must have
been impressed. And although Fletcher rejected any simple Lamarckian series, he did range the entire animal kingdom in ascending
order of complexity and degree of organization. Chambers found
this list highly suggestive, especially when juxtaposed with an outline of the appearance of various species in the geological record.
A chart illustrating this comparison became an important part of
Vestiges.[63]

Both Fletcher and Lord presented the well-known doctrine of
recapitulation. But Chambers was also aware of a rival, newer and
more sophisticated embryology, which had been developed by Von
Baer in Germany. This argued that embryos do not pass through
stages resembling the adult forms of simpler organisms; rather, the
process is one of differentiation, with the embryo starting as a gen-
eralized form and gradually exhibiting the special characters of the

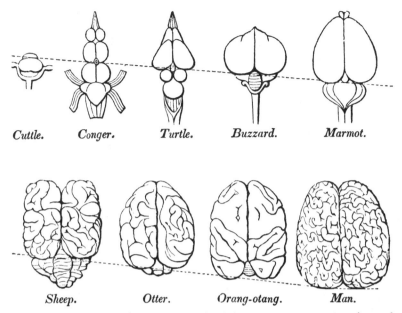

Cuttle.   *Conger.*   *Turtle.*   *Buzzard.*   *Marmot.*

Sheep.   *Otter.*   *Orang-otang.*   *Man.*

Comparative sections of the brains of adult organisms, ranging from the cuttlefish to man. The dotted line marks the posterior boundary of the cerebral hemispheres. (From John Fletcher, *Rudiments of Physiology* [Edinburgh: John Carfrae and Son, 1835–7], pp. 47–8. Reproduced by permission of the Syndics of Cambridge University Library.)

adult species. As Chambers noted, 'the resemblance is not to the adult fish or the adult reptile, but to the fish and reptile at a certain point in their foetal progress'. This idea, known as 'Von Baer's law', was just becoming known in Great Britain in the later 1830s, particularly through a series of articles by Martin Barry. Chambers may have read these (he did cite Barry's essay on fissiparous generation), but his most obvious source was the second edition of William Carpenter's *Principles of General and Comparative Anatomy*, which appeared in 1841.[64] Chambers was aware of this work by the summer of 1844, and sent *Vestiges* to John Churchill partly because he had published it. Moreover, Carpenter was doing a considerable amount of writing for the Chambers firm in the early 1840s, including a *Rudiments of Zoology* for the *Educational Course* and articles for an expanded edition of *Chambers's Information for the People*. Neither of these primers discussed embryology, but in editing them Chambers was more likely to consult the textbook than would otherwise have been the case. A summary of Carpenter's presentation and a revised version of his diagram appeared in *Vestiges* without acknowledgement.[65]

The text of the first edition of *Vestiges* strongly suggests that there were two stages in the prepublication development of Chambers' ideas on an evolutionary mechanism. Von Baer's law differed fundamentally from the concept of recapitulation, but it appeared only as an inserted qualification and not as the primary definition of a mechanism. The first edition thus seemed contradictory to many readers, and within a few weeks Chambers altered his discussion of recapitulation to bring it closer to Von Baer. By the third edition of January 1845, all references to embryos passing through the 'permanent forms' of adult organisms had been excised. This shift towards Von Baer continued right through the tenth edition of 1853, although it was never complete.[66]

It would appear, then, that the transmutation theory of *Vestiges* was initially constructed around the traditional concept of recapitulation available in the works of Lord and Fletcher. Only afterwards did Chambers become aware of the alternative being imported into Britain by Carpenter, and he added it to his hypothesis at some point in the early 1840s. Far from being old-fashioned by advocating a linear progression, *Vestiges* served more than any other book to bring novel concepts of branching and differentiation to general notice.[67]

The domestication of development

Beyond the specific sources of Chambers' theory, however, lies the basic problem that transmutation was a radical doctrine. It had been ridiculed in the *Journal* and had few if any supporters in Edinburgh phrenological circles. Even after *Vestiges* was published, neither Combe nor Nichol advocated the idea. How could Chambers have become willing to consider a subject previously seen as dangerous and degrading?

Important clues are provided by the unusual circumstances in which the book was written. Chambers composed *Vestiges* in the early 1840s, when he had left the urban bustle of Edinburgh, and moved to St Andrews in Fife. This move has usually been seen as a way of avoiding prying eyes that might pin the authorship of *Vestiges* upon him. But this is simply a myth. In fact Chambers had suffered a mental breakdown, and the move was undertaken to aid his recovery. For a decade he had written the larger part of the weekly *Journal* single-handedly. The strain was simply too much, and in 1842 he collapsed. As Anne Chambers noted, 'incessant mental exertion' and 'intense mental action' had left him with a mind in an 'unsound, or partly in a *diseased* state'.[68] To escape, the family moved to a villa on the outskirts of St Andrews. Chambers took up golfing, participated in the local literary culture, and wrote occasionally for the *Journal*. In August 1842 Anne had a miscarriage and in 1843 his mother died, but for the most part the Chamberses found their stay pleasant and relaxing.[69] For the first time in ten years, Robert could turn his full attention to something other than his weekly essays. In this quiet family retreat, removed from controversy and potential critics, Chambers wrote *Vestiges*. The book was, as he said at its close, 'composed in solitude, and almost without the cognizance of a single human being'.[70]

Chambers, of course, was not alone in St Andrews; he brought with him a large and growing family. In many respects, his familiarity with everyday processes of birth and development was much more important for his theories than his exposure to technical treatments of embryology. Chambers was an occasional and desultory reader of physiology texts, but a full-time and enthusiastic father. By 1843 he had nine children (six girls and three boys), and watching them mature was a fundamental part of the experience that went into the making of *Vestiges*. Chambers was particularly moved by the birth of his eldest son, Robert, and published a signed poem, 'To a Little Boy', in *Blackwood's Magazine* for July 1835. 'The feelings there expressed', he told a friend two years later, 'have suffered no change, for he himself, in advancing out of childhood, has lost none of its endearing qualities. What a history, though an incommunicable one, resides in my mind respecting the aspects, talk, doings, and traits of progressive intelligence of this dear boy in the course of his brief existence.'[71] It may not be too much to say that this 'incommunicable history' was finally expressed in the cosmic story of *Vestiges*, which pictured all nature engaged in gestation and development.

Images of pregnancy, birth, childhood and the family were deeply embedded in the structure and language of the book. In explaining the evidence for the nebular hypothesis, Chambers immediately appealed to an analogy based on the familiar details of human growth. Stars were visible in every stage of development; 'it may be presumed that all these are but stages in a progress, just as if, seeing a child, a boy, a youth, a middle-aged, and an old man together, we might presume that the whole were only variations of one being'.[72] And the process applied not just to stars, but to our own Solar System. *Vestiges* quoted a passage from John Herschel's *Astronomy* to show how the planets were joined with all the mutual bonds of a Victorian family gathered around the hearth:

> When we contemplate the constituents of the planetary system from the point of view which this relation affords us, it is no longer mere analogy which strikes us, no longer a general re-

semblance among them, as individuals independent of each
other, and circulating about the sun, each according to its own
peculiar nature, and connected with it by its own peculiar tie.
The resemblance is now perceived to be a true *family likeness*; they
are bound up in one chain – interwoven in one web of mutual
relation and harmonious agreement, subjected to one pervading
influence which extends from the centre to the farthest limits of
that great system, of which all of them, the earth included, must
henceforth be regarded as members.[73]

This appearance of happy domesticity, Chambers went on to explain, could be taken in a literal sense, for the nebular hypothesis
showed that the planets were 'children of the sun', generated from
the stellar body according to the same universal law that had produced the galaxies. Language of this kind would have been familiar
to Chambers from Nichol's *Architecture of the Heavens*, which spoke
of 'nebulous parentage' and the 'inexhaustible womb of the future'.[74]

What Chambers did in *Vestiges* was to use these same generative
images to bring the frightening notion of transmutation within the
realm of the familiar. The production of new species, he wrote, 'has
never been anything more than a new stage of progress in gestation,
an event as simply natural, and attended as little by any circumstances of a wonderful or startling kind, as the silent advance of an
ordinary mother from one week to another of her pregnancy'.[75] In
other words, the birth of a new species was no more or less to be
feared than the birth of a child. To disarm criticism, Chambers asked
his readers what it would be like 'to be acquainted for the first time
with the circumstances attending the production of an individual of
our race'. Faced by the facts of sex and reproduction, 'we might equally
think them degrading, and be eager to deny them, and exclude them
from the admitted truths of nature'. But these facts could not be
denied by 'a healthy and natural mind'. Neither, by implication,
could the analogous processes of gestation that gave birth to new
species.[76]

Chambers, as much as any writer of his time, was aware of the
expectations of his audience. For years he had spoken through his
essays for the *Journal*, assuming the character of an 'intimate acquaintance or friend' on a weekly visit to his readers.[77] *Vestiges* was
successful because it employed strategies learned during this long
apprenticeship. The most touchy subject the book treated, the creation of new species by transmutation, had never been allowed into
middle-class Victorian homes. But by building the generative model
of *Vestiges* around images of pregnancy, childhood, the family and
the hearth, Chambers was able to minimize the fears of his audience.
Transmutation, which has been associated in the public mind with

Robert Chambers in the summer of 1844, shortly before publication of *Vestiges*. This calotype was made by Hill and Adamson at the request of Alexander Ireland, Chambers' friend and intermediary in the book's publication (Chambers to Ireland, 16 July 1844, NLS, Dep. 341/111/18. Reproduced by permission of the Scottish National Portrait Gallery.)

radical revolutionaries and dissolute foreigners, became infused with all the domestic virtues. With *Vestiges*, evolution quite literally moved off the streets and into the home.

It is worth noting that the domestic connections of a transmutation theory based on 'the universal gestation of nature' must have had great appeal for Chambers himself. In this lies the solution to what has always seemed a great paradox about *Vestiges*: how a respectable middle-class paterfamilias came to write on this subject at all. There is a great irony here, too; for it was ultimately because of his family that Chambers kept his authorship hidden for so many years. His son-in-law asked him directly about the reasons for remaining anonymous:

> I can easily understand now why Robert Chambers shrouded himself in impenetrable mystery. The veil was raised to me a few years after I married his daughter. . . . He and I had been out for a walk together, and as we were returning home I said to him, "Tell me why you have never acknowledged your greatest work." For all answer he pointed to his house, in which he had eleven children, and then slowly added, "I have eleven reasons."[78]

The authorship had to remain secret. The business might be damaged, the 'saints' might attack, but the greatest threat was that his growing family – especially his eight daughters – might be placed beyond the pale of respectability. Even Chambers' well-honed literary skills could not defuse the social threat of transmutation.

Chambers in context

Chambers had neither the secure position in society nor the scientific reputation that allowed Darwin to put his name on the title page of the *Origin of Species*. But there are at least a few parallels between the two authors. Not least, they were both middle-class family men who wrote their books in relative isolation. Darwin, plagued by a mysterious illness, retreated to the privacy of Down;[79] Chambers suffered a nervous collapse, withdrew into his domestic circle at St Andrews and produced a text that brought evolution into the middle-class homes of Victorian England. Like the *Origin of Species*, however, *Vestiges* was not the product of one individual's idiosyncratic psychology. Rather, Chambers' use of his personal and domestic life can only be understood in the wider context of early Victorian religious, political and family history. He was well aware of his audience, especially the constraints that it imposed around public discussions of sex, development and transmutation. These limits created both opportunities and dangers, which he negotiated with considerable skill.

Vestiges, as we have seen, is most effectively viewed as the carefully crafted product of a leading journalist and author. A bestseller among the general public, the book was not always so successful in dealing with problems relating to specialist science. This aspect of the work has drawn particular attention from historians, who have often pictured Chambers as a bungling amateur naturalist, a failed Darwin.[80] The present essay has dealt at some length with the ways that Chambers may have elaborated his ideas relating to the organic world by using the specialist literature on embryology. In retrospect, these seem the most significant part of *Vestiges*, and even at the time transmutation created the most controversy. But problems in scientific physiology, like those relating to technical natural history, geology and astronomy, had only a minor part in the genesis of *Vestiges*.

Chambers' deployment of technical science, like his use of his domestic experience, can only be understood in relation to the wider context of the early Victorian era. In common with his associates Combe and Nichol, Chambers was intent upon using the natural sciences as part of a programme for gradual middle-class reform. The tendency of his book, despite its appeals to religion, was fundamentally secular. It advocated progress, hard work and the interests of the middle classes; it turned its back on antiquarian history, organized religion, conservative politics and the ancient universities. These were ideas and institutions that had been supported by science in the past. But in Chambers' view they had become barriers to progress, and needed to be replaced by a new order. In short, the natural sciences were being brought to the service of causes that Chambers had already advocated in hundreds of popular and readable essays. For that reason, the Reverend Adam Sedgwick of Cambridge was entirely correct to call *Vestiges* a 'rank pill of asafoetida and arsenic, covered with gold leaf'.[81] Thomas Henry Huxley hit the target even more precisely when he commented that *Vestiges* could have been written by anyone – anyone, that is, who was familiar with the contents of the most popular weekly periodical in the kingdom, *Chambers's Edinburgh Journal*.

Notes

Research for this essay was supported by a grant from the Royal Society of London. Peter Bowler, John Brooke, Adrian Desmond, David Kohn, Bernard Lightman, Jim Moore, Jack Morrell, Simon Schaffer and Anne Secord provided valuable criticisms of earlier versions. Permission to quote from the papers of Robert Chambers, deposited in the National Library of Scotland in Edinburgh, has been generously given by Mr A.S. Chambers. I am particularly grateful to Mr Chambers for making these papers available, and to Dr I.G. Brown and other members of the library staff for assistance in consulting them. All other manuscripts are quoted by permission of the Keeper of Manuscripts at the National Library of Scotland.

1. Quoted approvingly by Chambers from a work on the *Rights of Industry*, in 'Science and Labour', *Chambers's Edinburgh Journal*, 2 (1833), 375–6 (376).

2. [Robert Chambers], *Vestiges of the Natural History of Creation* (London: John Churchill, 1844; reprint edn, Leicester: Leicester University Press, 1969). In this essay 'Chambers' will refer to R. Chambers unless otherwise qualified, and '*Vestiges*' will refer to the first edition of 1844 unless otherwise indicated.

3. An example here is the debate between David Brewster and Whewell about the plurality of worlds. See John Hedley Brooke, 'Natural Theology and the Plurality of Worlds: Observations on the Brewster – Whewell Debate', *Annals of Science*, 34 (1977), 221–86 and Michael J. Crowe, *The Extraterrestrial Life Debate, 1750–1900: The Idea of a Plurality of Worlds from Kant to Lowell* (Cambridge: Cambridge University Press, 1986).

4. For the 'science of progress', see Simon Schaffer, this volume. Important works discussing the relations of the gentlemen of science with their audiences include Adrian Desmond, *Archetypes and Ancestors: Palaeontology in Victorian London, 1850–1875* (London: Blond and Briggs, 1982); Roger Cooter, *The Cultural Meaning of Popular Science: Phrenology and the Organization of Consent in Nineteenth-Century Britain* (Cambridge: Cambridge University Press, 1984); Jack Morrell and Arnold Thackray, *Gentlemen of Science: Early Years of the British Association for the Advancement of Science* (Oxford: Oxford University Press, 1981); and Robert M. Young, *Darwin's Metaphor: Nature's Place in Victorian Culture* (Cambridge: Cambridge University Press, 1985). For the confrontation between specialist sciences and the technologies of mass publication, see James A. Secord, 'Extraordinary Experiment: Electricity and the Creation of Life in Victorian England', in David Gooding, Trevor Pinch and Simon Schaffer, eds., *The Uses of Experiment: Studies in the Natural Sciences* (Cambridge: Cambridge University Press, 1989), pp. 337–83.

5. Chambers to A. Ireland, 'Friday night', National Library of Scotland (NLS), Dep. 341/110/41–2. For the frontispiece, reproduced here by permission of the Syndics of Cambridge University Library, see Chambers, *Illustrations of the Author of Waverley: Being Notices and Anecdotes of Real Characters, Scenes, and Incidents, supposed to be described in His Works*, 2nd edn (Edinburgh: John Anderson, 1825). In corresponding with Alexander Ireland of Manchester, who served as intermediary with the publisher, Chambers gave private code names to himself and his friends, so that the letters would not be identified if they fell into the wrong hands. *Vestiges* was 'opus', Ireland was 'Alexius', Chambers' wife Anne was 'Mrs Balderstone', George Combe was 'Jokum', and Chambers himself was 'Ignotus', 'Mr Balderstone', or 'The Unknown'.

6. Three partial exceptions include M.J.S. Hodge, 'The Universal Gest-

ation of Nature: Chambers' "Vestiges" and "Explanations"', *Journal of the History of Biology*, 5 (1972), 127–51; Milton Millhauser, *Just Before Darwin: Robert Chambers and Vestiges* (Middletown, Conn.: Wesleyan University Press, 1959); and Sondra Miley Cooney, 'Publishers for the People: W. & R. Chambers; The Early Years, 1832–1850' (Ph.D. dissertation, Ohio State Unversity, 1977). The last work contains much useful information about Chambers and deserves to be better known by historians of science.

7. Robert K. Merton, 'The Matthew Effect in Science', in *idem, The Sociology of Science: Theoretical and Empirical Investigations*, ed. Norman W. Storer (Chicago: University of Chicago Press, 1973), pp. 439–59.

8. [Chambers], *Vestiges*, 10th edn (London: John Churchill, 1853), pp. v–x; first discussed extensively in Hodge, 'Universal Gestation'.

9. William Chambers, *Memoir of Robert Chambers, with Autobiographic Reminiscences of William Chambers* (New York: Scribner, Armstrong, and Co., 1872), p. 256.

10. This important correspondence, which was preserved by Ireland despite Chambers' instructions, was first used extensively in Cooney, 'Publishers for the People', but has not (to my knowledge) been referred to by historians of science.

11. Frederic L. Holmes, 'The Fine Structure of Scientific Creativity', *History of Science*, 19 (1981), 60–70.

12. Chambers, 'General Preface', *Select Writings of Robert Chambers*, 7 vols. (Edinburgh: W. and R. Chambers, 1847), I, iv. For discussion of Chambers' editorial policies, see Cooney, 'Publishers for the People'.

13. Quoted in James Payn, *Some Literary Recollections* (London: Smith, Elder and Co., 1884), p. 143. The best source for Chambers' early life is the reminiscences he compiled in 1833; see 'Diary of Robert Chambers', NLS, Dep. 341/29 (typed transcript).

14. Chambers to W. Scott, 17 Aug. 1822, NLS, MS. 3895, fos. 51–2.

15. Chambers to W. Blackwood, [1830–32], NLS, MS. 4714, fos. 183–4. For 'Young Waverley', see James Grant Wilson, 'Robert Chambers', *Lippincott's Magazine*, 8 (1871), 17–26 (19).

16. Chambers to Scott, 30 Mar. 1830, NLS, MS. 3917, fos. 156–7.

17. Robert Chambers, *History of the Rebellions in Scotland, under the Marquis of Montrose, and Others, from 1638 till 1660*, 2 vols. (London: Constable, 1828), I, 10–11. The wider meanings of historical writing in this context are discussed in Stephen Bann, *The Clothing of Clio: A Study of the Representation of History in Nineteenth-Century Britain and France* (Cambridge: Cambridge University Press, 1984) and David Brown, *Walter Scott and the Historical Imagination* (London: Routledge and Kegan Paul, 1979).

18. Chambers, 'General Preface', pp. iii–iv.

19. Eliza Priestley, *The Story of a Lifetime* (London: Kegan Paul, 1908), p. 30. Chambers told a similar story to Charles Kingsley and A.K.H. Boyd in 1867; see [Boyd], *Twenty-five Years of St. Andrews, September 1865 to September 1890*, 2 vols. (London: Longmans, Green, and Co., 1890), I, 86. Boyd connects the incident with the first issue of the *Journal* in March 1832. Sales of the *Journal* are outlined by Scott Bennett, 'Revolutions in Thought: Serial Publication and

the Mass Market for Reading', in Joanne Shattock and Michael Wolff, eds., *The Victorian Periodical Press: Samplings and Soundings* (Leicester: Leicester University Press, 1982), pp. 225–57 (236).

20. David Dickson, *The Influence of Learning on Religion: A Sermon, preached before the Society in Scotland, (Incorporated by Royal Charter,) for Propagating Christian Knowledge* (Edinburgh: W. Creech, 1814). For Dickson, see [John Anderson, Jr], *Sketches of the Edinburgh Clergy of the Established Church of Scotland* (Edinburgh: John Anderson, 1832).

21. Chambers to Ireland, n.d., NLS, Dep. 341/110/243–4. For a similar case involving a move away from a presbyterian background, see the sensitive account of George Combe's career in Cooter, *Cultural Meaning*, pp. 101–33. A useful background on the Moderate faction in the Scottish church in which Chambers was raised is provided in Andrew L. Drummond and James Bulloch, *The Church in Victorian Scotland, 1843–1874* (Edinburgh: The Saint Andrew Press, 1975), esp. pp. 224–33, and J. David Hoeveler, Jr, *James McCosh and the Scottish Intellectual Tradition: From Glasgow to Princeton* (Princeton, N.J.: Princeton University Press, 1981), pp. 3–32.

22. James Payn, *Some Literary Recollections* (London: Smith, Elder, and Co., 1884), p. 142. Drummond and Bulloch, *Church in Victorian Scotland*, claim that Chambers continued his allegiance to the presbyterian church in his parish, although the source they cite does not support this contention. For Anne Chambers' comment, see A. Chambers to R. Chambers, 5 Sept. 1836, NLS, Dep. 341/90/19.

23. Hodge, 'Universal Gestation', pp. 131–3; James R. Moore, *The Post-Darwinian Controversies: A Study of the Protestant Struggle to come to terms with Darwin in Great Britain and America, 1870–1900* (Cambridge: Cambridge University Press, 1979), p. 344; Michael Ruse, *The Darwinian Revolution* (Chicago: University of Chicago Press, 1979), p. 112.

24. Chambers to Ireland, 30 June 1844, NLS, Dep. 341/110/32–3.

25. Chambers to Ireland [1844], NLS, Dep. 341/110/157. Of course, Chambers' project was not 'secular' in the militant sense associated with the Secularist movement of the later nineteenth century, although he was not a religious man at this stage of his life, and clearly hoped to remove the need for any ideas of direct divine intervention in the world.

26. There were, of course, other sciences on offer in Edinburgh than the phrenologically oriented 'science of progress', and evangelical presbyterians played important roles in these. For a discussion of the group (including John Fleming and Hugh Miller) of Free Church men of science, see David N. Livingstone, *Darwin's Forgotten Defenders: The Encounter between Evangelical Theology and Evolutionary Thought* (Grand Rapids, Mich.: William B. Eerdmans Publishing Co., 1987), pp. 7–16.

27. George Combe, *Of the Constitution of Man and Its Relation to External Objects* (Edinburgh: John Anderson, Jr, 1828). The literature on phrenology is extensive. See especially Roger Cooter, *Cultural Meaning*; Steven Shapin, 'Phrenological Knowledge and the Social Structure of Early Nineteenth-Century Edinburgh', *Annals of Science*, 32 (1975), 219–43; and *idem*, 'The Politics of Observation: Cerebral Anatomy and Social Interests in Edinburgh

Phrenology Disputes', in Roy Wallis, ed., *On the Margins of Science: The Social Construction of Rejected Knowledge*, Sociological Review Monographs, 27 (Keele: University of Keele, 1979).

28. Thus standard histories such as Peter J. Bowler, *Evolution: The History of an Idea* (Berkeley: University of California Press, 1984), Ernst Mayr, *The Growth of Biological Thought: Diversity, Evolution, and Inheritance* (Cambridge, Mass.: Belknap Press, Harvard University Press, 1982), and Ruse, *Darwinian Revolution* discuss *Vestiges* at considerable length, but barely mention phrenology, if they do so at all.

29. Surprisingly, one phrenological enthusiast who was not an intimate of the Chambers' household was Robert's brother William. From the late 1830s onwards, their relationship was confined to business matters. On a personal level, they were not close. The rift has hitherto remained a private secret masked by public displays of brotherly goodwill. For example, see R. Chambers to W. Chambers, 28 March 1859, NLS, Dep. 341/93, fos. 50–9 for evidence of longstanding mutual antipathy.

30. For these points, and a discussion of the *Educational Course*, see Cooney, 'Publishers for the People', pp. 153–213; see also [Chambers], *Introduction to the Sciences: For Use in Schools and for Private Instruction* (Edinburgh: W. and R. Chambers, 1836).

31. 'The brain has been already mentioned as the recognized seat of the mind and the centre of sensation. By this it is not meant that the brain is itself the mind, but merely a material organ, charged by Almighty power with the functions of the mind, which, in itself, we are taught to regard as a spiritual and distinct essence, destined to survive the dissolution of all matter whatever. In what manner the material and immaterial things are connected during life, no one can tell' ([Chambers], *Introduction to the Sciences*, 97).

32. Gov Hutchinson, 'Robert Chambers's Vision of Science: The Diffusion of Scientific Ideas to the General Reader in Early-Victorian Britain' (Ph.D. dissertation, Temple University, 1980), pp. 93–4.

33. Combe to Chambers, 26 Nov. 1835, NLS, MS. 7386, fos. 423–4 (copy).

34. Chambers to G. Combe, 25 Nov. 1835, NLS, MS. 7234, fos. 140–1. Alexander Duff (1806–78) was a leading presbyterian missionary on the Indian subcontinent.

35. Chambers to Ireland, [1845], NLS, Dep. 341/110/88–9. For further comments by Chambers on Combe and materialism, see 'George Combe', in Chambers, ed., *The Book of Days: A Miscellany of Popular Antiquities*, 2 vols. (Edinburgh: W. and R. Chambers, 1864), II, 213–14.

36. See *Testimonials on Behalf of George Combe as a Candidate for the Chair of Logic in the University of Edinburgh* (Edinburgh: John Anderson, 1836), p. 55.

37. Chambers to Ireland, [Nov. 1837], NLS, Dep. 341/110/9–10; also Chambers to Ireland, [1845], NLS, Dep. 341/113/164–5; and Wilson, 'Chambers', p. 22.

38. [Chambers], *Vestiges* (10th edn, 1853), p. v; see also Marilyn Bailey Ogilvie, 'Robert Chambers and the Nebular Hypothesis', *British Journal for the History of Science*, 8 (1975), 214–32.

39. [Chambers], 'Professor Nichol's Views of the Architecture of the Hea-

vens', *Chambers's Edinburgh Journal*, 6 (1837), 210–11.

40. As Schaffer notes (this volume), Nichol was often accused of authoring *Vestiges*, and privately he thought that Chambers had plagiarized from him. For Chambers' account of the Irish tour, see 'A Few More Days in Ireland', *Chambers's Edinburgh Journal*, 6 (1837), 289–90, 301–2, 309–10, 317–18, 325–6, 333–4.

41. Chambers to D.R. Rankine, 3 Sept. 1837, NLS, Dep. 341/109/1.

42. [Chambers], 'Gossip about Golf', *Chambers's Edinburgh Journal*, 11 (1842), 297–8; also in Chambers, *Select Writings*, II, 313–24.

43. [Chambers], 'Thoughts on Nations and Civilisation', *Chambers's Edinburgh Journal*, 11 (1842), 137–8 (138).

44. [*idem*], 'The Easily Convinced', *Chambers's Edinburgh Journal*, 11 (1842), 185–6; also in Chambers, *Select Writings*, IV, 14–22.

45. For the problems Chambers faced in this area after publication, see especially Richard Yeo, 'Science and Intellectual Authority in Mid-nineteenth Century Britain: Robert Chambers and "Vestiges of the Natural History of Creation"', *Victorian Studies*, 28 (1984), pp. 5–31; Ogilvie, 'Robert Chambers and the Nebular Hypothesis'; and Schaffer, this volume.

46. John Pringle Nichol, *Views of the Architecture of the Heavens, in a Series of Letters to a Lady* (Edinburgh: William Tait, 1837), p. 127.

47. [Chambers], *Vestiges* (10th edn, 1853), p. v.

48. 'Natural History: Animals with a Backbone', *Chambers's Edinburgh Journal*, 1 (1832), 337–8.

49. 'Popular Information on Science: Transmutation of Species', *Chambers's Edinburgh Journal*, 4 (1835), 273–4.

50. 'Sketches in Natural History: Monkeys', *Chambers's Edinburgh Journal*, 11 (1838), 251–2.

51. The dangerous associations of transmutation are well brought out in Adrian Desmond, 'Artisan Resistance and Evolution in Britain, 1819–1848', *Osiris*, 2nd ser., 3 (1987), 77–110; *idem*, 'Robert E. Grant: The Social Predicament of a Pre-Darwinian Transmutationist', *Journal of the History of Biology*, 17 (1984), 189–223; Michael Bartholomew, 'Lyell and Evolution: An Account of Lyell's Response to the Prospect of an Evolutionary Ancestry for Man', *British Journal for the History of Science*, 6 (1973), 276–303; and Pietro Corsi, 'The Importance of French Transformist Ideas for the Second Volume of Lyell's "Principles of Geology"', *British Journal for the History of Science*, 11 (1978), 221–44.

52. 'Jealousies of the Employed against Employers', *Chambers's Edinburgh Journal*, 7 (1838), 41–2; for similiar comments, see the speech by Chambers at the firm's annual soirée: 'Messrs Chambers's Soirée', *Chambers's Edinburgh Journal*, 12 (1843), 197–9.

53. In 1844 Chambers had tried to get his editorial assistant, David Page, to review *Vestiges*, but he refused; see *The Athenaeum*, no. 1414 (1854), 1463–4.

54. Chambers to Combe, 27 March 1834, NLS, MS. 7232, fos. 70–1. For Combe's own prudishness in phrenological discussions of sex, see Michael Shortland, 'Courting the Cerebellum: Early Organological and Phrenological Views of Sexuality', *British Journal for the History of Science*, 20 (1987), 173–99,

esp. 187–9. Some of the effects of taboos on discussions of sex in scientific work are brought out in John Farley, *Gametes & Spores: Ideas about Sexual Reproduction, 1750–1914* (Baltimore, Md.: Johns Hopkins University Press, 1982), pp. 110–28.

55. Cooney, 'Publishers for the People', p. 63.

56. [Chambers], 'Educability of Animals', *Chambers's Edinburgh Journal*, 22 (1842), 97–8; also in Chambers, *Select Writings*, IV, 154–62.

57. 'Popular Information on Science: Effects of Climate, &c. on Human Beings', *Chambers's Edinburgh Journal*, 12 (1843), 346–7.

58. [Chambers], *Vestiges* (10th edn, 1853), pp. v–vi; see also Hodge, 'Universal Gestation'.

59. See Adrian Desmond's essay, this volume.

60. For example, Chambers' reference to Tiedemann in *Vestiges*, p. 201 is drawn directly from Lord, while his comments on Lamarck on pp. 230–1 imply no further knowledge than is available in Lyell's *Principles*.

61. *Vestiges*, pp. 200–1, quoting (with minor alterations) from Perceval B. Lord, *Popular Physiology: Being a Familiar Explanation of the Most Interesting Facts connected with the Structure and Functions of Animals, and Particularly of Man* (London: John W. Parker, 1834).

62. Adrian Desmond, this volume; John Fletcher, *Rudiments of Physiology, in Three Parts* (Edinburgh: John Carfrae and Son, 1835–7), part i, esp. pp. 6–17.

63. *Vestiges*, pp. 226–7.

64. For these points, see the excellent discussion in Evelleen Richards, 'A Question of Property Rights: Richard Owen's Evolutionism Reassessed', *British Journal for the History of Science*, 20 (1987), 129–71 (133–9) and Dov Ospovat, 'The Influence of K.E. von Baer's Embryology, 1828–1859', *Journal of the History of Biology*, 9 (1976), 1–28.

65. *Vestiges*, p. 212. It is worth noting that the source was explicitly credited from the fifth edition of 1846, when Carpenter himself was actively contributing to the revision of *Vestiges*. For his participation in the other publications of the Chambers firm, see Cooney, 'Publishers for the People', p. 195.

66. Richards, 'Property Rights'. For examples of the revisions, see Marilyn Bailey Ogilvie, 'Robert Chambers and the Successive Revisions of the "Vestiges of the Natural History of Creation"' (Ph.D. dissertation, University of Oklahoma, 1973), esp. pp. 307–32.

67. Richards, 'Property Rights'; Ospovat, 'The Influence of K.E. von Baer's Embryology'. For an interpretation of Chambers that emphasizes the elements of linearity in the first edition, see Peter J. Bowler, *Fossils and Progress: Paleontology and the Idea of Progressive Evolution in the Nineteenth Century* (New York: Science History Publications, 1976), pp. 53–62. As Bowler has pointed out to me, *Vestiges* continued to use analogies with Babbage's calculating engine, which best fitted a linear model of progress, even after the embryological discussion had been revised to conform with Von Baer. Here, as in many other ways, Chambers was evidently more concerned with communicating a basic message about natural law than with analytical consistency. In fact, the embryology was revised towards Von Baer largely (although not

entirely) because the physiologist Carpenter assisted in improving the later editions.

68. A. Chambers to W. Chambers, n.d., NLS, Dep. 341/82/31; R. Chambers to [A. Chambers], 23 Nov. 1842, NLS, Dep. 341/82/32. Another important event leading to the move may have been the death of a daughter, Margaret, from scarler fever in March 1842. See [Chambers to Rankine], 14 Mar. 1842, NLS, Dep. 341/109/3. For the story that Chambers left St Andrews for secrecy, see Millhauser, *Just Before Darwin*, pp. 29–30.

69. [Chambers to Rankine], NLS, Dep. 341/109/5. For the death of Chambers' mother, see W. Chambers, *Memoir*, p. 243.

70. *Vestiges*, p. 387.

71. Chambers to Rankine, 3 Sept. 1837, NLS, Dep. 341/109/1. See also Chambers, 'To a Little Boy', *Blackwood's Edinburgh Magazine*, 38 (July 1835), 70. I am grateful to Mr A.S. Chambers for the following details concerning birth and death dates of the children of Anne and Robert Chambers: Jane (b. 1830), Robert (1832–88), Mary (b. 1833; unnamed twin died same day), Anne (b. 1835), Janet (1836–63), Eliza (1836–1909), Amelia (b. 1838), Margaret (1839–42), James (1841–1929), William (1842–3), William (b. 1843), Phoebe (1846–1918), Alice (1851–1925).

72. *Vestiges*, p. 8.

73. Ibid., pp. 11–12, quoting (with slight modifications) from John F.W. Herschel, *A Treatise on Astronomy* (London: Longman, Brown, Green and Longmans, 1830), p. 264. See the suggestive comments on this passage in Gillian Beer, *Darwin's Plots: Evolutionary Narrative in Darwin, George Eliot and Nineteenth-Century Fiction* (London: Routledge and Kegan Paul, 1983), p. 169.

74. Nichol, *Architecture*, pp. 194, 195.

75. *Vestiges*, p. 223.

76. Ibid., pp. 233–5. Because Chambers had prepared his readers so well, criticisms like those of Sedgwick tended to misfire.

77. Chambers to Combe, 17 Dec. 1840, NLS, MS. 7254, fos. 1–6.

78. R.C. Lehmann, *Memories of Half a Century* (London: Smith, Elder, and Co., 1908), p. 7, quoting memoirs of Frederick Lehman, who married one of Chambers' daughters in 1852.

79. James R. Moore, 'Darwin of Down: The Evolutionist as Squarson-Naturalist', in David Kohn, ed., *The Darwinian Heritage* (Princeton, N.J.: Princeton University Press, 1985), pp. 435–81.

80. A.O. Lovejoy goes to the opposite extreme, making Chambers *successful* on what are essentially Darwin's terms. See Lovejoy, 'The Argument for Organic Evolution before the *Origin of Species, 1830–1858'*, in Bentley Glass, Owsei Temkin and William L. Strauss, Jr, eds., *Forerunners of Darwin: 1745–1859* (Baltimore, Md.: Johns Hopkins Press, 1959), pp. 356–414. The dangers of these kinds of approaches were pointed out long ago in Hodge, 'Universal Gestation'.

81. A. Sedgwick to M. Napier, 10 April 1845, in Macvey Napier, ed., *Selections from the Correspondence of the Late Macvey Napier* (London: Macmillan, 1879), p. 492; [Thomas Henry Huxley], 'The Vestiges of Creation', *British and Foreign Medico-Chirurgical Review*, 13 (1854), 425–39 (438).

6

Of love and death: Why Darwin
'gave up Christianity'

JAMES R.MOORE

> One wonders . . . whether Darwin may not have left something
> out of the story, either deliberately or by lapse of memory.
>
> John C. Greene[1]

Victorian people lost faith in Christianity with a punctiliousness
equalled only by those who have found them out. No survey of Victorian literature is complete without a chapter on the 'crisis' of belief;
no biography of the period may omit the subject's *Sturm und Drang*
when an inherited creed is abandoned. Historians who have analysed
the so-called 'loss of faith' have reached the unsurprising conclusion
that Victorians renounced Christianity as much or more for moral
reasons as for intellectual ones. Those who have scrutinized individual crises have often found cause to agree. It is striking, therefore, that the reasons usually adduced for Charles Darwin's loss of
Christian faith are for the most part intellectual, while the distinction
between Christianity and theistic natural religion, which Darwin
himself maintained in writing of his beliefs, is largely neglected.

 Somehow latter-day punctiliousness may have failed. And it is not
hard to understand why. Unlike many Victorian doubters, Darwin
left an abundance of specific and highly detailed evidence for his
intellectual development. One is 'embarrassed by riches' when trying to explain his loss of Christian conviction. Comte on theology
and Hume on miracles, pain and purposelessness in nature, Fitz-
Roy's evangelicalism and Lyell's uniformitarianism, heterodox family
members and friends – the available reasons, intellectual and moral,
for the expiry of Darwin's original faith constitute an overkill.[2] To
discern the lethal from the merely local and limited factors, moreover, requires at least two additional insights. One must have a fair
idea of the kind of Christianity against which Darwin's various arguments and experiences would have counted; and one must be able

to tell when these arguments and experiences presented themselves severally to Darwin, and with what effect. Here again it is easy to see why historical punctiliousness has failed. Darwin's Paleyite Anglicanism, steeped in Unitarian nonconformity, is a brand of Christianity every bit as recondite now as it was outside a fairly rarified stratum of early Victorian intellectual culture. And the full sequence of events by which Darwin came to reject Christianity is not at all clear, let alone the differential impact of the arguments and experiences that confronted him at each stage.

This essay cannot analyse the many reasons for Darwin's fall from grace by reconstructing his initial state of belief and plotting its downward trajectory. To that extent punctiliousness will be wanting in what follows. But the end-point of the process, when Darwin said he 'gave up Christianity', can be determined with some precision, and this knowledge forms the basis of my argument. Through a close analysis of his *Autobiography*, probably the most widely familiar of all his works,[3] I shall establish a particular locus for his final apostasy. Then, by co-ordinating some of the chief events in Darwin's life within this context, I shall explain why, in a crucial moment of his existence, Darwin could no longer believe in Christianity, even as commended to him by his beloved wife. Finally, I shall suggest how this insight may shed light on episodes in Darwin's subsequent career, including his personal life and his science.

'I was very unwilling to give up my belief'

The most authoritative account of Darwin's loss of Christian faith appears in the *Autobiography*, a substantial manuscript written between 1876 and 1881 but not intended for publication. From this six-hundred-word account (85–7) it would appear that Darwin had come 'gradually' to distrust the Old Testament by the time he resettled in England after the *Beagle* voyage. The reasons given are both empirical and moral. During the subsequent period, probably from late 1836 to early 1839, Darwin seems also to have begun 'further reflecting' on the reliability of the New Testament, with the result that, as he states, 'I gradually came to disbelieve in Christianity as a divine revelation.' The reasons given here pertain chiefly to defects in historical evidence, although Darwin acknowledges a hermeneutic problem with New Testament morality. 'But', he hastens to add, 'I was very unwilling to give up my belief.' Very frequently – 'often and often' – he had tried to 'invent evidence' that would convince him of the historicity of the Gospels, and this prolonged his indecision. 'Thus disbelief crept over me at a very slow rate.' The process was 'so slow', however, that it caused him 'no distress', and when 'at last complete', it yielded a conviction so singular that, on publication of the *Auto-biography* in 1887, the family suppressed it (87): 'I have never since doubted even for a single second that my conclusion was correct.' Then, as if to leave no doubt in his reader's mind either, Darwin explains why there could be no turning back:

> I can indeed hardly see how anyone ought to wish Christianity to be true; for if so the plain language of the text seems to show that the men who do not believe, and this would include my Father, Brother and almost all my best friends, will be everlastingly punished.
>
>   And this is a damnable doctrine.

The family suppressed this remark as well. Seventy years would pass before a grand-daughter, Nora Barlow, revealed the strength and intensity of Darwin's disbelief in an unexpurgated text of the *Autobiography*.

The remainder of the three thousand words Darwin devotes to religious subjects in the *Autobiography* (87–96) concerns natural religion and morality. Christianity is completely set aside. This structural feature of the text has seldom received its due from historians, but even those who carefully distinguish Christianity from natural religion in Darwin's religious outlook have tended to conflate his growing doubts about the latter with his disbelief in the former. Since the publication of a series of so-called 'metaphysical' notes and notebooks that Darwin kept between 1838 and 1840, historians have been struck by the heterodoxy of his researches over the very period when, according to the *Autobiography*, the evidence for Christianity was being tried and found wanting. Add to this conjuncture the fact that in the middle of the same period Darwin courted and married his cousin Emma Wedgwood, whose simple heart-felt convictions served as the perfect foil for his sophisticated doubts, and the conclusion seems inescapable: his loss of faith, though possibly the precondition of his evolutionary speculations and dating from as early as 1831, was in all likelihood the product of his research. By 1839, when he was thirty years old – certainly no later than 1842 – Darwin had given up Christianity.[4]

I hold this prevailing view of Darwin's loss of faith to be wrong for a pair of reasons that underpin the rest of my analysis. First, it takes the *Autobiography* too seriously as a statement of causality and not seriously enough as an ascription of dynamics. Darwin's account of his gathering disbelief is an ideological reduction of a complex process to the terms of his mature views on evolution. In the intel-

lectual world as in the natural, empirical evidence is made to govern
the narrative; in neither realm does a constructive *saltus* take place.
Like the clerical career Darwin describes earlier in the text, his faith
dies a 'natural death' (57) – except for the *coup de grâce*. The slow,
gradual process of disbelief is made to end abruptly when Darwin
utters a conviction of singular certitude and clinches it with a reprobation. Evidential considerations surely played an important part
in this process – historians are right to emphasize the point – but
Darwin's clear intentions in ascribing its dynamics must not be ignored. The process was protracted because he was frankly reluctant
to give up Christianity; the process ended sharply and irrevocably
because he was outraged by 'the plain language of the text', which
demanded everlasting punishment for unbelievers. Therefore, despite
its ideological and possibly self-serving character, the *Autobiography*
points towards a different interpretation of Darwin's loss of faith from
that current among historians. Not only did Darwin reach a critical
stage in life beyond which Christian belief became impossible for him;
he also reached that stage perhaps a good deal later than has custom-
arily been allowed, and there met with rather more than an intellectual challenge that put paid to Christianity.

The second reason why I dissent from the prevailing view of Darwin's loss of faith is the uncontroverted testimony of Edward Aveling that Darwin did in fact finally relinquish Christianity at a period
approximately ten years later than the one usually assigned. Aveling
may have proved himself a plausible rascal, but in September 1881,
when he visited Darwin at Down House in the company of the German physiologist Ludwig Büchner, he was as yet merely an ingenuous
young Secularist bent on using Büchner's coattails to secure an endorsement of his own atheism from England's leading naturalist.[5] In the
event, Aveling got less than he wanted; and the account of the visit
he published after Darwin's death conveyed his host's lack of sympathy with sufficient candour to move an eyewitness of the encounter,
Francis Darwin, the first editor of the *Autobiography*, to remark that
Aveling had given 'quite fairly his impressions of my father's views'.
But this much was obtained from Darwin and published in Aveling's
account: the admission that 'I never gave up Christianity until I was
forty years of age'. As a deliberate, direct quotation the statement may
have special claim to represent Darwin's views 'quite fairly'.[6] It is surprising, therefore, that historians have made so little of it. Those who
refer to Aveling's narrative usually dwell on other aspects of the visit,
even when Darwin's statement is noted. The most recent study of
Darwin's religious views finds the statement 'curious, to say the least'.
In the light of Darwin's own account of his loss of Christian faith,

however, his admission makes ample good sense, quantifying the period of indecision and marking the point of no return. Although Darwin turned forty on 12 February 1849, I propose now to treat the age at which, he said, he 'gave up Christianity' as a round figure and to recall that period of his life through an analysis of the religious part of the *Autobiography*.[7]

'Queer conservato-grundiform feelings'

Like other self-revealing documents Darwin prepared – the first 'sketch' of his species theory and the metaphysical notebooks – the *Autobiography* was begun away from home. He started writing at the end of May 1876, while visiting Emma's brother and sister-in-law, Hensleigh and Fanny Wedgwood; he completed the main text at Down about two months later.[8] Over the next five years Darwin enlarged the manuscript by almost fifty per cent. The most important additions were a long memoir of his father (28–42); an equally long account of scientific society 'during the early part of our life in London' (99–114) – that is, just after his marriage in January 1839; a synopsis of his later publications (133–6); and a number of paragraphs on religion. The memoir of Dr Robert Darwin was written, according to Barlow, in 1878 or later; the account of London social life was added with the book descriptions in the month to 1 May 1881. The paragraphs on religion seem to have been incorporated at various times, but one and probably both of the longest paragraphs were 'written in 1879 – copied out Apl. 22 1881', according to a note by Darwin on the manuscript (95n). It was not his original intention to discuss religion separately in the *Autobiography*, as its table of contents reveals. The headings in the table do not exactly tally with the text (19). Two have been omitted, although the text contains their subject matter. And one heading appears in the text but not in the table: 'Religious Belief'.

The *Autobiography* was written for no one but the family, and Darwin seems to have had especially in mind the grandchildren he hoped would be born. A 'German editor' had asked him for 'an account of the development of my mind and character with some sketch of my autobiography', he wrote in the opening paragraph (21). This had probably occurred the previous autumn.[9] If Germans were curious about his personal life, he may have thought, how much more interesting a detailed account would be one day to his first grandchild, whose birth was expected in about four months to Francis and his wife Amy, who lived in Down village. 'I know', Darwin continued, 'that it would have interested me greatly to read even so short and

dull a sketch of the mind of my grandfather written by himself, and
what he thought and did and how he worked.' Besides, although
Darwin's health had improved considerably in recent years, he felt
that he could die at any time. There was urgency in his pen, and perhaps a trace of pain. 'I have attempted to write the following account
of myself, as if I were a dead man in another world looking back at
my own life. Nor have I found this difficult for life is nearly over with
me' (21).

Darwin wrote for an hour or so on most days and completed a
first draft of the *Autobiography* on 3 August 1876, a month before
Amy's confinement (145). In past years he 'used to dread the time
and hate it with all [his] . . . heart', but since 1850, when he first
administered chloroform to Emma, his mortal fear of childbirth had
diminished.[10] Still, old memories could revive. Twice before, while
awaiting a confinement, Darwin had tended dying loved ones. In
May 1848 he stayed with his father at Shrewsbury and eventually
became so ill that he feared for his own life.[11] Francis was born in
early August; his father did not die until November. In April 1851,
however, when he stayed at Malvern with Annie, his ten-year-old
daughter, the girl died just three weeks before Emma was delivered
of Horace, the youngest surviving son. Now, twenty-five years later,
while awaiting Amy's confinement and the birth of his first grandchild, Darwin felt his own life drawing to a close. He began the *Autobiography* on this note and echoed it near the end. After mentioning
his forthcoming book, *The Effects of Cross and Self Fertilisation in the
Vegetable Kingdom*, and a few other publications he hoped to complete, Darwin alluded to the canticle sung by Simeon, the devout
old priest, on being presented with the infant Jesus according to the
Gospel of Luke: 'My strength will then probably be exhausted, and
I shall be ready to exclaim "Nunc dimittis"' (133). And in the next
paragraph: 'My father lived to his eighty-third year with his mind
as lively as it ever was, and all his faculties undimmed; and I hope
that I may die before my mind fails to a sensible extent' (136).

Morbid Darwin's thoughts may be, but they were eminently Victorian and unexceptionable. When Francis, who had been specially
entrusted with the manuscript of the *Autobiography*, sought approval
from the family to publish the text in his father's *Life and Letters*, these
passages offended no one. What provoked Emma and, still more, the
elder daughter Henrietta and her husband, was the section headed
'Religious Belief'. For several months in 1885 the family bickered.
Litigation could not be ruled out if Francis, supported by William,
the eldest son, proceeded to publish the section intact, in its original
context. Eventually a 'compromise' was reached. The *Autobiography*

was dismembered, the section on religious belief was removed to a separate chapter in the *Life and Letters*, and only 'extracts, somewhat abbreviated, from a part ... written in 1876' were printed. On no account was Emma's name to be connected with the subject. Although Darwin's reply in 1879 to a young German enquirer, 'that the theory of Evolution is quite compatible with belief in a God', was penned by Emma, the correspondent, according to Francis in the chapter on religion, had been 'answered by a member of my father's family'.[12]

What caused a peaceable loving family suddenly to be riven by bitter controversy? Two reasons will suffice. Darwin seldom spoke on religious subjects. He gave his sons to believe that this was because he and Emma did not 'see eye to eye': 'These are very difficult matters. You boys must settle them for yourselves', he was remembered to say. Religion belonged to woman's realm in the Darwin household; nor does Emma seem to have discussed the subject with her adult sons.[13] Thus the controversy broke out, in part, because it had been systematically suppressed, and the depth of sentiment on the women's side startled Francis and his brothers. But the eruption of 'queer conservato-grundiform feelings', as Francis unkindly called them, was mainly provoked by passages in the religious section of the *Autobiography*, a section that the members of the family most sympathetic with their father's view of religion wished to see published intact, 'in the place he put it'.[14] The matter will be referred to at various stages of the argument, but for the present it is necessary to assess the placement of the section and the content of the main offending passages that were omitted from the *Autobiography* as it first appeared in the *Life and Letters* in 1887.

'It seems to me raw'

The section headed 'Religious Belief' in the unexpurgated *Autobiography* follows Darwin's account of his scientific work in London immediately after the *Beagle* voyage. 'During these two years', it begins, 'I was led to think much about religion' (85). Darwin refers either to the 28 months after the *Beagle* docked at Falmouth or to the 23 months after his own arrival in London to lodge in Great Marlborough Street. The *terminus ad quem* in both cases was his marriage, for when he finished the section on religion he resumed his account under the head, 'From my marriage, Jan. 29, 1839, and residence in Upper Gower Street to our leaving London and settling at Down, Sept. 14, 1842' (96). This new section begins with an intimate memoir of Darwin's family life, then picks up the story of his scientific work with the words, 'During the three years and eight months whilst we resided in London . . .' (98). Now when Francis Darwin edited the *Autobiography* in accordance with the family compromise worked out in 1885, everything from 'During these two years . . .' up to 'During the three years and eight months . . .' vanished from its original context. The section on religious belief, to which his father had not given a separate heading in his table of contents, as well as the intimate family memoir immediately following, were treated as parenthetical matter. Francis covered for the omission in the *Life and Letters* version of the *Autobiography* by a partial allusion to its contents: 'After speaking of his happy married life, and of his children, he continues. . . .' Two hundred and thirty-eight pages after this misleading reference, in a separate chapter of the *Life and Letters*, the section on religion reappeared, without the family memoir, as a part of the *Autobiography* 'not written with a view to publication'. It was severely censored, and the new context in which Francis placed it was the summer of 1842, when the Darwin household removed to Down.[15] Although Francis did not disguise the fact that his father's account referred to the development of his religious views beginning some five years earlier, he nevertheless, by skilful editing, dissociated the section on religious belief both from his parent's marriage and from the interior life of the family.

What then was so objectionable about the section on religious belief as it appeared in the original text of the *Autobiography*? The question may be answered by considering the section in its three sequential parts. First, Darwin disposes of Christianity (85–7); then he evaluates the two great doctrines of natural religion, the existence of God and personal immortality (87–94); finally, he describes the moral life of an agnostic (94–6).

The first part, dealing with Christianity, was written entirely in 1876. Two passages, one reflecting on the historicity of the Old Testament and its attribution to God of 'the feelings of a revengeful tyrant' (85), the other relativizing the morality of the New Testament (86), succumbed to the family compromise. Darwin's parting shot at the 'damnable doctrine' of eternal punishment, as quoted above, was expunged specifically at Emma's request. It is not hard to understand the immediate reason why she felt strongly about the passage. On 23 September 1882 the *Pall Mall Gazette* published a second reply from Darwin to the German enquirer whom Emma answered in 1879. Ernst Haeckel, a zoologist and bombastic supporter of Darwin, had got his hands on the second letter and published a German translation without permission. This text, in turn, fell into partisan hands and suffered translation back into English, with the result that Darwin's private thoughts on religion now stood naked before the self-same

[Margin annotation, handwritten:] Darwin's family was divided by his views on religion as expressed in his autobiography to such an extent that wasn't published in full until 70 yrs later. Demostrate radical nature of his views?

intelligentsia who had buried his body in Westminister Abbey just five months earlier. The family were not amused. In the letter Darwin denied 'that there ever has been any Revelation' and added, 'As for a future life every man must judge for himself between conflicting vague probabilities.'[16]

Emma probably learned of this publication very soon – whether she discovered that the Secularist *National Reformer* reprinted the letter a week later is not known[17] – and in October 1882, while Francis was working at home on the *Life and Letters*, she read his transcription of the *Autobiography* and annotated Darwin's bitter remark (87n):

> It seems to me raw. Nothing can be said too severe upon the doctrine of everlasting punishment for disbelief – but very few now would call that 'Christianity', (tho' the words are there). There is the question of verbal inspiration comes in too.

After the public disclosure of Darwin's temporizing on the problem of a future life, Emma, it seems, was unwilling that his views on this sensitive subject should be further exposed. Since 1878 Archdeacon F.W. Farrar, in *Eternal Hope*, had convulsed ecclesiastical opinion with his denial of the doctrine of endless punishment.[18] But even devout Unitarians had long believed that, if future punishment were a reality, it would be reformatory, not retributive and eternal. Somehow Darwin had stuck at an outmoded version of Christianity.

Most of the second part of the section on religious belief was also written in 1876. Here Darwin dealt with arguments from reason and intuition respecting the existence of God and personal immortality. Two subsequent paragraphs stating the conditions under which 'I deserve to be called a Theist' (92–3) and a final paragraph explaining why 'I for one must be content to remain an Agnostic' were most likely later accretions. Apart from a badly worded sentence on the sufferings of animals (90), two main passages vanished in the *Life and Letters* edition. The first depreciates theistic intuitions by comparing them to the beliefs of primitive religions and of 'barbaric tribes' (90–1). The second, in the final paragraph, undercuts both reason and intuition with an evolutionary argument (93):

> May not these be the result of the connection between cause and effect which strikes us as a necessary one, but probably depends merely on inherited experience? Nor must we overlook the probability of the constant inculcation in a belief in God on the minds of children producing so strong and perhaps an inherited effect on their brains not yet fully developed, that it would be as difficult for them to throw off their belief in God, as for a monkey to throw off its instinctive fear and hatred of a snake.

Again Emma's personal objection was decisive. It arose 'partly', she

told Francis, 'because your father's opinion that all morality has
grown up by evolution is painful to me', but also, it seems, because
the close juxtaposition of the final sentence with the statement, 'I for
one must be content to remain an Agnostic', might suggest, 'however
unjustly, that he considered all spiritual beliefs no higher than hereditary aversions or likings, such as the fear of monkeys towards
snakes' (93n). Emma had, after all, practised 'the constant inculcation
in a belief in God' on the minds of her own children, and her husband
had not interfered.[19]

The last and most personal part of Darwin's statement on religious belief was completely suppressed. This consisted of the two
long paragraphs, one and probably both of which had been 'written
in 1879 – copied out Apl. 22, 1881'. The date of composition is significant because it coincided with a set of circumstances that made Darwin
reflect again on the nature of his religious convictions. In 1879 he set
aside his scientific work for the first time to give his full attention to
another discipline, a biographical sketch of his grandfather, Erasmus
Darwin. During March and April that year he collected primary
sources and felt as though he were 'having communion with the
dead'.[20] In May and June he wrote up the account, stressing his
grandfather's intellectual and moral qualities and their hereditary
transmission to his father and himself. One of the attributes Darwin
dwelt on was his grandfather's religious heterodoxy. Although 'certainly a theist in the ordinary acceptation of the term, he disbelieved
in any revelation' – the rumour that he called for Jesus on his death-
bed was plainly 'incredible'.[21] But while Darwin was assembling
these thoughts, he was three times asked about his own religious
beliefs: twice, in April and in June, by the young German, and in May
by the writer John Fordyce, who wanted to know whether a theist
could also be an evolutionist. Darwin, it will be recalled, told the
German that he did not believe 'there ever has been any Revelation'.
To Fordyce he replied in words which would suggest that the paragraphs on theism and agnosticism in the Autobiography, discussed
above, were written in 1879 as well: 'Whether a man deserves to be
called a Theist depends on the definition of the term. . . . In my most
extreme fluctuations I have never been an atheist in denying the
existence of a God, I think that generally (and more and more as I
grow older), but not always, that an Agnostic would be the more
correct description of my state of mind.'[22]

Thus in 1879, when Darwin came to write the last two paragraphs
of his statement on religious belief, he was immersed in his grandfather's life and engaged with religious interlocutors. His replies to
the latter show clearly that he identified with the heterodoxy of the

former. The identity was not merely conventional, however, for he saw a hereditary linkage between his grandfather's attitudes and dispositions, and his own. The penultimate paragraph bears this out. Darwin allowed that 'a man who has no assured and ever present belief in the existence of a personal God or of a future existence with retribution and reward' (94) can find a basis for morality apart from religion in the cultivation of hereditary 'social instincts'. A few sentences later, echoing his tribute in the biographical sketch to the moral qualities of his grandfather, whose social instincts had led him to be 'highly benevolent' and respected of men, he declared abruptly, 'As for myself, I believe that I have acted rightly in steadily following and devoting my life to science. I feel no remorse from having committed any great sin ...' (95). The 'constant inculcation' of disbelief in the Darwin family, from grandfather down to grandson, had produced neither moral obliquity nor guilt.

Here again the family dug their heels in. Henrietta, whom Darwin had allowed to excise about twelve per cent of the sketch of his grandfather, including a final paean to his character, objected 'insanely', according to Francis, to the publication of the last, unrepentant remark. This was one reason why the entire 1879 addendum in which it appeared was expunged.[23] Another reason – possibly the real source of Henrietta's wrath – could be found in the next paragraph, where Darwin concluded the section on religious belief by discussing the particular sentiments of women (95):

> Before I was engaged to be married, my father advised me to conceal carefully my doubts, for he said that he had known extreme misery thus caused with married persons. Things went on pretty well until the wife or husband became out of health, and then some women suffered miserably by doubting about the salvation of their husbands, thus making them likewise to suffer.

No one in the family questioned to whom this passage referred. And the reference was distressingly close to Darwin's self-exculpatory remark. Henrietta may well have been incensed on this account alone.

The paragraph continues for several sentences recalling Dr Darwin's views on marriage and religion. This suggests that the 1879 addendum and Darwin's lengthy memoir of his father were written at the same time, while he was preparing his grandfather's biography. The memoir of Dr Darwin, inserted earlier in the *Autobiography*, not only discusses the resolution of 'family quarrels' (31) in his medical practice (a portion judiciously excised in the *Life and Letters* edition), but gives various anecdotes related to bleeding, morbidity and death.

'Many patients, especially ladies', wrote Darwin of his father, 'consulted him when suffering from any misery' (31); 'ladies often cried while telling him their troubles' (37). Darwin then recounts two cases of dying gentlemen whose wives suffered needlessly from lack of hope. 'My father said he had often since seen the paramount importance, for the sake of the patient, of keeping up the hope and with it the strength of the nurse in charge' (38–9). The parallel here with the passage quoted above is almost exact, the wives' misery in both instances arising from fear of death. Darwin may have incorporated these statements, with the rest of the memoir of his father, into the *Autobiography* on 22 April 1881, when he inserted the paragraph on marriage and religion. This paragraph ends by quoting the 'unanswerable argument', as heard by Dr Darwin, by which an 'old lady' assured herself of personal salvation in the hereafter: 'Doctor, I know that sugar is sweet in my mouth, and I know that my Redeemer liveth' (96). The old lady, suspecting Dr Darwin of 'unorthodoxy', had 'hoped to convert him' thus by force of religious intuition. A new heading follows and the first sentence begins, 'You all know well your mother....'[24]

'(Mem: her beautiful letter to myself . . .)'

The new heading, 'From my marriage, Jan. 29, 1839, and residence in Upper Gower Street to our leaving London and settling at Down, Sep. 14, 1842', differs slightly from its counterpart in the table of contents. It is not known whether this heading is original or when it was added to the manuscript. Nor can it be known, until a critical text of the *Autobiography* is established, the editorial status of the first two paragraphs, which begin with the words quoted above.[25] These paragraphs seem curiously disjoined from the subsequent text, where Darwin resumes the account of his scientific work in London with the words, 'During the three years and eight months....' Indeed, they say nothing – nor does the *Autobiography* say anything – specifically about Darwin's marriage and domestic life in London. But the paragraphs, containing an intimate memoir of Emma and the children, do relate closely to the preceding section on religious belief. And like the last two paragraphs of that section, they were suppressed.

Again, it is not hard to understand why. Darwin's moving tribute to Emma as a helpmeet, mother and nurse, 'my wise adviser and cheerful comforter', was too personal for publication in her lifetime.[26] His remarks on their 'most pleasant, sympathetic and affectionate' children (97) could scarcely have been published by themselves. But there may well have been a further reason for suppressing the family memoir. In suggesting it I want to bring my extended analysis of the *Autobiography* to a focus and cast new light on the period at which Darwin 'gave up Christianity'.

We must return to Darwin, writing his *Autobiography* in the summer of 1876. He is an expectant grandfather for the first time. He also expects to die. Thoughts of birth and death mingle painfully in his mind as he remembers awaiting Emma's confinement in 1848, when his own father, 'whose memory', he wrote, 'I love with all my heart' (28), was dying at Shrewsbury. The grandchild born was not his father's first, but, like himself, his father was an unbeliever. If Christianity were true he should even now be 'looking back' at his own life, as though dead, from beside his father in Hell, where his brother Erasmus would join them. The doctrine of eternal damnation is itself 'damnable', not those whom it would threaten. Darwin cannot see 'how anyone ought to wish Christianity to be true', although his wife did so fervently. Emma's, of course, was a reasonable faith, based firmly on intuition as well as revelation. She did not believe in eternal punishment – at least not now. She had powerful 'inward convictions and feelings' about God and personal immortality, which, however, Darwin coldly compared with primitive superstitions and dismissed as 'evidence of what really exists' (91). Yet – and here the subsequent paragraphs on theism and agnosticism, and on morality and marriage, had not been written – 'You all know well your mother', he immediately assures his children, '. . . my greatest blessing, . . . so infinitely my superior in every single moral quality, . . . my wise adviser and cheerful comforter. . . .' Then an inserted phrase, parenthetical, like so many of Darwin's strong emotional statements (97): '(Mem: her beautiful letter to myself preserved, shortly after our marriage.)'[27]

In this letter, written in 1839, which Barlow republished with the unexpurgated *Autobiography*, Emma addressed 'the question' that had already begun to divide them. She referred to 'that dread & fear which the feeling of doubting first gives' and to 'a danger in giving up revelation, . . . that is the fear of ingratitude in casting off what has been done for your benefit as well as for that of all the world'. The 'dread & fear', which her husband then apparently lacked, concerned his personal salvation. The 'danger', she believed, was that he might forego his share in Christ's redemption of the sins of humankind. Thus 'the question' that divided them pertained to nothing less than his eternal destiny. 'I should be most unhappy', Emma wrote towards the end of her letter, 'if I thought we did not belong to each other forever.' Here undoubtedly is the Christianity at which Darwin finally stuck, the simple evangelicalism of his young wife, and her manner of expression, using words such as 'dread & fear', 'danger' and 'forever', would appear to indicate that she then accepted the doctrine of eternal punishment. But the date was 1839, ten years short of the period when Darwin reportedly 'gave up Christianity'. And as Emma herself believed, according to her letter, 'you do not consider your opinion as formed'.[28]

So we return again to 1876. Darwin has remembered his wife and her 'beautiful letter' to him immediately after uttering the harshest judgement on her beliefs he has ever penned. And he has done the deed for his own children, as well as theirs, children who had long been shielded from this deep domestic conflict. Now he remembers them affectionately, until his affection turns to tears (97–8). 'I have been most happy in my family. . . . When you were very young it was my delight to play with you all, and I think with a sigh that such days can never return. . . .'

> We have suffered only one very severe grief in the death of An-
> nie at Malvern on April 24th, 1851, when she was just over ten
> years old. She was a sweet and most affectionate child, and I
> feel sure would have grown into a delightful woman. But I need
> say nothing here of her character, as I wrote a short sketch of
> it shortly after her death. Tears still sometimes come into my
> eyes, when I think of her sweet ways.

The date of death is wrong. Annie died on 23 April. Darwin knew that his father had possessed an 'extraordinary memory' for dates, which often 'annoyed' him, according to the memoir Darwin later added to the *Autobiography*, because 'the deaths of many friends were often recalled to his mind' (39). In this deliberate, tender passage any annoyance in the son is turned to grief. The mistake was not for revising, even if, as seems likely, Darwin miscopied it from his journal.[29] Indeed, the only other date of death in the *Autobiography* is also wrong, which Barlow thought 'curious'. Writing again in the summer of 1876, Darwin gives 'November 13th, 1847' as the day 'when my dear father died' (117). Here the diary is unambiguous and Darwin has simply blundered by a year. Later he would assail Anna Seward's biography of his grandfather for being 'full of in-accuracies even to his age when he died'.[30] But in writing of his own father, the son was not preserved from error either. Nor, given ample opportunity, did he ever revise the text.

I believe, however, that neither mistake is merely 'curious'. Darwin has just recalled his 'one very severe grief', a worse grief than even that of his father's death. The remembrance of his daughter's death, like his father's, brings back painful memories of Emma, whose sim-ple heart-felt faith he has just denied: memories of awaiting her confinement in 1851, and of their separation, as he kept vigil at Malvern with Annie. Thus by 3 August, when the first draft of the *Autobiography* was complete, the expectant grandfather, who expects to die at any time, has bracketed, in effect, the period at which his religious beliefs diverged categorically from his wife's. On the one hand, he had drawn the line in the first part of the section on religious belief by angrily rejecting Christianity for its 'damnable doctrine' that would put his father in hell. Dr Darwin died on 13 November 1848. On the other hand, he has drawn a line of demarcation in the family memoir, which closely follows the section on religious belief, by recalling poignantly his wife's 'beautiful letter' to him on the subject of his own eternal salvation, and by then remembering the 'very severe grief' they suffered and the 'short sketch' he wrote to commemorate their deceased daughter. Annie died on 23 April 1851. Two strong emotions, anger and grief, in the *Autobiography* mark off the years from 1848 to 1851 as the period when Darwin finally renounced his faith. Two deaths, and two births, lie at its extremes. When he wrote and revised the manuscript, which gives many other dates accurately, he suppressed his critical interest in the two deaths to such an extent that he failed to recognize his errors. And when the *Autobiography* came up for publication, the family suppressed any reference to the deaths that would indicate their emotional or religious significance. Whether this was done purposefully or not, the effect was the same: to obscure the dynamics of Darwin's loss of faith in relation to his life with Emma.

Now, however, that Emma's hidden agenda has been brought to light in the context of the *Autobiography*, a locus has been established for investigating not merely when, but why Darwin 'gave up Christianity'. This locus consists of his domestic and family existence at mid-century. I turn therefore to consider the chief events of this place and period, which culminated in Darwin's final apostasy.

'A feather-bed to catch a falling Christian'

The pre-history of the 'beautiful letter' Emma gave her husband in 1839 was as significant as its aftermath. From early adulthood the two cousins encountered death and interpreted it in different ways.[31] In 1832 Emma was marked permanently by the sudden loss of her dearest sister Fanny at the age of twenty-six. They had been inseparable companions, Emma two years younger; ever after she governed her life by the solemn hope, central to her Unitarian upbringing, that loved ones would be reunited beyond the grave.[32] In a large and close-knit family this hope was often put to the test. It was not found wanting. Emma lost an infant and both parents in the 1840s, two more children and two aunts in the 1850s, a sister, another aunt and a nephew in the 1860s, and thereafter, in the years before Darwin's death in 1882, her last sister, a brother and a remaining aunt. Her one 'unhealed' grief was for Annie, her eldest daughter; she 'very rarely' spoke of the child. But although the vividness of her first faith may have waned, overshadowed by this second tragic loss, she retained through all these trials the essential Christian hope of her Wedgwood heritage. In 1893, three years before her own death, she expressed it by transcribing lines from Tennyson's 'In Memoriam' into a note-book: 'Yet less of sorrow lives in me/For days of happy commune dead,/Less yearning for the friendship fled/Than some strong bond which still shall be.'[33]

Darwin's belief in the Christian afterlife was neither so well rooted nor so amply proved. His mother's death in 1817, when he was eight, came too early to affect his religious attitudes, and after persuading himself as a young man, by reading John Pearson's *Exposition of the Creed*, that 'the resurrection of the body' and 'the life everlasting' must, as the *Autobiography* puts it (57), be 'fully accepted', he lost no one near and dear to him until his father's death in 1848.[34] Only a Cambridge friend and three shipmates on the *Beagle* had died by the time Darwin wrote in his diary of primeval forests, 'No one can stand unmoved in these solitudes, without feeling that there is more in man than the mere breath of his body.'[35] Nor does it appear that he departed farther from his mother's Unitarianism than to sketch a post-Priestleyan evolutionary eschatology by the time his infant daughter, Mary Eleanor, died in 1842. As part of his species research, however, Darwin had begun to speculate in his metaphysical notebooks on the heritability of religious and moral sentiments. A chief source of information on the subject was his father, who also advised him, probably in 1838, on the differences between Darwins and Wedgwoods and on the religious propensities of women. Darwin had decided to marry a Wedgwood, just as his father had. In religion the Wedgwoods were as warm-hearted as the Darwins tended to be hard-headed. Dr Darwin and his elder son were already unbelievers. The younger son could now appreciate the remark with which his one grandfather, Erasmus Darwin, had twitted the other, Josiah Wedgwood I: Unitarianism is 'only a feather-bed to catch a falling Christian'.[36] Heritable religious sentiments were clearly manifest in the families, and Darwin no doubt married with the expectation that Emma's beliefs would diverge from his.

But nothing, I believe, forced Darwin to face up to their differences until the time of Dr Darwin's death. Emma's hidden agenda – hidden, that is, first from the family, then from historians – had been tabled in her 'beautiful letter' shortly after their marriage. Life went on thereafter with the usual anxieties about childbearing and the usual grief, in Emma's case, at the loss of aged parents. Her husband meanwhile underwent a painful gestation of his own, 'confessing a murder' by revealing his secret thoughts on species, then entrusting the manuscript of his theory to Emma herself, 'in case of my sudden death'.[37] Darwin never felt quite well as he struggled in the face of conventional beliefs to articulate a new vision of nature. He remembered Emma's concern for their eternal life together and feared for his own demise. In June 1846, while 'watching the thunderstorms' in the summer-house at the end of the Sandwalk at Down, he composed a loving note to Emma, stating how 'sure I am that I can now say so and shall say so on my death-bed, bless you, my dear wife'. Further assurance of his own religious convictions was apparently not to be had, and two years later, in May 1848, he wrote even less auspiciously to his trusted colleague J.D. Hooker about the repugnant reproductive features of the barnacles he had been dissecting. After talk of ill-health, dying and resting in Down churchyard, Darwin struck a note of defiance: '. . . you will perhaps wish my barnacles and species theory al Diavolo together. But I don't care what you say, my species theory is all gospel.'[38]

A week later Darwin was on his way to Shrewsbury to keep vigil with his father, who was expected to die at any time. Emma remained at Down, six months pregnant. The fortnight in his old home was harrowing, with 'continual anxiety' and attacks of illness. Darwin wrote Emma, 'I do long to be with you & under your protection for then I feel safe[.] God bless you.' The illness worsened during the summer, and in October, when Darwin saw his father for the last time during another fortnight's visit, he remained very unwell. Dr Darwin died in mid-November. The news reached Down while Emma was away. She rushed home to comfort her stricken husband before he went again to Shrewsbury, fearing he had worn her with his 'unwellness & complaints'. Nothing is known of this visit except Darwin's recollection in the *Autobiography* (117) that illness kept him from attending the funeral or acting as one of his father's executors.[39] Afterwards he remained sick and depressed. 'All the autumn & winter I have been much dispirited and inclined to do nothing but what I was forced to', he told his cousin W.D. Fox, an Anglican clergyman, the week before his fortieth birthday. With 'health very bad with much sickness & failure of power', Darwin turned to hydropathy for the first time. From March to the end of June 1849 he recuperated with the family at the spa town of Malvern under the care of Dr James

Gully. It was his longest absence from Down. On returning he began self-treatment in accordance with Dr Gully's methods and started a health diary, which he kept for about six years. His condition began to improve.[40]

'The dreadful doctrine of the Eternal Hell'

During this period of unprecedented ill-health, when Darwin alternately explored the guts of 'vile molluscous animals' and fought off wretched vomiting; when he first contemplated the universe devoid of the wisdom, benevolence and medical counsel of his father, a rank unbeliever; indeed, when he himself became convinced that he was about to die – during this period Darwin consulted a number of books, the contents of which would suggest that his life had now reached the critical stage in which Christianity was being scrutinized for the last time. Since 1840 he had shown an interest in various titles that, on the assumption he was already a confirmed unbeliever, should perhaps not have engaged him.[41] Now in April 1848, according to his reading notebook, Darwin exposed himself to the cold erudition of Andrews Norton, the late professor of sacred history at Harvard University, in his commanding two-volume work, *The Evidences of the Genuineness of the Gospels*. Norton, dubbed the 'Unitarian Pope' by Thomas Carlyle for his *ex cathedra* hostility to German critical scholarship, argued that the Gospels 'remain essentially the same as they were originally composed' and that 'they have been ascribed to their true authors'. Darwin, who had re-read William Paley's *Evidences* in 1841, could as yet find this sort of argument 'good'.[42]

Then in July, with the onset of 'swimming head, depression, trembling, many bad attacks of sickness', Darwin finished Julius Hare's memoir of John Sterling, the victim of an enfeebled constitution, who died at the age of thirty-eight.[43] The parallels between Sterling's life and his own were unmistakable. Sterling, born in 1806, left Trinity College, Cambridge, the year before Darwin arrived at Christ's, having lacked the stimulus to take a degree. He fell under the influence of Samuel Taylor Coleridge, the Wedgwoods' friend, who loosed him from conventional pieties, and William Wordsworth, whose 'Excursion' Darwin twice enjoyed. After marrying in 1830 he spent several years abroad and there formed the idea, which Darwin had shared, of entering the Church. He was ordained a deacon in 1834 and became curate of a rural parish, where, like Darwin's mentor, the Reverend John Stevens Henslow, he conducted a reforming ministry. Illness soon caused him to remove to London, and thence, as the *Beagle* sailed home, to more salubrious resorts. While in London

Sterling became acquainted with Carlyle, the friend of Darwins and Wedgwoods, and began research for treatises on revelation and ethics. It was, like Darwin's London years, the 'period of his greatest moral and intellectual energy'. As he read deeper into German neology, however, Sterling lost sympathy with biblical religion and ecclesiastical authority. He now looked for a 'scientific theory of the Bible' and found it – alas for his faith – in David Strauss's ruthless Hegelianizing of the Gospels. By 1840 Sterling, the essayist, the poet and the playright by turns, was as certain as Darwin would ever be 'of doing some good by shewing the absence of all coherence and life in the prevalent English notions, and my own faith in the possibility of deep and systematic knowledge on the laws and first principles of our existence'. But time was short, health precarious at best. In 1843 Sterling was profoundly shaken by the death of his mother on Easter Sunday and of his wife during childbirth two days later. He himself succumbed in September 1844, firmly beliving, as he told his friends, that 'we shall meet again. . . . Christianity is a great comfort and blessing to me, although I am quite unable to believe all its original documents.'[44]

Darwin, the erstwhile ordinand, anxious about his own health, found the memoir 'modestly good', according to his reading notebook. It seemed no better perhaps because of Sterling's muddled adherence to Christianity and the repeated explanation offered by his clerical biographer that 'withdrawal from the practical ministerial life' had left Sterling to 'employ himself in speculation', resulting in the growth of 'negative convictions'.[45] This was too close for comfort. The memoir was, after all, a tract. But, having instructed himself from the life of a fallen curate, Darwin turned in subsequent months to the writings of one of Sterling's closest friends, an Anglican missioner-turned-freethinking theist.

Between 1847 and 1850 Francis Newman, the younger brother of a recent celebrated convert to the Church of Rome, published three books. Darwin read them all. He began in September 1849 with *The Soul*, an impassioned plea for universal religion based on spiritual intuitions.[46] The 'affections of Awe, Wonder, Admiration', leading through 'perceptions of Order, Design, Goodness and Wisdom' to a feeling of 'Reverence', a 'sense of sin', and the experience of a 'personal relation to God' were all intuitions that Darwin either came to distrust or simply could not share. The *Autobiography*, indeed, reads in places like a rebuttal of Newman's so-called 'natural history' of the soul. But in chapters discussing the 'future life' and the 'prospects of Christianity', Newman may well have secured Darwin's assent. 'Logical demonstration of human immortality' in the manner of Paley and

Joseph Priestley was, he insisted, inaccessible to 'the great mass of mankind', whose salvation is supposed to depend on it.[47] And what of the alternative that faced them, 'the dreadful doctrine of the Eternal Hell'?

> Are not men ... driven into a Self-righteous belief, that they in some sense deserve heavenly glory, merely because they cannot feel that they deserve the awful alternative which alone is treated as possible?
> ... If it be said that the fixed doctrine comforts us on the loss of pious relatives, is it forgotten what distress it inflicts on those whose near kinsfolk die without clear marks of piety?[48]

The only grounds for belief in a future life, according to Newman, lay in 'a full sympathy of our spirit with God's Spirit'. The book-religion of a Christianity based on 'Evidences' must give way to a heart-religion for all. 'Unless the appeal can be made directly to the Conscience and the Soul, faith in Christianity once lost by the vulgar is lost forever.'[49]

Unfortunately, it is not known whether Darwin confronted this stark alternative, although Emma, who read *The Soul* about the same time, may well have done so on his behalf. And he left no opinion about the book. Of Newman's *History of the Hebrew Monarchy*, which he opened next, in August 1850, there can be no doubt as to his verdict. Darwin called it 'poor'.[50] This may seem surprising in view of the praise the work has subsequently received. Newman offered a radical view of Israelite history, informed by German scholarship; it was advanced, insightful and lucidly written.[51] What Darwin objected to, however, was surely not Newman's historical method, but its premise: his belief in a static religious instinct. The 'obvious presumption' this entailed, that 'the relations between the divine and the human mind are still substantially the same as ever', seemed plainly irreconcilable with the long and bloody annals of Hebrew political culture. Indeed, Newman himself had to acknowledge from time to time that the record of atrocities committed or commanded by Jehovah God was one with which 'it is hard for Christian or modern feelings to sympathize'. But to him these acts were merely the perverting circumstances that attended the distillation of 'religious wisdom'. Amid death, famine, rapine and the war of Semitic tribes had arisen the highest spiritual faith, a belief in 'the Holiness of God and his Sympathy with his chosen servants'.[52]

To Darwin the material circumstances of life were essential, not merely accidental, to the development of religion. From barbarity nothing but barbarity. The God of a primitive people could be none other than – as the *Autobiography* would put it (85) – a 'revengeful tyrant'. Like everything else, however, the Jewish religious consciousness had evolved, and Christianity was the proof. Now, with Newman, Darwin had to confront his own identification with this other ancient faith. He continued to dissect barnacles; he began preparing his description of them for the press. Gastric flatulence beset him day and night, although uncontrolled vomiting did not return. On 16 March 1851, according to his reading notebook, he recorded his highest accolade, 'excellent', for Newman's spiritual autobiography, *Phases of Faith*.[53]

The modest volume was not beyond perusal in a single sitting. It reconstructed, step by step, the passage of a young man, destined at Oxford for holy orders, from Calvinistic evangelicalism, through Unitarianism, to the fringes of free religion. Intellectual and moral considerations, being integral elements of his faith, received equal emphasis in the narrative. Scruples about the Thirty-nine Articles, missionary problems among Mohammedans, and doubts about scriptural authority came at the start. Almost the first distinctly moral difficulty was presented by the doctrine of eternal punishment. If a finite being deserves infinite punishment because of an offence committed against an infinite being, then 'the fretfulness of a child is an infinite evil!' Efforts to 'pare away the vehement words of the Scripture' on the subject proved unsuccessful, and with 'moral sentiment' and Scripture 'no longer in full harmony', the entire Calvinist scheme of redemption gave way. Newman's heart was opened to the Unitarians, although, he said, 'I distinctly believed that English Unitarianism could never afford me a half-hour's resting-place.'

And he was right. The Bible straightaway began to disintegrate in his hands: the Old Testament with its creation legends and moral monstrosities, the Synoptic Gospels and the New Testament epistles with their inconsistencies and superstitions. '"Of course then you gave up Christianity?"' Newman asked of himself rhetorically. 'Far from it', he replied. 'I gave up all that was clearly untenable, and clung the firmer to all that appeared sound.' The history of the rise and spread of Christianity and of its moral effects did not, however, commend it to him above other religions as a divine revelation; the evidences of miracle and fulfilled prophecy only served to show the folly of faith that rests on the credulity of ancient witnesses. The Gospel of John long remained for Newman the 'impregnable fortress of Christianity', but after criticism showed that Jesus may never have spoken the discourses attributed to him there, the founder of Christianity 'gradually melted into dimness' and receded from his practical faith. Even the last resource of Unitarianism, the moral teaching of Jesus, fell away, convicted of expression in 'enigmatical and pretentious' parables. 'I felt no convulsion of mind, no emptiness of soul, no inward practical change', Newman averred. The transition to a pure theistic faith was as anodyne as it was complete.[54]

Thus *Phases of Faith* could seem 'excellent' to Darwin for several reasons.[55] It told the story of a young man like himself, but unlike Sterling, who had the fortitude to pursue an enquiry to its logical conclusion. It offered a compendium of objections, most of which Darwin had already encountered if not actually entertained, to the entire range of Christian apologetic arguments. It portrayed Unitarianism as a soft option like his grandfather had, a half-way house on the way to a purely theistic religion which, as Newman foresaw, 'shall combine the tenderness, humility, and disinterestedness, that are the glory of Christianity, with that activity of intellect, untiring pursuit of truth, and strict adherence to impartial principle, which the schools of modern science embody'.[56] Finally, the book was a model of spiritual autobiography conceived as the outgrowth of one 'phase' of faith from another, forming a natural progression in which the abandonment of Christianity appears at the end of a plausible, gradualistic narrative. Darwin would reach the same destination in his *Autobiography* using a similar, though more compressed, technique. In real life, however, as in his account, he would not do so without a pang. Emma's agenda was to be recalled once more with particular brutality. On the eighth day after completing *Phases of Faith*, Darwin set off for Malvern with Annie.

'He will be longer getting over it than Emma'

Since his father's death in 1848, Annie, now ten years old, had become Darwin's favourite child. They enjoyed a close and affectionate relationship. Dr Darwin had had a favourite daughter, Susan, who nursed him until the end; afterwards, when Darwin himself became severely ill and feared for his life, he may have thought of Annie as his own nurse. If he survived, she would be his 'solace' in old age. But Annie became generally unwell towards the end of 1850, and it was Darwin, fearful that her vomiting in early March meant that she had inherited his constitution, who undertook to place her in the care of his own physician, Dr Gully.[57] At Malvern Darwin left her with a nurse, her governess and her little sister Henrietta in a big boarding house across the road from where the family had stayed in 1849, high on the slope beneath the Worcester Beacon, dominating the town. He returned to Down via London, where he was reported looking very well, and at home resumed work on his barnacles. A fortnight later, on 15 April, came word that Annie was feverish and vomiting again. Darwin insisted that Emma, now eight months pregnant, should stay behind. He rushed to Malvern, arriving on the 17th, Maundy Thursday.

What transpired in the following week is best revealed in the correspondence between Darwin, Emma and their cousin Fanny Wedgwood, who arrived in Malvern on the Saturday to give comfort and support. It is the most touching sequence in the voluminous family archive.[58] Nothing before or after would so expose the sacred interiors of Darwin's and Emma's relationship. Their situation was unspeakably tragic. Darwin, the trained clinical observer, who had given up medicine after witnessing pre-anaesthetic surgery performed on a child, now compelled to watch the slow torture and death of his most precious girl while recording every change – the vomiting, the diarrhoea, the pathetic struggles against the catheter – every wretched alternation of his own spirits, hour-by-hour, night and day, for the consolation of his anxious wife, who for once could not attend him. Emma, in her ninth pregnancy at the age of forty-three, helpless now except to save the life within her, paralysed at the coming of each post, breathing prayers along with medical advice in daily letters, and writing selflessly at the last, when she knew her husband faced the graveside alone, 'I feel so full of fears about you. They are not reasonable fears but my power of hoping seems gone.... I can't bear to think of you by yourself.'[59]

It was a three-fold struggle for existence, and Emma, who had been supported at home by a godly aunt and her eldest sister Elizabeth, must have sensed it. The unborn infant, the beloved daughter, Darwin himself on the Malvern hillside – all suspended between life and death, earth and heaven, in Easter-week. And there was no resurrection. The old fear that she and her husband would be separated, not merely for a few days, but 'forever', could only have been renewed. His tender words to her after reporting Annie's death on 23 April, 'We must be more and more to each other my dear wife', simply made the thought more poignant. When Darwin returned to Down – he did not attend the burial – it was 'some sort of consolation to weep bitterly together'. Time and again he had broken down as Annie's life ebbed away; afterwards his crying with Fanny, then Emma, seems to have helped prevent any serious relapse of his illness. But 'God knows', he told his brother Erasmus, 'we can neither see on any side a gleam of comfort.'[60] Emma bore the destruction of her hopes 'gently & sweetly', crying 'without violence', and her pregnancy was never at risk. She struggled to 'attain some feeling of submission to the will of Heaven'. For Darwin the loss was much worse. His cousin Elizabeth saw him 'sadly knocked down' on arriving home. 'He will be longer getting over it than Emma', she observed.[61]

In subsequent days Emma jotted down a few precious memories, just as she would do after her husband's death thirty years later. Darwin meanwhile informed his cousin Fox, the clergyman, of their 'bitter & cruel loss', ascribing it to a 'low & dreadful fever'. 'Thank God she suffered hardly at all', he added, '& expired as tranquilly as a little angel.'[62] The tension in these remarks, with their gratitude and reprobation, indicate perhaps something of the conflict Darwin experienced as he sought to reconcile Annie's death with his view of life and such faith as he could muster. Indeed, all through the anguished week's letters to Emma, the most frank and uncontrived religious expressions of gratitude, hope and blessing had jostled uncomfortably against the bleakest utterances of grief and despair. 'Oh my own it is very bitter indeed –', he had written on Good Friday, after a night in which Annie had been expected to die; 'God preserve & cherish you.'[63] But now, exactly a week after she was gone, the day after writing to Fox, Darwin struck a different note. In fifteen hundred words, written solely for Emma and himself 'in after years, if we live', he set down a stunning vindication of the child. Full of vivid and precise details, spontaneous but controlled in its expression of grief and love, free of all taint of bitterness, the memoir was perhaps the most beautiful and certainly the most intensely emotional piece he would ever write.

The memoir can be read at several levels and does not bear minute analysis at any one.[64] But as a waymark in Darwin's religious life, a single feature is paramount. Annie appears as a type-specimen of all the highest and best in human nature. Physically, intellectually and morally she was all but perfect: her movements 'elastic & full of life & vigour'; her 'whole mind ... pure & transparent'; her conduct 'generous, handsome & unsuspicious; ... free from envy & jealousy; good tempered & never passionate'. Darwin recalled that 'she hardly ever required to be found fault with, & was never punished in any way whatever'. 'A single glance of my eye, not of displeasure (for I thank God I hardly ever cast one on her) but of want of sympathy would for some minutes alter her whole countenance.' It was this fine sensitivity that left her 'crying bitterly ... on parting with Emma even for the shortest interval' and that made her exclaim when very young, '"Oh Mamma, what should we do, if you were to die."' The perfection of Annie's character – her sheer 'joyousness' – was, however, completed not only by her sensitivity but by her physically affectionate ways. From infancy she would 'fondle' her parents, much to their delight. And 'she liked being kissed: indeed every expression in her countenance beamed with affection & kindness, & all her habits were influenced by her loving disposition.' It was Annie's face above all that Darwin remembered in the memoir, her

ANNE ELIZABETH
DARWIN
BORN MARCH 2.1841.
DIED APRIL 23.1851.
A DEAR AND GOOD CHILD

Memorial to Darwin's 'insufferable grief': the sturdy stone, its surrounding palisades long since weathered away, still stands prominently beneath a cedar of Lebanon in the Priory Churchyard, Great Malvern.

tears and kisses, her 'sparkling eyes & brindled smiles', her 'dear lips'. Again and again he summoned the innocent features before his mind, comparing them with the daguerrotype taken two years earlier, probably then beside him. 'Oh that she could now know how deeply, how tenderly we do still and shall ever love her dear joyous face', he ended. 'Blessings on her. –'[65]

By this account, then, Annie did not deserve to die; she did not even deserve to be punished – in this world, let alone the next. 'Formed to live a life of happiness', as Darwin observed, she stumbled on ill-health and nature's check fell upon her, crushing her remorselessly. This was 'bitter & cruel' enough without the added prospect of retribution. Whatever her final destiny – Darwin clearly longed that she might somehow survive – it could not be merely contingent. He thanked God, as if defiantly, that he himself had never shown displeasure in her; he expected as much from God in return. Was her conduct, even during the final illness, not 'in simple truth – angelic'? Was her body not interred beneath a gravestone warning that 'a dear and good child' lay there? Annie of all people had no need – as his *Autobiography* would put it – to feel 'remorse from having committed any great sin'.[66] Since Darwin had, I believe, virtually reached this conclusion on behalf of his dead father and for himself in the preceding three years, aided by judicious reading that consolidated his doubts, the application to Annie was now straightforward. If contemplation of Dr Darwin's eternal destiny had spiked Christianity – Emma's Christianity, the only living faith he really knew – Annie's death clinched the matter *a fortiori*. Darwin was forty-two years old. Thereafter he would worry about God and pain and immortality, unencumbered by the Christian plan of redemption. Annie's face, however, would haunt him to the end, with her loving kisses and her tears when leaving Emma.

'Horrid wretches like me'

The circumstances under which Darwin came at last to reject Christianity were full of pain. Intellectual considerations weighed heavily with him, but his decisive objection was moral. Retribution to him was a natural fact, not a supernatural threat. Rewards and penalties for conduct took place in the present life, not another. Christianity posed an ultimatum – Emma had tabled an agenda – and Darwin found its terms of reference inadequate for living in this world. The loss of Dr Darwin in 1848 concentrated his mind on the problem and so helped to break his health, which in turn caused Emma to fear for his life, making the problem worse. 'Some women', Dr Darwin had said, 'suffered miserably by doubting about the salvation of their husbands, thus making them likewise to suffer.' The loss of Annie in 1851 was the point of no return. His feelings about her death became, like a scar, part of his personality.[67] And 'inward convictions and feelings', as the *Autobiography* says (91), were always central to his capacity for religious experience. The perfect child, a vengeful God – Christianity broke its back on the conundrum. Eventually, his former senses dulled by emotional scar-tissue, theism and immortality became unreal, leaving nothing certain but the remembrance of 'insufferable grief'.[68]

Historians have concentrated until now on the impact on Darwin's career of the years between 1837 and 1842, when he formulated the theory of natural selection. If my argument in this essay is accepted, the formative influence of another critical period in Darwin's life will have to be traced as well. During the years of his routine descriptive and classificatory work on barnacles, when the principle of divergence was forming in his mind; midway between the completion of the 'essay' version of his theory and the start of work on his 'big book' – in this relatively quiescent interval of note-taking and fact-gathering, Darwin underwent an upheaval that marked him permanently. He began a doubter; he ended a resolute unbeliever. While living the life of an incumbent in a pleasant rural parish, his non-Christian self-identity became established. Emma alone, who had been affected in quite the opposite way by the death of her beloved sister Fanny, understood the significance of the period between 1848 and 1851. But as the years passed, what 'much pains all one's female relations', as Darwin would put it, was gradually exposed to the world.[69]

Further research will show how far the events surrounding Darwin's loss of faith subsequently influenced his personal life and his science. Certainly, parts of the *Origin of Species* and *The Expression of the Emotions in Man and Animals* will need to be interpreted afresh. And Darwin's relations with Charles Lyell, Alfred Russel Wallace and George Romanes – not to speak of Emma – will acquire a new dimension when seen in the light of the complex problem of love, death and immortality. I may conclude, however, by offering early evidence of Darwin's changed self-identity and of his altered affective response to nature.

Twice in as many months, immediately after starting his full-scale work on species, Darwin placed himself beyond the pale of Christianity. In May 1856 he warned a young entomologist about compromising on the subject of transmutation:

> I have heard Unitarianism called a feather-bed to catch a falling Christian; & I think you are now on just such a feather bed, but I believe you will fall much lower & lower. Do you not feel that "your little exceptions" are getting pretty numerous? It is a funny argument of yours that I (& other horrid wretches like me) may be right, because we are in a very poor minority! anyhow it is a comfort to believe that *some others* will soon be with me.[70]

The imagery is unmistakable. Darwin sees himself among the damned – 'horrid wretches' – consigned to nether regions. The argument he finds 'funny' is nevertheless addressed. It comes from the Gospels: 'Wide is the gate, and broad is the way, that leadeth to destruction, and many there be which go in there at: . . . strait is the gate, and narrow is the way, which leadeth unto life, and few there be that find it.'[71] On this reasoning, the damned, being few, must have found the way of life, and Darwin is comforted to be among them. Then a month later, writing to Hooker, he again evoked his new identity according to the Christian scheme of things. Reacting to the statement of T.H. Huxley, that the 'indecency' of the method of fertilization in ciliograde acalephae shows 'nature becoming very *low* in all senses amongst those creatures', he exclaimed, 'What a book a Devil's Chaplain might write on the clumsy, wasteful, blundering low & horribly cruel works of nature'![27] 'Low' and 'cruel' were words he had used to describe the manner of Annie's death. Darwin had personally been through the kind of hell to which Christians would consign him. Nature now was infernal enough, future punishment damnable. One need not be a 'Devil's Chaplain' to be a confirmed unbeliever.

When Darwin condensed the *Origin of Species* from the manuscript of his larger work, he portrayed nature in a still more revealing way. 'We see the face of nature bright with gladness', he begins in the chapter on the struggle for existence, drafted and revised between 19 March and 10 May 1859;[73] we forget, he adds subsequently, that 'heavy destruction inevitably falls either on the young or old during each generation or at recurrent intervals'. Then, immediately, 'an image of uncontrollably intense and repellent anthropomorphism': 'The face of nature may be compared to a yielding surface, with ten thousand wedges packed close together and driven inwards by incessant blows.' Darwin here has constructed a figure that corresponds emotionally to his 'most sombre sense of the individual within the natural order'.[74] Where did it come from? He ends the chapter in search of a palliative: 'We may console ourselves with the full belief, that the war of nature is not incessant, that no fear is felt, that death is generally prompt, and that the vigorous, the healthy, and the happy survive and multiply.'[75] But the words ring hollow. Why should consolation be sought unless some *one* has been bereaved? And can bereavement be so readily assuaged? Nature is the victim in Darwin's figure, but nature is also given a 'face', a face 'bright with gladness', to which death comes promptly without fear. Only one face in Darwin's experience ever did that. He could recall it 'with *much distinctness*' – 'her eyes sparkled brightly; she often smiled' – and he had the imaginative ability with bygone faces to make them 'do anything I like'.[76] Here, then, nature may be tortured that health and happiness should prevail, but the face is also sacrificed for the redemption of the world.

The bereavement is finally his own; the real victim, tragically, a child already perfect. Annie, who died at Easter, became the paschal lamb of Darwin's post-Christian evolutionary soteriology.

Notes

Many people have contributed generously in different ways to this essay, and I wish to thank especially Fred Burkhardt, Peter Gautrey, David Kohn, Anne Secord, and, above all, Ralph Colp, Sue Davies and Jessica Drader. Grateful acknowledgement is made to the University Library and the Library of Christ's College, Cambridge; to the American Philosophical Society, Philadelphia; to Messrs Josiah Wedgwood and Sons Ltd, Barlston, Stoke-on-Trent, and the Keele University Library, where the Wedgwood-Mosley manuscript collection is deposited; and to all other holders of unpublished materials employed in this essay, for permission to read and quote therefrom.

1. John C. Greene, *Darwin and the Modern World View* (Baton Rouge: Louisiana State University Press, 1961), p. 10.
2. John Hedley Brooke, 'The Relations between Darwin's Science and His Religion', in John Durant, ed., *Darwinism and Divinity: Essays on Evolution and Religious Belief* (Oxford: Basil Blackwell, 1985), p. 61.
3. Nora Barlow, ed., *The Autobiography of Charles Darwin, 1809–1882, with Original Omissions Restored* (London: Collins, 1958). Page numbers given in the text refer to this edition. See Ralph Colp, Jr, 'Notes on Charles Darwin's "Autobiography"', *Journal of the History of Biology*, 18 (1985), 357–401.
4. The more important discussions of Darwin's personal religious views are found in: Benjamin Breckinridge Warfield, 'Charles Darwin's Religious Life: A Sketch in Spiritual Biography' (1888), in *idem, Studies in Theology* (New York: Oxford University Press, 1932), pp. 541–82; John Addington Symonds, 'Darwin's Thoughts about God' (1890), in *idem, Essays Speculative and Suggestive*, 3rd edn (London: Smith, Elder and Co., 1907), pp. 399–402; Robert E.D. Clark, *Darwin: Before and After; The Story of Evolution* (London: Paternoster Press, 1948), pp. 81–90; Arthur Keith, *Darwin Revalued* (London: Watts and Co., 1955), pp. 233–49; Maurice Mandelbaum, 'Darwin's Religious Views', *Journal of the History of Ideas*, 19 (1958), 363–78; Gertrude Himmelfarb, *Darwin and the Darwinian Revolution*, 1959 (New York: W.W. Norton and Co., 1968), pp. 381–7; Michael T. Ghiselin, 'The Individual in the Darwinian Revolution', *New Literary History*, 3 (1972), 113–34; Howard E. Gruber and Paul H. Barrett, *Darwin on Man: A Psychological Study of Scientific Creativity* (London: Wildwood House, 1974); Sandra Herbert, 'The Place of Man in the Development of Darwin's Theory of Transmutation: Part I. To July 1837', *Journal of the History of Biology*, 7 (1974), 217–58; *idem*, 'The Place of Man in the Development of Darwin's Theory of Transmutation: Part II', *Journal of the History of Biology*, 10 (1977), 155–227; Silvan S. Schweber, 'The Origin of the "Origin" Revisited', *Journal of the History of Biology*, 10 (1977), 229–316; *idem*, 'The Genesis of Natural Selection – 1838: Some Further Insights', *Bioscience*, 18 (1978), 321–6;

Edward Manier, *The Young Darwin and His Cultural Circle: A Study of the Influences which helped shape the Language and Logic of the First Drafts of the Theory of Natural Selection* (Dordrecht, Holland: D. Reidel, 1978); Neal C. Gillespie, *Charles Darwin and the Problem of Creation* (Chicago: University of Chicago Press, 1979); Michael Ruse, *The Darwinian Revolution* (Chicago: University of Chicago Press, 1979), pp. 180–4; Phil Diamond, 'The Natural Theologians and Darwin: A Case of Divergent Evolution in the History of Ideas', *Australian Journal of Politics and History*, 26 (1980), 204–11; Peter Brent, *Charles Darwin: 'A Man of Enlarged Curiosity'* (London: Heinemann, 1981), pp. 255–6, 313–15, 450–7; Dov Ospovat, *The Development of Darwin's Theory: Natural History, Natural Theology, and Natural Selection, 1838–1859* (Cambridge: Cambridge University Press, 1981); Brooke, 'Relations between Darwin's Science and His Religion'; and Frank Burch Brown, *The Evolution of Darwin's Religious Views* (Macon, Ga.: Mercer University Press, 1986).

No one until Himmelfarb took account of the suppressed portions of the *Autobiography*. The analyses of Warfield and Mandelbaum are, however, exemplary. Symonds, Himmelfarb and Ospovat make no effort to distinguish Darwin's Christianity from his general religious convictions, while Keith, Ghiselin, Gruber, Schweber, Manier and Diamond only deal with it implicitly in the context of Darwin's growing agnosticism, which they understand to have reached a mature or atheistic form any time from within a few years after 1831 (Ghiselin) to as late as 1842 (Manier). Only Jonathan Miller and Borin Van Loon in *Darwin for Beginners* (London: Writers and Readers, 1982) place Darwin's loss of faith in 'orthodox Christianity' as early as 'the time he set foot on the Beagle' (p. 117). Among those who discuss Darwin's Christianity *per se*, Ruse places the loss of faith shortly after 1836 – cf. his *Darwinism Defended: A Guide to the Evolution Controversies* (Reading, Mass.: Addison-Wesley, 1982), p. 29 – Brooke conjectures that it soon followed Darwin's engagement in 1838, Brent moves the date as late as 1840, and Brown strongly suggests that Darwin's Christianity had been pared down to a 'minimal' theism by 1842. Herbert may incline towards this view, although her opinion is more guarded. Only Gillespie suggests a period as late as 1850 for the 'earliest possible' abandonment of Christianity. 'A firm date', he states, 'is impossible to establish' (pp. 136–8).

5. For views of Aveling, see Lewis S. Feuer, 'Marxian Tragedians: A Death in the Family', *Encounter*, no. 110 (Nov. 1962), 23–32 and the less slanted account in Edward Royle, *Radicals, Secularists, and Republicans: Popular Freethought in Britain, 1866–1915* (Manchester: Manchester University Press, 1980), pp. 105–6.

6. Edward B. Aveling, 'A Visit to Charles Darwin', *National Reformer*, new ser., 40 (22 Oct. 1882), 273–4; (29 Oct. 1882), 293–4: reprinted verbatim by Aveling as *The Religious Views of Charles Darwin* (London: Freethought Publishing Co., 1883), where Darwin's admission appears on p. 5. For Büchner's recollection, which does not feature Darwin's admission, see 'The Unknowable' in his *Last Words on Materialism and Kindred Subjects*, translated by Joseph McCabe (London: Watts and Co., 1901), p. 147. For Francis Darwin's assessment, see his *The Life and Letters of Charles Darwin, including an Autobiographical*

Chapter, 3 vols. (London: John Murray, 1887), I, 317n (hereafter *LL*) and cf.
F. Darwin to W.E. Darwin, [1884–5], in the Darwin Archive, Cambridge
University Library (DAR), 210.8.

7. Brown, *Evolution*, p. 21. Among those who refer to Aveling's account
(see n. 4 above), only Himmelfarb, p. 319 omits Darwin's age. Others mention it in passing or otherwise attach little significance to it: Warfield, pp.
549–52; Clark, pp. 83–4; Mandelbaum, p. 366n; Gillespie, p. 138; and Brent,
p. 455. See also G.W. Foote, *Darwin on God* (London: Progressive Publishing
Co., 1889), pp. 27–9 and Geoffrey West, *Charles Darwin: A Portrait* (New
Haven, Conn.: Yale University Press, 1938), p. 310 for further notable but
uninteresting references.

8. Gavin De Beer, ed., 'Darwin's Journal', *Bulletin of the British Museum
(Natural History)*, Historical Series, 2 (1959), 8, 10, 20 (hereafter 'Journal').

9. Cf. Ralph Colp, Jr, '"I Never Wrote so Much About Myself": Charles
Darwin's 1861–1870 Autobiographical Notes', in E. Geissler and W. Scheler,
eds., *Darwin Today: The 8th Kühlungsborn Colloquium on Philosophical and Ethical
Problems of Biosciences* ..., Abhandlungen der Akademie der Wissenschaften
der DDR, 1983, no. 1 (Berlin: Akademie-Verlag, 1983), pp. 37–51, where the
editor is identified as Wilhelm Preyer, with whom Darwin corresponded in
1869–70, and *idem*, 'Notes', pp. 359–61, which quotes a letter to Darwin,
dated 20 Sept. 1875, from Ernst von Hesse Wartegg, requesting autobiographical information for a newspaper article he was to write. See Frederick
Burkhardt and Sydney Smith, eds., *A Calendar of the Correspondence of Charles
Darwin, 1821–1882* (New York: Garland Publishing, 1985), 10162 (hereafter
Calendar).

10. Darwin to J.D. Hooker, 6 June [1868], DAR 94:69–70. See Ralph C.
Colp, Jr, *To Be an Invalid: The Illness of Charles Darwin* (Chicago: University of
Chicago Press, 1977), pp. 27, 68–9, 167–9.

11. Colp, *To Be an Invalid*, p. 38.

12. *LL*, I, 69, 307; *Calendar*, 11971, 11981. The controversy may be followed
by comparing the correspondence and memoranda in DAR 210.8 with the
manuscript of the *Autobiography* in DAR 26:74 and the edited fair copy in
Francis Darwin's hand, held in DAR 149.

13. Margaret Keynes, *Leonard Darwin, 1850–1943* (Cambridge: Cambridge
University Press, 1943), p. 42. For the context in which the Darwin women
became religious specialists, see Evelleen Richards, 'Darwin and the Descent
of Woman', in David Oldroyd and Ian Langham, eds., *The Wider Domain of
Evolutionary Thought* (Dordrecht, Holland: D. Reidel, 1983), pp. 81–2, 85.

14. F. Darwin to W.E. Darwin, [1885], DAR 210.8.

15. *LL*, I, 69, 307; cf. p. 26, where it is acknowledged that other omissions
have been made without notice. Five years later, when Emma was 84 years
old, Francis published an abbreviated, one-volume edition of the *Life and
Letters*, entitled *Charles Darwin* ... (London: John Murray, 1892). There the
section on religious belief is removed back to a chapter immediately following
the *Autobiography*.

16. Darwin to N.A. von Mengden, 5 June 1879, DAR 139.12. Only a copy
of Darwin's reply survives. It is not known whether the original was written

by him or by an amanuensis.

17. 'A Hitherto Unpublished Letter of Charles Darwin', *National Reformer*, new ser., 40 (1 Oct. 1882), 235.

18. See Geoffrey Rowell, *Hell and the Victorians: A Study of Nineteenth-Century Theological Controversies concerning Eternal Punishment and the Future Life* (Oxford: Clarendon Press, 1974), ch.7.

19. For Darwin's experiments with monkeys and snakes in the Zoological Gardens, London, see Charles Darwin, *The Descent of Man and Selection in Relation to Sex*, 1871, 2nd edn (London; John Murray, 1874), pp. 71–2 and *idem*, *The Expression of the Emotions in Man and Animals* (London: John Murray, 1872), p. 144.

20. Darwin to [Reginald Darwin], 4 April 1879, quoted in Ralph C. Colp, Jr, 'The Relationship of Charles Darwin to the Ideas of His Grandfather, Dr. Erasmus Darwin', *Biography*, 9 (1986), 11 (also *LL*, III, 219).

21. I quote from galleys 47–8 of Darwin's 'preliminary notice' to Ernst Krause, *Erasmus Darwin* . . . , translated by W.S. Dallas (London: John Murray, 1879), contained in DAR 210.20. The deathbed episode was excised by Henrietta (see below).

22. Darwin to J. Fordyce, 9 May 1879, in Gavin De Beer, 'Further Unpublished Letters of Charles Darwin', *Annals of Science*, 14 (1958), 88.

23. Colp, 'Relationship', p. 15; F. Darwin to W.E. Darwin, [1885], DAR 210.8. The full remark was, however, insinuated on the last page of the *Life and Letters* and attributed to a portion of the *Autobiography* added to the manuscript in 1879.

24. The biblical text to which the old lady alluded, Job 19:25–6, would have been familiar to Darwin and Emma from the libretto of Handel's *Messiah*. Darwin may also have encountered it, italicized, when he read 'with care' John Pearson's *Exposition of the Creed*, Article XI, 'The Resurrection of the Body', before going up to Cambridge (*Autobiography*, p. 57).

25. Evidence that the *Autobiography* may have been written in more than a single draft is described by P.J. Gautrey, 'A Previously Unidentified Darwin Manuscript in Cambridge University Library', *Cambridge University Libraries Information Bulletin*, 4 (Oct.–Nov. 1980), 2–3.

26. The tribute first appeared in Francis Darwin and A.C. Seward, eds., *More Letters of Charles Darwin: A Record of His Work in a Series of Hitherto Unpublished Letters*, 2 vols. (London: John Murray, 1903), I, 30 (hereafter *ML*).

27. Colp, 'Notes', p. 382n.

28. Emma Darwin to Charles Darwin, [*c.* Feb. 1839], in Frederick Burkhardt and Sydney Smith, eds., *The Correspondence of Charles Darwin* (Cambridge: Cambridge University Press, 1985), II, 171–2 (hereafter *Correspondence*). The letter first appeared in the privately printed 1904 edition of *Emma Darwin* . . . (pp. 187–9) by Henrietta [Darwin] Litchfield; it was later published in *idem*, *Emma Darwin: A Century of Family Letters, 1792–1896*, 2 vols. (London: John Murray, 1915), II, 173–4 (hereafter *ED*). Barlow restored an important annotation in her transcription that accompanied the *Autobiography*.

29. 'Journal', p. 13. Annie's death date had appeared correctly on her gravestone and, at Darwin's request, in her obituary notice in *The Times*, 28

April 1851. See Darwin to E.A. Darwin, [25 April 1851], in the Wedgwood-Mosley Collection, Keele University Library, Keele, Staffordshire (W/M), 310.

30. Darwin to E. Krause, 19 March 1879, quoted in Colp, 'Relationship', p. 10.

31. This paragraph and the following one contain a summary of the author's unpublished research.

32. *ED*, I, 250–1. On Unitarianism, see Rowell, *Hell and the Victorians*, p. 38.

33. *ED*, II, 137; cf. p. 286. Emma's 'book of extracts' remains in the family; it is quoted by the nephew of a favourite niece of Emma's, Hugh Edmund Langton Montgomery, in 'Emma Darwin', *The Month*, 29 (1963), 288–94, where he also offers a different view of her religious heritage.

34. See Ralph Colp, Jr, 'The Evolution of Charles Darwin's Thoughts about Death', *Journal of Thanatology*, 3 (1975), 191–206 and Robert T. Keegan and Howard E. Gruber, 'Love, Death, and Continuity in Darwin's Thinking', *Journal of the History of the Behavioral Sciences*, 19 (1983), 15–30.

35. Nora Barlow, ed., *Charles Darwin's Diary of the Voyage of H.M.S. 'Beagle'* (Cambridge: Cambridge University Press, 1933), p. 427 (24. Sept.–2 Oct. 1836). The same passage appears substantially unaltered in Darwin's *Journal of Researches*, which went to press in August 1837. Darwin recalls it in the religious part of the *Autobiography*, p. 91.

36. The saying was first published in Darwin's 'preliminary notice' to Krause, *Erasmus Darwin*, p. 45, although it had been current in the family for many years. I quote from the galleys of the notice, corrected in Darwin's hand, where the word 'only' is inserted (DAR 210.20). See also Darwin to Hooker, 11 May [1859], *LL*, II, 158.

37. See Ralph Colp, Jr, '"Confessing a Murder": Darwin's First Revelations about Transmutation', *Isis*, 77 (1986), 9–32.

38. Darwin to Emma Darwin, [June 1846], *ED*, II, 104; Darwin to Hooker, 10 May 1848, *ML*, I, 63–5. Cf. Colp, '"Confessing a Murder"', p. 31.

39. Darwin to Emma Darwin, [25 May, 27 May, 17 Nov. 1848], DAR 210.19; Darwin to [?], [13 or 20 Oct. 1848], *Calendar*, 1204. Cf. Colp, *To Be an Invalid*, pp. 38–9.

40. Darwin to W.D. Fox, 6 Feb. [1849], Darwin-Fox Correspondence, Christ's College Library, Cambridge (CCLC), 71; 'Journal', p. 12 (1 Jan.–10 Mar. 1849); Colp, *To be an Invalid*, pp. 43–53.

41. On 'vile molluscous animals', see Gavin De Beer, ed., 'Darwin's Notebooks on Transmutation of Species Part III: Third Notebook, July 15th 1838–October 2nd 1838', *Bulletin of the British Museum (Natural History)*, Historical Series, 2 (1960), 132 (*D* 37) for the original remark and Ralph Colp, Jr, 'Charles Darwin's Reprobation of Nature: "Clumsy, wasteful, blundering low & horribly cruel"', *New York State Journal of Medicine*, 81 (1981), 1116–19. On Darwin's health, see Darwin to Hooker, 28 Mar. 1849, *LL*, I, 373; Darwin to J.S. Henslow, 6 May 1849, *ML*, I, 66; and Darwin to J. Herschel, 13 June [1849], in Gavin De Beer, 'Some Unpublished Letters of Charles Darwin', *Notes and Records of the Royal Society of London*, 14 (1959), 34. For Darwin's reading, see

Peter J. Vorzimmer, 'The Darwin Reading Notebooks (1838–1860)', *Journal of the History of Biology*, 10 (1977), 107–53 (hereafter 'Reading Notebooks').

42. Andrews Norton, *The Evidences of the Genuineness of the Gospels*, 2nd edn, 2 vols. (London: John Chapman, 1847), I, 11; 'Reading Notebooks', p. 140.

43. 'Journal', p. 12 ('from July to end of year'). Julius Charles Hare edited the *Essays and Tales by John Sterling*, 2 vols. (London: John W. Parker, 1848) to forestall his fellow executor, Thomas Carlyle, who took a different view of Sterling's career. There is no evidence that Darwin, who had followed Carlyle closely, ever read his *Life of Sterling* (1851).

44. Hare, *Essays and Tales*, I, lxxxii, cvi, cl, ccxiv.

45. 'Reading Notebooks', p. 140; Hare, *Essays and Tales*, I, cxxv, cxxx; cf. ccxxv.

46. 'Reading Notebooks', p. 142. See Rowell, *Hell and the Victorians*, pp. 57–60.

47. Francis William Newman, *The Soul, Her Sorrows and Her Aspirations: An Essay towards the Natural History of the Soul, as the True Basis of Theology*, 1848, 2nd edn (London: John Chapman, 1849), pp. 227, 258.

48. Ibid., p. 230.

49. Ibid., pp. 233, 249.

50. 'Reading Notebooks', p. 142. For Emma's reading of *The Soul*, see *ED*, II, 125.

51. J. Estlin Carpenter, *The Bible in the Nineteenth Century* (London: Longmans, Green, and Co., 1903), p. 126; John Rogerson, *Old Testament Criticism in the Nineteenth Century* (London: SPCK, 1984), pp. 192–3.

52. [Francis William Newman], *A History of the Hebrew Monarchy from the Administration of Samuel to the Babylonish Captivity* (London: John Chapman, 1847), pp. iv, 67, 370. Although published annoymously, the book had been attributed to Newman in the publisher's list by May 1850.

53. Colp, *To Be an Invalid*, pp. 51–2; 'Reading Notebooks', p. 143.

54. Francis William Newman, *Phases of Faith; or, Passages from the History of My Creed*, 1850, 6th edn (London: Trübner and Co., 1865), pp. 47, 49, 50, 62, 115, 116, 125, 133, 168.

55. For Emma's probable assessment, see her aunt's reaction in *ED*, II, 125–6.

56. Newman, *Phases of Faith*, p. 175.

57. *ED*, II, 184; Darwin to Fox, 29 April 1851, CCLC 79; Darwin's memoir of Annie, DAR 210.13; Darwin to Fox, [27 Mar. 1851], CCLC 78a.

58. The full correspondence, including numerous condolences from family and friends, is in DAR 148, 210.13 and 210.19; W/M 310 and 939.1–5; and CCLC 79. Ralph Colp, Jr, has recounted the week's events and placed them in the wider context of Darwin's emotional life in 'Charles Darwin's "insufferable grief"', *Free Associations*, 9 (1987), 7–44. I am grateful to Dr Colp for advice and encouragement over the years in which we have worked on this episode, both individually and together.

59. Emma Darwin to Charles Darwin, [24 April 1851], DAR 210.13.

60. Darwin to Emma Darwin, [23 April 1851], DAR 210.13; Darwin to E.A. Darwin, [25 April 1851], W/M 310. The only interruption to Darwin's health

diary (see the text above at n.40) took place during this period.

61. S.E. Wedgwood to Darwin, [24 April 1851], DAR 210.13, Emma Darwin to F.E. Wedgwood, [24 April 1851], and S.E. Wedgwood to F.E. Wedgwood, [27 April 1851], both in W/M 310.

62. Darwin to Fox, 29 April 1851, CCLC 79. Emma's recollections are in DAR 210.13.

63. Charles Darwin to Emma Darwin, [18 April 1851], DAR 210.13.

64. Except to discover how more than half of it was suppressed when published after Emma's death: cf. the original in DAR 210.13 with *ED*, II, 147–8, which is identical to the version published in the 1904 edition of *Emma Darwin* (see n.28 above).

65. The daguerrotype, discovered in 1987 with other family artefacts in a vault at the Royal College of Surgeons, has been published in volume four of the Darwin *Correspondence*.

66. In the note reporting Annie's death to Emma, written on the same afternoon ([23 April 1851], DAR 210.13), Darwin felt moved to remark, 'I cannot remember ever seeing the dear child naughty'.

67. Colp, 'Charles Darwin's "insufferable grief"', p. 32. Annie's death was the real sting for Darwin, not his father's. See Darwin to Hooker, [28 Sept. 1865], DAR 115:275 and Darwin to Hooker, 30 Aug. [1881], DAR 95:530–1.

68. Darwin to Hooker, [28 Sept. 1865], DAR 115:275.

69. Darwin to W.B. Carpenter, 3 Dec. [1859], *LL*, II, 239. See James R. Moore, 'Darwin of Down: The Evolutionist as Squarson-Naturalist', in David Kohn, ed., *The Darwinian Heritage* (Princeton, N.J.: Princeton University Press, 1985), pp. 435–81.

70. Darwin to T.V. Wollaston, 6 June [1856], Edinburgh University Library, Gen. 1999/1/30. Wollaston's communication, which does not survive, came as a note attached to a letter from Charles Lyell.

71. Matthew 7:13–14 (KJV).

72. Darwin to Hooker, 13 July [1856], DAR 114:169. For a supplementary analysis, see Colp, 'Charles Darwin's Reprobation of Nature'.

73. 'Journal', p. 15; Darwin to J. Murray, 5 April and 10 May [1859], *Calendar*, 2447, 2460.

74. Charles Darwin, *On the Origin of Species by means of Natural Selection, or the Preservation of Favoured Races in the Struggle for Life* (London: John Murray, 1859), pp. 62, 66. The interpretation is Gillian Beer's in ' "The Face of Nature": Anthropomorphic Elements in the Language of the "Origin of Species"', in L.J. Jordanova, ed., *Languages of Nature: Critical Essays on Science and Literature* (London: Free Association Books, 1986), p. 237. On the wedging imagery, see the provocative article by Ralph Colp, Jr, 'Charles Darwin's Vision of Organic Nature: "A force like a hundred thousand wedges"', *New York State Journal of Medicine*, 79 (1979), 1622–9, 2136.

75. Darwin, *Origin of Species*, p. 79.

76. Darwin's memoir of Annie, DAR 210.13; Darwin's replies to Francis Galton's 'Questions on the Faculty of Visualizing', 1879, in *LL*, III, 238, 239.

7

Encounters with Adam, or at least the hyaenas: Nineteenth-century visual representations of the deep past

MARTIN RUDWICK

Adam is indeed dead in the flesh – at least until the final resurrection – but in visual representations he is alive and well and on view in every major museum of natural history.[1] Of course he is not named Adam; but *ha'adam*, The Man, the origin and fount of us all, is presented to our view under the form of *Homo habilis* or *H. erectus*, standing upright (more or less), knapping a flint or carving a carcase, or gazing out from a cave mouth to survey an artist's impression of the East African landscape of the early Pleistocene.

Often this transmuted Garden of Eden is but the last in a sequence of dioramas that depict various phases in the history of life, far back beyond The Man. For example, *Tyrannosaurus rex*, that perennial children's favourite, lords it over the lesser dinosaurs of the Wild West of the Cretaceous. Or our view dips underwater, like visitors to a dolphinarium, into the Jurassic seas of Dorset or Württemburg, to watch a school of ichthyosaurs darting after their squid-like prey. Far further back in time, a steamy swamp scene shows us the Everglades of the Carboniferous, with gigantic tree ferns in place of mangroves. And still further back we peer, like scuba divers in the Caribbean, at the coral reefs of Silurian Shropshire or Gotland, and see the trilobites and nautiloids going about their unspectacular business among the coral clumps. Reversing our stroll through the museum gallery, and following these scenes in their real geohistorical order, we are transported on a Wellsian time machine, or perhaps more appropriately like Dr Who, through an updated Genesis One. It is a creation narrative as powerful – and as mythopoeic – as its original:

> And Evolution said, "Let the waters bring forth swarms of living creatures." So Evolution created the great sea monsters and every living creature that moves. And Evolution said, "Let the

earth bring forth creeping things and beasts of the earth." And it
was so. Then Evolution said, "Let us make humankind in our
image." So Evolution created humankind in her own image.[2]
Evolution has merely replaced *'elohim*. In the implicit message of
the dioramas the plural actions of natural selection melt into an agency
as overarching, and imaginatively almost as personal, as the regally
plural God of the Genesis narrative. Such sequences of reconstructed
scenes do not logically entail a theory of evolution; but, to borrow
a term from the physical sciences, they certainly help to give that
theory *Anschaulichkeit*, conceivability or picturability. They make
the prehuman and barely-human past, its initial otherness and its
gradual approach towards familiarity, conceivable and imaginable.
They thereby create the imaginative groundwork for an evolutionary
interpretation of the scenes to become, if not logically compelling,
at least as plausible and persuasive as any scientific theory can be.

These sequences of dioramas in our modern museums, with their
trompe d'oeil backdrops merging forwards into fully three-dimensional
reconstructions of extinct animals and plants, have themselves descended from ancestors that were equally ambitious in conception
although more modest in their means. Those ancestors are to be
found not in museums but in books, as sequences of two-dimensional
pictures of the prehistoric past.[3] The changing style and iconography
of these visual representations reveals more than their designers
intended, precisely because their form was (as it still is) so greatly
underdetermined by the available evidence. It is customary to refer
disparagingly to what is 'an artist's impression' of the scene. But it is
the scientist who must decide how to extrapolate beyond the fragmentary evidence, to create a scene that will carry conviction; the
artist merely translates that vision on to paper or into plaster and
plastic, in an operation that is no less creative for being under the
constraints of the scientist's instructions. The genre of the reconstructed scene demands explicitness; it forces the scientist to reveal
judgements of likelihood, or mere hunches and prejudices, that can
discreetly be left implicit in more formal media of publication. Above
all, therefore, reconstructed scenes give us a unique insight into how
successive generations of scientists (and their collaborating artists)
have visualized the long aeons of 'deep time' that lie beyond human
history or even the origins of our humanity. In this brief essay I want
to explore the historical origins of this visual genre, which has become
such a powerful means of rendering the remote past imaginable, and
its evolutionary explanation persuasive.[4] I start at a point at which
the genre can be seen to have been already well established, and then
trace it backwards in search of its origins.

The World before the Deluge

One of the earliest full sequences of scenes from the time-machine, and certainly one of the most influential, was due to the prolific scientific popularizer Guillaume Louis Figuier (1819–94). Figuier was the author of profusely illustrated books on an amazing variety of topics: among others, *Great Discoveries of Modern Science* (1851), *Marvels of Modern Times* (1860), *The Earth and the Seas* (1864), *The Vegetable World* (1865), *Fish, Reptiles and Birds* (1868), *The Human Race* (1872), and so on and on to *Metropolitan Railways* (1886), not to mention *The Day after Death, or Our Future Life According to Science* (1871). Most of these books went through several editions in French; they were also translated into English, German and other languages, and thereby reached a literate audience that was almost global.

The World before the Deluge (1863), which contains Figuier's time-machine series, was no exception. Already in its sixth edition by 1872, it had by then appeared in English (published in both London and New York), German, Spanish (published in Mexico), Danish and probably other languages too. With such a spread of editions, it was immensely influential in conveying to a wide reading public – not least the young people to whom it was explicitly directed – a sense of the spectacular implications of the burgeoning sciences of geology and palaeontology.[5] Figuier borrowed most of his illustrations – some 300 engravings of fossils – from a respectable academic source, namely the two-volume *Elementary Course on Palaeontology and Stratigraphical Geology* (1849–52) by Alcide d'Orbigny (1802–57), who until his death had been professor of palaeontology at the Natural History Museum in Paris.[6] But Figuier added new illustrations of his own, to bring d'Orbigny's fossils to life: a magnificent sequence of twenty-five (in later editions, thirty) full-page 'ideal views' of the past history of the earth and its living inhabitants. These were drawn under his direction by the young painter and illustrator Edouard Riou (1833–1900), who, at the very same period, was also taking his imaginative time-machine and his artistic style in the other direction, in his illustrations for the science fiction of Jules Verne.

Figuier's great temporal panorama began even before the beginnings of life, with a view of the original condensation of water on to the surface of the hot primitive earth, a scene of torrential rain, lightning and boiling seas. It continued with a view of the Sun breaking through clouds over a Silurian seascape, with trilobites and nautiloid shells thrown up on the shore. After a similar Devonian scene came a view of Carboniferous marine life that at first glance could have been of a crowded Victorian aquarium, together with a view of a steamy

Carboniferous coal swamp in a style that has remained canonical for such reconstructions to the present day. And so the panorama continued, still in a style that is unmistakably continuous with that of our modern museums, through Triassic and Cretaceous, Eocene and Pliocene. The last 'ideal view' in that style was a chilly scene of Europe during the Quaternary Ice Age, with mammoths and cave-bears, woolly rhinoceros and the so-called Irish elk.

After that scene, however, came one entitled 'The Deluge of the North of Europe'. This, and the title of the book itself, are a reminder that the 'Deluge', although long abandoned by geologists as a global event or major agent of change, remained vivid in the imagination of the reading public even in the later nineteenth century. But it was Figuier's business as a popularizer to keep abreast of the latest science. So what he depicted in visual form (and described in his text) was an interpretation of the Deluge that was quite widely current among scientists in the 1860s and even the 1870s. A mass of swirling waters was shown submerging a sub-Arctic landscape of conifers, with huge icebergs being swept along, dropping their erratic boulders – unseen – as the ice melted, scratching the bedrock with their embedded stones and churning up the surface of the ground. A localized 'Deluge of the North of Europe' could thus explain the problematical Diluvium or Drift deposits of that region; and it did so with impeccable naturalism, since such a Deluge seemed a plausible physical consequence of the Ice Age, if it had indeed ended abruptly.

That Deluge, however, was clearly *not* the one that was recalled in Genesis and other ancient literatures. On this point Figuier adopted a standard compromise of the period. He affirmed the historicity of the Deluge of Genesis, but restricted its effects to Mesopotamia; he attributed it naturalistically to a sudden elevation of the land to the north, accompanied by a catastrophic volcanic eruption that had produced Mount Ararat. Riou illustrated this 'Asiatic Deluge' with an apocalyptic scene in the style of John Martin's celebrated painting of *The Deluge*, which had recently been exhibited in Paris: he showed an imaginary prehistoric city (and a couple of elephants) being overwhelmed by swirling waters and torrential rain.

Between these two distinct and contrasted Deluges, the origins of humankind had somehow to be accommodated. In his later editions, Figuier claimed that the discovery of a human jaw in the gravels near Abbeville in 1863 had at last precipitated a consensus among the scientists, that the human beings who had crafted the well-known prehistoric stone axes must have lived among the equally well-known extinct mammals of the Pleistocene.[7] So he got Riou to draw a new 'ideal view' that would embody that latest discovery.

'The Appearance of Man': an engraving by Edouard Riou from a later edition of Louis Figuier's *World before the Deluge* (1866), in which the co-existence of early human beings and the mammals of the Pleistocene was first depicted.

'The Appearance of Man' depicted a tribal group dressed in the skins of animals. The women were grouped domestically in a cave mouth (and were as topless as on a modern French beach); the men were equipped with stout flint-headed axes, and stood ready to ward off a hostile nature and hunt it down for food and clothing. For with the incongruous exception of a fine thoroughbred horse and a solitary deer, nature was not only wild but hostile: bears, hyaenas, rhinoceros and mammoths threatened the very survival of humanity. A deep ditch separated the tribal group from this hostile environment; more symbolically, it was a chasm that divided the human world from the natural. For Figuier's human beings, although primitive in time, and simple in tools, clothing and shelter, were no primitives in any other sense: they were unmistakably white and European, and wholly modern in physical appearance.[8]

When Figuier's book was first published, however, this co-existence of early human beings with the mammals of the Pleistocene was still highly controversial. In the early editions, his interpretation of the origin of humankind was even more traditional, as its visual representation clearly showed. The original version of the engraving just described was strikingly different in character: it depicted not a tribal group outside a cave, but a primal human family in a verdant temperate landscape.[9] Man, woman and child were surrounded by the sheep, goats, cattle and horses that they would soon domesticate, with only a distant deer to suggest the wilder nature that might need hunting rather than herding. To match this pastoral economy, the man was equipped not with a flint axe, but with no more than a staff. The scene could almost have come from the hand of a Poussin or a Claude; in style and tone, this Garden of Eden was in striking contrast to the long sequence of strange scenes from the deeper past, which so vividly illuminated the body of the book. The Deluge of Figuier's title stood here as a symbolic barrier between the immense panorama of deep time and its brief human epilogue, between 'the ancient world' of his subtitle and the civilized human world described in so many of his other books. The Arcadian scene of his original 'Appearance of Man' portrayed the not-so-distant origins of that world of humanity; the whole of the preceding history of the earth, despite its manifest diversity as portrayed in all the other 'ideal views', could in this perspective be lumped together as *the* 'ancient world'.

This contrast between Figuier's earlier and later editions neatly captures the historical moment at which the beginning of humanity came to be depicted not only as the final episode of a long sequence that stretched back beyond the beginning of life, but also as an episode not wholly different in character from the deeper past. Figuier may not have been persuaded by the evolutionary theories that a certain Englishman had just brought back to the forefront of public debate; but whether he knew it or not, and whether he willed it or not, his great series of 'landscapes of the ancient world' must surely have helped to make such theories literally *anschaulich*, picturable, and therefore potentially plausible to his vast and almost worldwide audience.

Landscape representations

Needless to say, however, Figuier was not the author of the genre. Fortunately for our phylogenetic task, he himself named the direct ancestor of his visual representations: it was *The Primitive World in Its Different Periods of Formation* (1847), by the Austrian palaeobotanist Franz Unger (1800–70).[10] By the time Figuier borrowed from his work, Unger had become professor of botany in Vienna, but when he first published his series of 'landscape representations' he held a similar but provincial post in Graz. Just as Figuier's 'ideal views' were published in a popular book, so likewise Unger's work had its origin in a lecture course to a group of amateurs in Graz. As a professional scientist, Unger claimed he had had serious scruples about 'crossing from strictly scientific research into fantasy'; but the work of his illustrator, the topographical artist Josef Kuwasseg (1799–1859) of Graz, convinced him that the exercise need be no more speculative than any other portrayal of the unobservable past. The landscapes that formed the *raison d'être* of the book thus embodied – and, for the first time, explicitly – the partnership between artist and scientist that has remained characteristic of the genre to the present day. Unger ensured that their work would become widely known in the scientific world, by publishing it with the text in both German and French; and the list of subscribers, who had enough confidence in Unger's abilities to risk their money beforehand, included naturalists from most of the European countries and even one (Asa Gray) from North America.

Unger, more cautiously than Figuier, began his series of fourteen scenes not with the pre-biotic world – for which there was still no unquestioned evidence in the stratigraphical record – but with a scene from the Transition period, using the older name for what had only recently been differentiated into Silurian and Devonian.[11] The series continued in a sequence that Figuier was to borrow with little substantive change, through the Carboniferous and the New Red Sandstone periods, the Oolite and the Chalk, and so into the Tertiary with scenes of Eocene and Miocene life. But Kuwasseg's lithographs were far superior to Riou's engravings, not only artistically but also in how they portrayed the fauna and flora of each period. Working under Unger's close supervision, Kuwasseg produced scenes that Unger could proudly claim were more authentic than any earlier 'landscape representations' had been. Rather than crowding into one scene a set of reconstructions of all the characteristic fossils found in a given group of strata, Unger got his artist to portray an ecologically plausible view of the *community* of organisms that had lived at each period in the past. Being a botanist, it is not surprising that his scenes gave the plant life more prominence than the animals; but the animals were shown in their likely ecological relations, as predators and prey, carnivores and herbivores, and so on.

Unger rounded off his series with two scenes that brought the immensely long history of life into the human epoch. Unlike Figuier's series, there was no portrayal of any Deluge, even a prehuman and periglacial one. The penultimate scene was of the glacial period on the edge of the Alps – it might have been a scene near Graz itself. In the foreground were cave bears, one with its paw on a pile of huge mammoth bones; in the middle distance, a herd of bison stood peacefully at the edge of a lake not far from the snout of a glacier. That scene, like Riou's later adaptation of it, was of a world still without human beings. By contrast, Unger's final scene depicted the culmi-

The Ice Age in Europe: a lithograph by Joseph Kuwasseg, forming the penultimate 'landscape representation' in Franz Unger's *The Primitive World in Its Different Periods of Formation* (1847). (Reproduced by permission of the Syndics of Cambridge University Library.)

The origin of humankind, as depicted by Kuwasseg for the final scene in Unger's *Primitive World*. (Reproduced by permission of the Syndics of Cambridge University Library.)

nation of the history of life in a style similar to Riou's *original* drawing
for Figuier. As the sun rose on the dawn of humankind, a family
group in primal nakedness – the man with no more than a staff in
his hand – surveyed a subtropical Eden in which only some playful
horses were visible to represent the untamed animal world, and the
most prominent natural object was a large palm tree – as it were, a
secular Tree of Life.

Unger claimed that his 'landscape representations' were superior
to their forerunners; in fact, such forerunners are hard to find. As
we push our search for the origins of the genre back into the earlier
nineteenth century, the evidence becomes fragmentary and problematic. There was certainly a tradition of such visual representations
before Unger's were published, but his book may well represent
the first time such scenes were published as a *sequence* stretching all
the way from the beginnings of life to the beginnings of humanity.
The significance of that step, for creating the imaginative infrastruc-
ture for a developmental – and later, an evolutionary – view of the
living world, needs no emphasis.[12]

Still, even the reconstruction of a scene from a *single* period in the
prehuman past represents an imaginative achievement, the magnitude of which is only masked from us by its familiarity. One of
Unger's likely models decorated the title-page of a short monograph
(1836) on the then recently discovered 'colossal skull' of the *Dinotherium*, an extinct elephant from the Tertiary of Hesse. The authors
were two Hessian geologists: August Wilhelm von Klipstein (1801–
94), who had recently been appointed to the chair of mineralogy at
Giessen (thereby becoming a colleague of the more famous professor
of chemistry, Justus von Liebig), and Johann Jakob Kaup (1803–73), a
curator at the grand-ducal museum at Darmstadt. Their monograph
consisted of a short descriptive text, with seven plates of maps, geological sections and drawings of the fossils themselves. Outside this
conventional series of illustrations, however, were two others: a
sketch of the workmen, supervised by a geologist, excavating the
gigantic fossils; and a single reconstructed landscape, showing the
living *Dinotherium* in its inferred original conditions of life, browsing
peacefully by the edge of a river, unconcerned at a lion's attack on a
herd of horses in the background or by the volcano in full eruption
in the far distance. This pair of vignettes neatly symbolized the act
of reconstruction: the 'before' of careful excavation and the 'after' of
imaginative reconstruction; or, conversely, the 'before' of the reality
of the deep past, and the 'after' of its fragmentary survival into the
present. That Klipstein and Kaup included a reconstructed scene in
their otherwise conventional monograph is a sign of how they hoped

A lithograph of the *Dinotherium* in its natural habitat, drawn to decorate the title page of the monograph on its fossil remains by August von Klipstein and Johann Kaup (1836). (Reproduced by permission of the Syndics of Cambridge University Library.)

to transcend the goals of merely descriptive palaeontology; but that the scene was printed as a decoration to the title-page, and not on one of the full-page lithographed plates, is equally a sign of the tentative and experimental character of that act of reconstruction.[13]

Caricatures of the deep past

This scene by Klipstein and Kaup is, at least at present, the earliest known reconstruction of the prehuman past to be published, albeit marginally, as part of a standard scientific report.[14] But at least one other example of the genre must have been circulating informally in German scientific circles a few years earlier. In a letter written in 1831, the Oxford geologist William Buckland (1784–1856) told his fellow geologist Henry De la Beche (1796–1855) that 'a German parody of your Duria Antiquior' had just reached England.[15] The reference was to a recent lithograph in which De la Beche – an accomplished artist as well as a fine geologist – had portrayed a scene of 'a more ancient Dorset', namely the Dorset of the period when the famous ichthyo-saurs, plesiosaurs and other Liassic fossils of Lyme Regis had been alive. De la Beche had reconstructed a scene which, although rather implausibly crowded with all the more spectacular denizens of Liassic Dorset, vividly brought those animals to life.[16] Almost every animal was shown eating, or being eaten by, another. Such habits had been inferred in the preceding years from a wealth of sober scientific evidence, notably by Buckland himself. The slightly vulgar showmanship of Buckland's lectures, in which the professor himself frequently acted the part of the extinct animals he was describing, had delighted most members of his audiences as much as it had been frowned upon by a minority of his more staid colleagues. Above all, however, Buckland's highly ecological approach to the study of fossils had made the deep past *anschaulich* as never before. De la Beche's half-humorous, half-serious portrayal of *Duria Antiquior* merely translated that vision on to lithographic stone and thence on to paper. In that form it circulated widely among the circle of gentlemanly geologists in England. At least in Buckland's opinion, a copy must then have been sent to Germany, to become the inspiration for a similar scene (as yet unlocated), perhaps based on the equally fine Liassic fossils of Württemburg. It may have been that German version that became the inspiration for Klipstein and Kaup's single reconstruction of a quite different period in the history of life, and hence, indirectly, for Unger's complete sequence of 'landscape representations'.

However that may be, Buckland's letter to De la Beche records more than that possible link from England to Germany. Like the happy accident of a Solnhofen slate, it preserves in unpublished form the earliest known 'fossil record' of the *idea* of constructing a whole sequence of scenes from the deep past.[17]

Buckland implored De la Beche not to stop with *Duria Antiquior*, but 'to put on the stocks in your best style 2 or 3 more restorations of scenes in the ancient world'. In fact he proposed three, which together with the existing one would have made a sequence of four scenes. His first suggestion ran as follows:

I. The Period immediately preceding the formation of Diluvium

'A more ancient Dorset' (*Duria Antiquior*), the half-humorous lithograph drawn by Henry De la Beche in about 1830 to illustrate the life habits of the ichthyosaurs, pleisiosaurs and other animals found as fossils in the Liassic strata of Dorset. (Reproduced by permission of the Department of Geology, National Museum of Wales.)

> – a Land Piece – with only rivers plains & mountains – as in Pale-
> strina Pavement – exhibiting the gamboled Battles of Elephants
> & Rhinoceros & Mastodons – Hippopotami jumping into the
> Rivers – Megatherium sitting on his Haunches with one fore
> Paw against the trunk of a Tree and the other reaching down an
> enormous Branch – Horse Ox and Elk scampering before a Pack
> of Wolves and falling headlong into fissures – Hyaenas in their
> Den or dragging into its Mouth their Prey – Tigers crouching to
> spring on Deer.[18]

This lively verbal representation of the fauna of the late Tertiary – with the animals of the Old and New Worlds mixed up somewhat implausibly but to good dramatic effect – was followed by a scene from the early Tertiary, based on the celebrated Parisian research of Georges Cuvier (1769–1832) earlier in the century:

> II. A lake scene from the F[resh] water Period. Ponds full of
> Palaeotherium Anoplotherium and Chaeropotamus and all the
> Paris Pigs of those Days. Dogs and Sarigues at Montmartre.
> Birds and Reptiles, snakes and water Rats in Auvergne Tortoises
> Beavers Crocodiles – Volcanoes in the Distance.[19]

With *Duria Antiquior* already available to represent the much earlier period of the Lias, Buckland then moved for his final scene still further back in time:

> III. A sea scene – Sea with tropical Islands of Carboniferous and Transition Periods. Land very short of animals, but glorious hypertropical Vegetation – Sternberg's Lepidodendron, stems of gigantic Cactus. Sea full of Tropical islands – covered with Coal Plants – under water, encrinites, Corals, chain Corals – Orthoceratites and Nautili at Surface – Spirifer, Producta – Trilobites – and a few fish. The Trilobites wd caricature well.[20]

Buckland's sequence of verbal sketches was scribbled down with scant regard for formal punctuation, and in the characteristic handwriting that led his friends to claim that he employed a trained spider as amanuensis. But it expressed perfectly his vivid ecological imagination, and the lively reality with which in his mind's eye he transported himself back into the remote prehuman past. His verbal sketches could almost have been a rough draft of Unger's instructions to *his* artistic collaborator, but for one striking difference. Unger arranged his scenes in their sequence in real time, from ancient to recent; by contrast, back near the origins of the genre, Buckland's were listed in retrospective order. Like most geologists who summarized stratigraphical knowledge in the 1820s and 1830s – including Charles Lyell (1797–1875), in the third volume (1833) of his *Principles of Geology*[21] – Buckland considered it natural to probe backwards, from the relatively well-known recent past towards the more obscure periods of the deepest past. Nevertheless, despite that important difference in presentation, Buckland's letter to De la Beche marks the potential start of the genre of a *sequence* of reconstructed scenes. De la Beche seemed not to have acted on the proposal, and the idea may therefore have remained for the time being in the private realm. Knowingly or not, however, it was eventually to be realized in Unger's great series of 'landscape representations.'

Spy-holes into the past

How then did Buckland and his circle come to conceive of the idea of depicting even a *single* scene from the deep past? A clue is provided by the apparently casual comment at the end of Buckland's verbal proposal; 'the Trilobites would caricature well'. The scenes he hoped De la Beche would draw would be not only serious attempts to depict the past, but also, at the same time, light-hearted *caricatures* of the results of scientific research. De la Beche was the obvious man to approach, as he was the best-known geological caricaturist. In addition to *Duria Antiquior* he had, for example, just lithographed a caricature showing a 'Professor Ichthyosaurus' lecturing on a human skull: this was an imagined view into the post-human *future*, which ridiculed the implausible cyclicity in the history of life that Lyell had postulated in the first volume (1830) of his *Principles*.[22]

Caricatures such as these were privately – but widely – circulated among the gentlemanly geologists in the research triangle of London, Oxford and Cambridge. The earliest and most significant of them, however, dates from a decade before either *Duria Antiquior* or Professor Ichthyosaurus. It was drawn by Buckland's close friend and former Oxford colleague William Conybeare (1787–1857), and it reflects the immense impact of Buckland's celebrated analysis of the bone-bearing cave at Kirkdale in Yorkshire. This research won him the Royal Society's Copley Medal, and formed the centrepiece of his *Reliquiae Diluvianae* (1823). That title was a striking misnomer, though doubtless it helped the sales of the book.[23] The organic relics that Buckland described and analysed were not themselves relics of the Deluge at all, though he did argue that a transient diluvial event had sealed them into the cave and thus preserved them. They were relics of the period immediately *before* the Deluge, a period of tranquility and normality, or at least a period marked by ecological relations of predator and prey, scavengers and scavenged, closely analogous to those of wilder regions in the modern world.

The cave at Kirkdale, although far smaller and less spectacular than other bone-bearing caves already known on the continent, gave Buckland the evidence with which to reconstruct a den of prediluvial hyaenas, whose scavenging had concentrated and preserved a sample of an entire fauna. It gave him an opportunity to exercise his scientific imagination in a reconstruction not just of the bodies of single animal species, as Cuvier had done more authoritatively before him, but of an entire fauna interlocked in ecological relations with one another.[24] His analysis was so persuasive and his verbal reconstruction so vivid that it seemed to his contemporaries to disclose the reality of the prehuman past as never before.

This was the achievement that Conybeare translated into visual form. His caricature was not a sober reconstruction of the kind that came to fruition in Unger's work, but it was no less decisive historically. It depicted Buckland himself crawling into Kirkdale cave, candle in hand, and there encountering the hyaenas of the prediluvial period, very much alive among their scavenged bones. The geologist became in caricature a *participant* in the scene he had soberly reconstructed in words. The entrance of the cave became symbolically a passage through the epistemic barrier that separated the observable present from the prehuman past; the candle became symbolically the illumi-

'The Hyaena's Den at Kirkdale', the lithographed caricature drawn by William Conybeare to celebrate William Buckland's analysis (1822) of the bone-bearing cave in Yorkshire.

nation that the geologist could bring to that otherwise inaccessible past.

That point was made by Conybeare himself, at the end of some sixty lines of dreadful doggerel appended to the caricature:

Mystic Cavern, thy chasms sublime,
All the chasms of History supply;
What was done ere the birth-day of time,
Thro' one other such hole I could spy.[25]

The aspiration was of course exaggerated for poetic effect. Conybeare, who had just completed his great *Outlines of the Geology of England and Wales* (1822), knew perfectly well that the 'birth-day of time' was unimaginably remote, and that far more than *one* 'other such hole' into the deep past would be needed in order to understand the history that had filled it.[26] But the principle had been established. Buckland's cave research demonstrated, more vividly than ever before, the feasibility of penetrating the epistemic and imaginative barrier between present and past, and of rendering the past *anschaulich*. If Buckland could encounter the hyaenas through a spy-hole at Kirkdale, he and others could hope by careful research to construct other windows on to the past, and so to encounter the palaeotheria, the ichthyosaurs, the lepidodendrons, and so on back to the trilobites.

Back to the Ark

Conybeare's caricature of Buckland with the hyaenas seems to represent the very origin of the genre of reconstructed scenes from the prehuman past. Significantly, it is preceded only by representations that belong to a quite different tradition. A late example, dating from less than twenty years before Conybeare's caricature, formed the frontispiece to the first volume (1804) of the first substantial illustrated book on fossils to be published in England, the *Organic Remains of a Former World* by James Parkinson (1755–1824).[27] It depicted a rocky shoreline and a view out to sea, with the sun breaking through stormclouds. On the shore were the drifted shells of some of the fossils illustrated in the lavish hand-coloured engravings that embellished the book. But although the shells included ammonites, this was no view of ancient Oolitic or Liassic times. For the sky was marked by a rainbow; and at the rainbow's end, stranded on a rocky islet, was a distant but unmistakable Ark. Parkinson regarded *all* his fossils as the remains of a *single* former world; more precisely, it was 'the antediluvian world' that had been brought to an end by an event that was at least congruent with the Genesis narrative. Such a belief had long been abandoned by those who studied fossils on the continent – as Parkinson realized in time for his later volumes – but his original frontispiece can stand as a late example of an iconographical tradition that stretches back at least as far as the encyclopaedic *Noah's Ark* (1675) by the Jesuit polymath Athanasius Kircher (1601–80).[28]

The collective memory of that tradition may perhaps explain the hesitancy with which the newer genre came to be established. The scientific geologists may well have been concerned to distance their project from the earlier tradition, for theirs was self-consciously based on reasoning from natural evidence alone. Certainly, it is striking how 'marginal' the scientific genre remained in its early years. Born with an explicit caricature, lithographed for the entertainment of a close-knit circle of scientific friends, it only slowly shed its gentle humour and its semi-private circulation. The first fully public and fully realistic examples were still printed in a 'marginal' manner, separately from more conventional scientific illustrations. Even the first full series of scenes was presented somewhat apologetically, with the excuse that it had originally been devised for the benefit of amateurs; and the most influential of later series was explicitly aimed at a juvenile readership.

Yet it may be no coincidence that this scientific genre of reconstructed scenes first arose within precisely the circle of geologists who were most concerned to maintain at least the *religious* authority of the biblical records, while conceding or even welcoming the interpretive insights of the new biblical criticism.[29] It is not fanciful to conjecture that the genuine piety of men like Conybeare enabled them to *imagine* the unitary biblical narrative with a greater sense of concrete reality than their less religious counterparts; and that that distinctive vision of the human past facilitated their first attempts to project a similar concrete reality into the deep time revealed by their geology.

On to Darwin

'He who can read Sir Charles Lyell's grand work on the Principles of Geology', wrote Charles Darwin in the *Origin of Species*, 'yet does not admit how incomprehensibly vast have been the past periods of time, may at once close this volume.'[30] Darwin did indeed need Lyell's extremely long time-scale – significantly longer, incidentally, than modern estimates based on radiometric methods – in order to validate his conception of extremely slow evolutionary change by means of natural selection.[31] But the fame of the intellectual lineage that runs through Lyell to Darwin should not blind us to its limitations. The evolutionary view of the natural world, which in its organic aspect we have come to associate so crucially with Darwin, needed far more than the mechanism of natural selection to lend it plausibility. It needed more than a Lyellian vision of vast *time*, within which natural selection could operate effectively. It needed equally, or perhaps even more, a concrete vision of an unimaginably lengthy prehuman *history*.

It is one of the ironies of the 'Darwinian Revolution', to the understanding of which John Greene has contributed with such distinction, that this sense of prehuman history was fed into the stream of nineteenth-century thought, not by the 'apostolic succession' of heroes that led through Lyell to Darwin, but by those who in an earlier historiography were cast as the villains, the opponents of 'progressive' evolutionary thinking. The reason, however, is not hard to find. Lyell and Darwin were ultimately concerned to discover the generalities of the natural causal *laws*, both inorganic and organic, by which the world maintained itself as a natural system. Others, like Conybeare and Buckland, who were formerly reviled as catastrophists or dismissed as diehard opponents of evolution, were ultimately more concerned to reconstruct the particularities of the *history* that had brought the world to its present state. In the jargon of historiographical theory, Lyell and Darwin set themselves objectives that were above all *nomothetic*; their opponents, by contrast, set themselves primarily *idiographic* objectives.

The two sets of goals were not, and are not, mutually exclusive; on the contrary, they are blended together in the modern evolutionary understanding of the earth and its biosphere. In our enthusiasm for Darwin's own role in his 'revolution', therefore, we should not overlook the historical importance of those who first made *some* kind of evolutionary theory seem a plausible possibility, whether intentionally or not, by displaying the vast scale and rich diversity of the *history* that needed causal explanation. In particular, we should recognize the achievements of those who made the long history of the earth and of its fauna and flora visually imaginable, *anschaulich*: first and tentatively to their fellow geologists, then more boldly to amateurs of the science, and ultimately to an even wider public – the public that read *The World before the Deluge* a century ago and that now throngs our museums of natural history.

Notes

This essay is based upon work supported by the National Science Foundation under grant no. SES-88-96206.

1. The main allusion is to John C. Greene, *The Death of Adam: Evolution and Its Impact on Western Thought* (Ames: Iowa State University Press, 1959), the first and perhaps most influential book in the *oeuvre* honoured by this volume. The other side of the allusion is a reminder – which may be needed by some historians of science, though not by John Greene – that the figure of Adam remains 'alive' in religious practice, because it is so rich in symbolic and therefore practical meaning; from this perspective, it is simply irrelevant that it is indeed 'dead' in modern natural-scientific practice.

2. The parody is adapted from Genesis 1:20–7 (RSV).

3. Visual reconstructions in books remain, of course, as important as those in museums. For some recent, artistically attractive, and scientifically authoritative examples, see Anthony J. Sutcliffe, *On the Tracks of Ice Age Mammals* (Cambridge, Mass.: Harvard University Press, 1985), especially the Welsh hyaena den on p. 119.

4. Constraints beyond my control limit severely the number of 'visual quotations' that can be included in this essay, although images, and not words, are its subject matter. The early history of other, more technical, visual genres in geology is explored in Martin J.S. Rudwick, 'The Emergence of a Visual Language for Geological Science, 1760–1840', *History of Science*, 14 (1976), 149–95. The felicitous phrase 'deep time' is borrowed from John McPhee,

Basin and Range (New York: Farrar, Strauss, Giroux, 1981).

5. Louis Figuier, *La terre avant le Déluge: ouvrage contenant 25 vues idéales de paysages de l'ancien monde* (Paris: Hachette, 1863 and later edns); *The World before the Deluge* (London; 1865 and later edns; also New York: D. Appleton, 1869). Other translations are listed in standard major library catalogues. Figuier's *La terre* was the first volume in his series entitled 'Tableau de la nature: ouvrage illustré à l'usage de la jeunesse'.

6. Alcide d'Orbigny, *Cours élémentaire de paléontologie et de géologie stratigraphique*, 2 vols. (Paris: Victor Masson, 1849–52).

7. *La terre*, 6th edn, p. 419n. Ironically, the authenticity of the Moulin-Quignon mandible was soon rejected by the experts; but a consensus on the main point, based on a much wider range of evidence, did indeed congeal at just this time: see Donald K. Grayson, *The Establishment of Human Antiquity* (London: Academic Press, 1983), ch. 9. Charles Lyell's book on *The Geological Evidences of the Antiquity of Man* (London: John Murray, 1863) was but one popularization of the new view.

8. *La terre*, 6th edn, fig. 322. The strikingly subtropical character of the vegetation reflects Figuier's claim that 'the appearance of Man' had taken place in its traditional location somewhere in Asia, *not* in the evidently harsher climate of Pleistocene Europe.

9. *La terre*, 1st edn, 1863, fig. 310. In some later editions this engraving was retained, incongruously, in addition to the one that was designed to supersede it; in the New York edition (1869) it was even elevated in position to form the frontispiece of the book!

10. F. Unger, *Die Urwelt in ihren verschiedenen Bildungsperioden: 14 landschaftliche Darstellungen mit erläuterndem Texte* (Vienna, 1847). I am preparing a modern edition of this important work.

11. In the second edition (Vienna, 1858), Unger added two new views, of the Silurian and Devonian periods, while adding the qualifier 'moderne' to his original view of the 'Epoque de Transition'.

12. Unger himself did believe in some kind of evolutionary explanation for the changing character of the organic world; but his visual representations did not require, let alone compel, such a conclusion.

13. A. von Klipstein and J.J. Kaup, *Beschreibung und Abbildung von dem in Rheinhessen aufgefundenen colossalen Schadel der Dinotherii gigantei, mit geognostischen Mittheilungen über die Knochenführenden Bildungen des mittelrheinischen Tertiärbeckens* (Darmstadt, 1836). The vignette was lithographed by R. Hoffmann and L. Becker, whom I have been unable to identify further.

14. Other reconstructions from the same period are all found in popular books, or in works that in some other way are not 'standard' scientific reports; e.g. John Martin's highly imaginative (and inaccurate) apocalyptic frontispieces for Gideon Mantell's *The Wonders of Geology; or, A Familiar Exposition of Geological Phenomena* (London: Rolfe and Fletcher, 1838), and for *The Book of the Great Sea-Dragons, Ichthyosauri and Plesiosauri, Gedolim Tanimim of Moses, Extinct Monsters of the Ancient Earth* (London: William Pickering, 1840), by the eccentric fossil collector Thomas Hawkins.

15. W. Buckland to H.T. De la Beche, 14 Oct. 1831, De la Beche papers,

National Museum of Wales, Cardiff. I am indebted to the Keeper of the Museum for permission to quote from this letter.

16. *Duria Antiquior* is analysed in Paul J. McCartney, *Henry De la Beche: Observations on an Observer* (Cardiff: Friends of the National Museum of Wales, 1977), pp. 44–7. See also James A. Secord, 'The Geological Survey of Great Britain as a Research School, 1839–1855', *History of Science*, 24 (1986), 223–75 (esp. 241–7). A fine, water-colour version, perhaps the drawing on which the lithograph was based, is in the National Museum of Wales, and is reproduced in colour on the cover of S.R. Howe, T. Sharpe and H.S. Torrens, *Ichthyosaurs: A History of Fossil 'Sea-Dragons'* (Cardiff: National Museum of Wales, 1981).

17. The Solnhofen 'slate' (in fact a fine-grained limestone), which was later to become famous for its *Archaeopteryx*, was already yielding exceptionally fine specimens of a wide variety of Jurassic organisms that would not have been preserved under more normal circumstances. Their discovery was due to the exploitation of the stone for use in lithography, a technique that came into general use only in the 1820s.

18. See n.15 above. The 'Palestrina pavement' was the Hellenistic 'Barberini mosaic' in the Italian town (the birthplace of the composer), which depicts the landscape of the Nile with its animal (and human) inhabitants.

19. The latest collection of Cuvier's work was in his *Recherches sur les ossemens fossiles, où l'on rétablit les caractères de plusieurs animaux dont les révolutions du globe ont détruit les espèces*, 3rd edn, 5 vols. (Paris and Amsterdam: G. Dufour and E. Ocagne, 1825). A *sarique* is an opossum. Buckland noted that De la Beche's own recently published *Geological Manual* (London: Treuttel and Würtz, Treuttel Jun. and Richter, 1831) would provide him with the data for sketching 'the appropriate vegetation' for that scene.

20. The reference in this last passage was to Kaspar, Graf von Sternberg, *Versuch einer geognostisch-botanische Darstellung der Flora der Vorwelt* (Leipzig, Prague and Regensburg, 1820–38).

21. Charles Lyell, *Principles of Geology, Being an Attempt to explain the Former Changes of the Earth's Surface, by reference to Causes now in Operation*, 3 vols. (London: John Murray, 1830–3).

22. This caricature, and the sketches that led to it, are analysed in Martin J.S. Rudwick, 'Caricature as a Source for the History of Science: De la Beche's Anti-Lyellian Sketches of 1831', *Isis*, 66 (1975), 534–60. See also McCartney, *Henry De la Beche*, pp. 50–3. Another of his caricatures, which circulated widely at the time although it was never even lithographed, is his telling visual comment (1834) on the nascent Devonian controversy: see Martin J.S. Rudwick, *The Great Devonian Controversy: The Shaping of Scientific Knowledge among Gentlemanly Specialists* (Chicago: University of Chicago Press, 1985), p. 104; also McCartney, *Henry De la Beche*, p. 31.

23. William Buckland, *Reliquiae Diluvianae; or, Observations on the Organic Remains contained in Caves, Fissures, and Diluvial Gravel, and on Other Geological Phenomena, attesting the Action of an Universal Deluge* (London: John Murray, 1823). That the title was a misnomer is noted by Nicolaas A. Rupke, *The Great Chain of History: William Buckland and the English School of Geology (1814–1849)*

(Oxford: Clarendon Press, 1983), p. 39. Buckland's book was enlarged from a long paper read to the Royal Society in 1822 and published in its *Philosophical Transactions*, vol. for 1822, pp. 171–236.

24. Cuvier seems never to have carried his reconstructions further than mere outlines of the bodies of extinct mammals: see his lively MS. sketch of a *Palaeotherium* reproduced in William Coleman, *Georges Cuvier, Zoologist: A Study in the History of Evolution Theory* (Cambridge, Mass.: Harvard University Press, 1964), p. 122; also the drawings published in Cuvier, *Recherches*, 3rd edn, III, pl. 66.

25. [W.D. Conybeare], 'The Hyaena's Den at Kirkdale, near Kirby Moorside in Yorkshire, discovered A.D. 1821', lithographed broadsheet, anonymous and undated. The poem is reprinted in C.G.B. Daubeny, *Fugitive Poems connected with Natural History and Physical Science* (Oxford: James Parker, 1869), pp. 92–4, where it is attributed to Conybeare and dated 1822.

26. Conybeare revised and greatly enlarged an earlier book by his nominal co-author (and publisher) into what became internationally a standard reference work on stratigraphical geology: William Daniel Conybeare and William Phillips, *Outlines of the Geology of England and Wales, with an Introductory Compendium of the General Principles of that Science, and Comparative Views of the Structure of Foreign Countries. Part I* [all issued] (London: William Phillips, 1822).

27. James Parkinson, *Organic Remains of a Former World: An Examination of the Mineralized Remains of the Vegetables and Animals of the Antediluvian World; Generally termed Extraneous Fossils,* 3 vols. (London: Sherwood, Neely and Jones, 1804–11). See J.C. Thackray, 'James Parkinson's *Organic Remains of a Former World* (1804–1811),' *Journal of the Society for the Bibliography of Natural History,* 7 (1976), 451–66.

28. Athanasius Kircher, *Arca Noë in Tres Libros Digesta, sive De Rebus Ante Diluvium, De Diluvio, et De Rebus Post Diluvium a Noemo Gestis* (Amsterdam: Johann Jansson, 1675).

29. See, for example, W.D. Conybeare, *An Elementary Course of Lectures on the Criticism, Interpretation, and Leading Doctrines of the Bible, delivered at Bristol College in the Years 1832, 1833* (London: John Murray, 1834).

30. Charles Darwin, *On the Origin of Species by Means of Natural Selection, or the Preservation of Favoured Races in the Struggle for Life* (London: John Murray, 1859), p. 282. The comment was somewhat overdone, because the reader would already have got through more than half the book before encountering it!

31. See Joe D. Burchfield, 'Darwin and the Dilemma of Geological Time', *Isis,* 65 (1974), 300–21.

8

Huxley and woman's place in science: The 'woman question' and the control of Victorian anthropology

EVELLEEN RICHARDS

> If then we are not to speak of grave or scientific things in "society", & are shut out from almost all scientific "Societies", how are we to learn?
>
> Eliza Linton to T.H. Huxley (1868)[1]

As John Greene has stressed, it was Thomas Henry Huxley rather than Herbert Spencer who became the 'chief expounder and champion of Darwinism' as a 'world view' after the publication of the *Origin of Species*. It was Huxley who in 1860 undertook to depict 'the picture which science draws of the world' and throughout the decade he elaborated a Darwinian world view of 'harmonious order governing eternally continuous progress' in which Nature was always 'fair, just, and patient' but 'without remorse' in enforcing the universal struggle for existence. One of Huxley's more colourful metaphors pictured the world as 'Nature's university' in which all 'mankind' are enrolled. Those who will not learn and obey her laws are ruthlessly 'plucked', while those who learn to live in harmony with Nature are the liberally educated: 'They will get on together rarely; she as his ever beneficient mother; he as her mouth-piece, her conscious self, her minister and interpreter.'[2]

In the 1860s, therefore, Huxley constituted himself as Nature's leading Darwinian 'mouth-piece', and his better-known role as the interpreter of 'man's place in nature' was subsidiary to his overarching 'moralizing naturalism'. The picture of the world that Huxley projected in his popular essays and lectures of this period was one in which 'natural knowledge' would lay the foundations of a 'new morality'; a world where the 'man of science' through his access to reliable natural knowledge would guide the conduct and organization of society. Those men of science best fitted to bring about the 'New Reformation' envisaged by Huxley were of course the 'young

253

guard' Darwinians, and throughout the 1860s he worked tirelessly towards their interrelated professional and social advancement.[3] Much of this effort was channelled into the socially sensitive science most closely concerned with the study of 'man' – anthropology – and its divided and divisive theoretical models and institutions. For in this crucial period of the Huxley-led drive for Darwinian dominance of nineteenth-century science, professional anthropology was riven by the conflict between the older-established and religious-oriented Ethnological Society and the short-lived, but extremely influential, Anthropological Society of London (1863–71), the institutional stronghold of the physical anthropologists who were led by the charismatic racist, James Hunt. Although, as I shall show, Huxley's anthropological position was more congruent with that of the naturalistic and anti-clerical oriented Anthropologicals, he and the leading Darwinians allied themselves with the Ethnological traditionalists, and Hunt and Huxley became locked in an extended and acrimonious struggle for the control of Victorian anthropology.[4] It is my intention in this essay to explore the significance of the 'woman question' to their conflict and its resolution.

Although it has received scant attention from historians, Hunt himself pointed up the decision of the Ethnologicals to follow the example of the Royal Geographical Society and admit 'Ladies' to their meetings, as one of the major reasons for his secession from the Ethnological Society in 1863 and the foundation of the Anthropological Society. In addition to its racism, Hunt's Anthropological Society was characterized by an overt anti-feminism, and it played a key role in the 'scientific' refutation of the claims by nineteenth-century liberal feminists for social and intellectual equality.[5] By contrast with Hunt, Huxley has been conventionally depicted as adopting an 'enlightened' stand on women's issues.[6] However, closer scrutiny makes his publicly reiterated support for female education and entry to the professions somewhat problematic. Moreover, Huxley played a little-known but leading role in actively excluding women from scientific societies, specifically the Ethnological Society. Huxley and Hunt, it would seem, were at one in their opposition to female admission to the Ethnological Society.

Most accounts of the Huxley/Hunt conflict have argued the basic intellectual incompatability of the Darwinians and the Anthropologicals, but as John Greene has reminded us: 'The lines between science, ideology, and world view are seldom tightly drawn.'[7] In this essay I offer a reinterpretation that stresses the crucial role of ideological and social factors in the Huxley/Hunt dispute. It thus provides a framework within which Huxley's position on the woman question and its implications for the professional and scientific aspirations of the Darwinians may be clarified.

Finally, I have sought to relate the politics of female admission to the complexities of nineteenth-century feminism through the controversial figure of Eliza Lynn Linton, one-time radical, successful woman journalist, and latterly, committed Darwinian. Lynn Linton offered the only documented resistance to Huxley's exclusion of women from the Ethnological Society, and her passionate eight-page petition on behalf of the women 'visitors' stands as a unique (and symbolically neglected) testament to the relations of women to mid-Victorian organized science and Darwinism in particular.

'I perceive . . . that Ladies *are* to come – More's the pity'

The admission of women to the Ethnological Society was first formally proposed at the council Meeting of 17 October 1860, when John Crawfurd, the recently elected president, gave notice that at the next meeting he would move a resolution that 'Ladies be admitted as visitors'.[8] Crawfurd's move was in line with recent liberal developments in female education, where, following the establishment of Queen's College and the Ladies College in Bedford Square, women had shown an increasing interest in attending public lectures on science at a time when these constituted almost the only form of scientific education available to them. However, Crawfurd seems to have been less motivated by a liberal concern with furthering female education than by his desire, as president, to increase attendance at the meetings of the moribund Society by making them, as Hunt sneeringly put it, 'fashionable and popular'.[9]

The attendance of 'the fair sex' at public scientific lectures had been generally endorsed by their male patrons on the grounds that it 'doubled the enjoyment' of the men by providing a sort of decorative backdrop to the occasion.[10] Crawfurd and a 'large and powerful section' of the Ethnological Society evidently wished to capitalize on this feminine drawing-power in the same way the Geographical Society had recently done. However, while it was socially acceptable for women to decorate the more frivolous scientific occasions, when it came to their actual membership of learned societies and attendance at society meetings, even avowed liberal 'advocates' of female education like Huxley drew the line. Some months previously he had forcefully stated and explained his opposition to female admission to Charles Lyell.

Lyell had written to Huxley stating that the admission of women to

the Geological Society might aid the cause of 'geology' by exposing
them to the non-creationist viewpoint and counteracting the unfavourable influence of organized religion. That 'jesuit in disguise', the
Bishop of Oxford, was going about and 'inculcating the doctrine that
no woman should ever be allowed in any of the authorised places of
education to hear both sides discussed'. But Lyell saw no reason why
'the power of the tongue & its influence on one half of society'
should be left exclusively to those '60,000 sworn teachers of endorsed
opinions'.[11]

However, by return of post, Huxley quashed Lyell's notion of
recruiting the forgotten half of society to the Darwinian cause. He
reacted against it as vehemently as the Bishop of Oxford, Samuel
Wilberforce, if for somewhat different reasons:

> [T]he Geological Society is not, to my mind, a place of education
> for students but a place of discussion for adepts: and the more it
> is applied to the former purpose the less competent it must
> become to fulfil the latter – its primary and most important
> object.

Women were necessarily amateurs, and their presence at serious
scientific discussions would jeopardize the professional status of the
Society. It was not, Huxley hastened to assure Lyell, that he wished
to place 'any obstacle' in the way of the intellectual advancement and
development of women. On the contrary, he did not see how society
could progress as long as one half of the human race remained sunk
in the ignorant superstitions inculcated by parsondom. But he did not
believe that others would follow his plans for educating his own daughters in basic science to the extent that the next generation of women
might become 'fit ... companions of men in all their pursuits' (not
that Huxley thought men had anything to fear from their
'competition'):

> [Y]ou know as well as I do that other people won't do the like,
> and five sixths of women will stop in the doll stage of evolution,
> to be the stronghold of parsondom, the drag on civilization,
> the degradation of every important pursuit with which they
> mix themselves – "intrigues" in politics and "friponnes" in
> science.[12]

If Huxley's 'claws and beak' were 'good for anything', he assured
Lyell that such 'dolls' would be kept from hindering the progress of
any science he had 'to do' with.

Even by early 1860, as this exchange makes explicit, Huxley had
put a Darwinian gloss on the woman question and prescribed its
limitations. A minority of women, suitably educated, might become
the 'fit companions' of men, but not their 'competitors'. Like Henrietta Huxley (that paragon of scientific wives), they might assist

their husbands – exhibit an intelligent interest in their work, illustrate or proof-read their manuscripts, even occasionally accompany them to the more popular scientific meetings. Their proper role was to be more concerned with the scientist than his science. For careerists like Huxley, who had a 'good deal of fighting to do in the external world', it was essential to have 'light and warmth and confidence within the four walls of home'.[13] It was inconceivable that women might actually engage in the 'fighting' that Huxley found so invigorating, and advance the professional status and rewards of science. On the contrary, their inexpert and unprofessional presence would 'hinder [its] progress'. As for the great majority of women, the 'five sixths' whose lot was to remain stunted at the 'doll stage' of evolution in the thrall of parsondom, their reactionary and frivolous presence was to be excluded at all costs from the forums of professional science.

Huxley's prejudices were founded on his own experiences and expectations. Mid-Victorian science was an all-male preserve, which women entered, if they entered at all, only as spectators – at the most as fashionable dabblers, not to be taken seriously. Thus Huxley could be jovially impressed by Mrs Buckland's newly discovered fossil 'Echinoderm', but even the 'very jolly' Mrs Buckland brought out his only semi-facetious 'unutterable fear of scientific women'.[14] The majority of women who dabbled in natural history were 'Naturalists of the Boudoir' who kept a shell or mineral collection, a fern case or aquarium, and rarely ventured into the serious study of their hobby. The exclusion from the learned societies of those few women who went beyond the dilettante pursuit of their interest was in itself a strong disincentive to research, for it was usually only through the various *Transactions* that findings could be published.[15]

Victorian feminine fragility also obstructed serious research. Girls were discouraged from real intellectual application on the grounds that it was unhealthy and fatiguing. While Huxley persistently over-worked himself to the point of mental exhaustion, churning out publications and lectures until he was persuaded to recoup his forces by energetically clambering up and down various European peaks, Henrietta carefully inculcated the Victorian canon of feminine intellectual and physical frailty. 'I am sure you are right', wrote her acquaintance the Countess of Portsmouth, 'in not allowing your girl to do much work – I am grown very nervous about any forcing of the brain with young growing girls. I think it horribly capable of addling the brain instead of filling it. . . .'[16]

Another major obstacle to the feminine pursuit of the natural sciences was the crippling 'delicacy' of the age, which made it necessary for middle-class women to shy in prudish alarm from any publicly expressed hint of sex or reproduction. It took a strong motivation indeed for women to flout this Victorian sensibility. The few male naturalists who consented to address women on such issues, expurgated their lectures to such an extent that it is doubtful whether their audience grasped the essential physiology. Publications likely to be read by women were also suitably censored, or the offensive matter might be rendered in Latin. In 1856 Richard Owen negotiated carefully with John Murray over the respectability of his proposed inclusion of the 'reproductive economy and apparatus' of a bee in his article on parthenogenesis for the Tory *Quarterly Review*;[17] while around the same time, his arch-rival, Huxley, sniggeringly refused an invitation to lecture to ladies at London University: 'What on earth should I do among the virgins, young and old in Bedford Square? ... I should be turned out ... for some forgetful excursions into the theory of Parthenogenesis or worse.'[18]

It was not only on sexual matters that Huxley had to guard his tongue in female company. The overwhelming majority of Victorian women did not share Henrietta Huxley's faith in the ennobling revelations of science and her open-mindedness on religious issues. Huxley could not speak without repercussion even to his fellow naturalists' wives, using the same openness with which he evidently discussed his agnosticism with Henrietta. Some years earlier, Huxley's 'unusually plain manner [of speaking] of his want to faith' had so 'alarmed' Andrew Ramsay's young wife Louisa that she 'worked herself into a fever' after an 'intellectual evening' at the Huxley's. On this occasion, as Ramsay had feared, Huxley's plain speaking broke up the projected joint visit of the Ramsays and the Huxleys to Switzerland.[19] Such incidents undoubtedly reinforced Huxley's conventional Victorian view of the pious frailties of middle-class women: they were plainly unequal to the give and take of robust 'intellectual' discussion and thus better excluded from it.

It is fascinating to note how this female stereotype manifested itself on the legendary occasion of Huxley's celebrated confrontation with Wilberforce at the Oxford Meeting of the British Association for the Advancement of Science, only a few months after his exchange of views with Lyell. This was the highpoint of the Darwinian debates of the 1860s (now regarded by historians as more apocryphal than apocalyptic), when Huxley supposedly routed the reactionary and anti-scientific forces of parsondom in the person of Wilberforce with his devastating and apposite response to the bishop's fatal quip about Huxley's simian ancestry: 'Was it through his grandfather or his grandmother that he claimed his descent from a monkey?' Huxley's son Leonard (who had not then been born) authoritatively describes the 'ladies' who packed the windows of the lecture room to urge on the bishop's attack on natural selection with a dainty 'waving and fluttering' of their white handkerchiefs.[20] The full horror of the social solecism of Huxley's unprecedented public counter-attack on a man of the cloth is epitomized in the 'lady' who fainted and had to be carried from the room. This often-recounted incident also, of course, encapsulates the intellectual inappropriateness of the response and, more generally, of the presence of women at the debate. In one contemporary version, the bishop's slur on Huxley's ancestry is interpreted as a misguided attempt to trade on the antipathy to degrading *woman* (presumably in the person of Huxley's grandmother) to the level of the quadrumana. Huxley's reply is then represented as a scholarly and dignified eschewing of such vulgarity, and Wilberforce, abashed, is forced to recognize that he 'had forgotten to behave like a perfect gentleman'.[21] The piquancy of this account lies in its reversal of roles, with the bishop emerging as no gentleman, and the upstart marginal middle-class Huxley winning the day through his higher moral tone and implicit defence of Victorian values and Victorian womanhood. It gains some support from historians who have pointed out the extent to which the leading Darwinians capitalized upon their collective gentlemanly image – their solid financial, political and sexual respectability and general Victorian conventionality – in the promotion of unconventional scientific opinion.[22]

Whatever the reality behind the mythology that has accreted around the Oxford debate, it serves as a cliché of the conventional relation of women to mid-Victorian science and to the Darwinian debates in particular. It was not a convention that Huxley cared to flout, and indeed, in a number of ways, he subscribed to it. He was, moreover, as I shall show, not averse to deploying it for institutional and social ends, and to the detriment of his liberal views. In contradiction of his public stance, Huxley's personal antipathy to the presence of women at scientific lectures was made explicit in a note he wrote in 1862 to Edward Perceval Wright, who had invited him to address the Dublin University Association on the question of the 'common origin of men and apes'. Wright had expressed some concern that Huxley's controversial topic might provoke a religious backlash, but Huxley in reply made it clear that he was far more concerned at the emasculating prospect of female attendance than by any mere 'blackguarding' by 'your Irish Holy Willies': 'I perceive I misunderstood the tenor of your former note – and that Ladies *are* to come – More's the pity – I shall have to emasculate my discourse or else be unintelligible – I think I prefer the latter alternative.'[23]

By 1864 Huxley had found anatomical evidence of the feminine inferiority that he and his Victorian contemporaries took for granted.

In his Hunterian Lectures for that year he described the structural differences he had supposedly observed between the brains of men and women: 'On the whole [the cerebral convolutions] are simpler in women than in men, and in the lower races the convolutions have a greater simplicity and symmetry than in the higher.'[24] It was this anatomical and intellectual ranking of women and blacks below white European males that, for Huxley, made it 'simply incredible' that women and blacks could ever endanger the supremacy of men like himself, and it was this scientific certainty that underpinned the reassuring message of his essay, 'Emancipation – Black and White', published the following year. Here Huxley eschewed the 'new woman-worship' on scientific grounds and confronted the 'irrepressible' woman question with the unshakable Victorian conviction that 'in every excellent character, whether mental or physical, the average woman is inferior to the average man, in the sense of having that character less in quantity and lower in quality'. History proves that man is more intelligent, responsible, passionate, artistic, and even more beautiful than woman. But although Nature has not made men and women equal, these 'facts' do not afford the 'smallest ground' for refusing to educate women as well as men, or giving them the same civil and political rights. In the name of justice, moralized Huxley, law and custom should not add to the biological burdens that weigh women down in the 'race of life': 'The duty of man is to see that not a grain is piled upon that load beyond what Nature imposes; that injustice is not added to inequality.' Let women achieve their liberal rights: let them compete with men; give them a fair field but no favour, and let Nature judge the outcome. 'So far from imposing artificial restrictions upon the acquirement of knowledge by women, throw every facility in their way'; let us have 'sweet girl graduates'; let women even become merchants, barristers, politicians; it would make no difference to the status quo:

> Nature's old salique law will not be repealed, and no change of dynasty will be effected. The big chests, the massive brains, the vigorous muscles and stout frames of the best men will carry the day, whenever it is worth their while to contest the prizes of life with the best women. . . . The most Darwinian of theorists will not venture to propound the doctrine, that the physical disabilities under which women have hitherto laboured in the struggle for existence with men are likely to be removed by even the most skilfully conducted process of educational selection.[25]

But not even such newly erected Darwinian biological barriers were strong enough to keep women in their proper place and out of serious scientific discussions. Although as secretary and then president,

Huxley had vigilantly kept the 'friponnes' from the door of the Geological Society, he had not been able to perform the same service to ethnology, for he was not then a member of the Ethnological Society. In 1860, under the aegis of Crawfurd and in Huxley's absence, women had been formally admitted to the meetings of the Society. In 1868, when Huxley was elected president, he immediately flew in the face of his own liberal platitudes by initiating the move to exclude them. Huxley was now the acknowledged self-constituted Darwinian spokesman and anatomical expert on the central and most contentious issue of the Darwinian debates: 'man's place in nature'. His popular lectures to working men on the relations of man to the lower animals, and his 1863 book with this title, had assured him of professional and public recognition.[26] But this 'question of questions', which precipitated Huxley into the fight for Darwinian control of Victorian anthropology, was not to be debated by women. The new science of 'man' might pronounce upon 'woman' (indeed, this was shortly to become one of its major concerns), but it was defined and applied by Huxley specifically to exclude the very object's participation. The symbolic Huxleyean 'beak and claws' were not only used to rend parsondom in defence of the Darwinian programme, but also, as he had forewarned Lyell, turned against women's scientific aspirations. The Ethnological Society, which had so briefly (and disastrously in Huxley's view), opened its doors to women, was to be professionally closed to them. No longer would it be viewed disparagingly as a 'ladies' Society', but henceforth, like its institutional competitor, the all-male Anthropological Society, it would be relieved of the emasculating presence of women. At a stroke, Huxley thereby sought to upgrade the professional status of the Darwinian-led Ethnologicals and to remove one of the major impediments to their amalgamation with the recalcitrant and vociferous Anthropologicals. By 1868, the continuing schism between the two Societies had become a serious obstacle to Darwinian dominance of this key discipline, and Huxley was intent upon their unification under the 'proper direction' of the Darwinians.[27]

'We are the students and interpreters of nature's laws'

In 1860, when the contentious issue of female admission was first broached in the Ethnological Society, ethnology had not been admitted to the Darwinian pantheon. It was not one of the sciences with which Huxley had as yet 'to do', and the Society was not, generally speaking, a grouping likely to attract his iconoclastic attention. The ethnologists, who had their roots in Quaker and evangelical philanthropy, conducted their researches within a religious framework and sought to account for racial diversity in terms consistent with the Bible. They were primarily 'monogenists' who accepted some modification over time as races diverged from their original unity of type.

However, in recent years, the religious conservatism of the Society had been challenged by a small but growing membership of physical anthropologists, including James Hunt. The physical anthropologists were primarily 'polygenists' who advocated the ultimate diversity of the human races and opposed the theological concern of the ethnologists to derive all races from a single stock. They were generally men like Hunt himself, with a background in medicine, and their method was strictly anatomical. They placed great emphasis on describing, measuring and classifying the physical types of humanity, forming rigid categories that maximized racial differences and justified the polygenist belief in essential human diversity and inequality. On the whole, they lacked the benevolent, protectionist racial attitudes of the ethnologists, and they were inflexibly determinist in their interpretation of racial differences. In Hunt's case, racial determinism extended to an extreme and virulent racism that eventually precipitated his final break with the Ethnologicals.[28] But by that stage Hunt had established the Anthropological Society, which met for the first time on 6 January 1863. And as Hunt and others represented it, the issue of female admission was crucial to the formation of the new Society.

Hunt conceived the Anthropological Society as a platform for his racial/political opinions, which he could not voice within the confines of the Ethnological Society. He and his supporters wanted a forum where they could pursue their version of anthropology, untrammelled by theological or social restraints. And the admission of women to meetings of the Ethnological Society was viewed by Hunt as the single greatest threat to the objects and duties of a truly scientific society. The 'grave, erudite, and purely scientific study' of anthropology required the 'most free and serious discussion, especially on anatomical and physiological topics', and this was totally at odds with the admission of women. As a dedicated anthropologist, Hunt had been the most vigorous opponent of this 'fatal mistake', but in vain:

> You will, doubtless, smile at the strange idea of admitting
> females to a discussion of all Ethnological subjects. However,
> the supporters of the "fair sex" won the day, and females have
> been regularly admitted to the meetings of the Ethnological
> Society during the past three years. Even now the advocates of
> this measure do not admit their error, nor do they perceive how

they are practically hindering the promotion of those scientific objects which they continue to claim for their society.[29]

Hunt's account is supported by the sparse evidence of the Ethnological Society minutes. As they record, Crawfurd's proposal that women should be admitted as 'visitors' was discussed at the meeting of 27 November 1860 and clearly provoked considerable opposition, not only from Hunt and his followers. 'After some discussion', the motion was amended to read that 'Ladies be admitted to the Meetings on all occasions specified by the Council.' 'This amendment', the minutes tersely state, 'was carried', and the secretaries were authorized to invite 'strangers' to the meetings. Hunt was then joint secretary, and such was his indignation and chagrin at the decision that at the meeting following his defeat, he tendered his resignation on the ostensible grounds of 'health' and 'too much Society business'.[30] This, however, he was persuaded to withdraw, and he continued as secretary until his final rupture with the Ethnological Society, when he could devote the whole of his considerable energy and talents to strengthening his flourishing male-only stronghold, the Anthropological Society. One of the earliest motions adopted by the new society, with Hunt in the chair, was that 'Ladies' might become financial members of the Anthropological Society, but 'shall on no occasion whatever be allowed to attend any of the meetings of the Society'.[31] Their cash was welcome, but not their persons.

From the start, Hunt made clear that he was not merely founding a new society, but a 'new science', and that the overwhelming significance of the new anthropology devolved upon its political implications. Race was for Hunt, as it had been for his mentor, the controversial transcendental anatomist and racial determinist Robert Knox, the key to 'scientific' political legislation and social procedure.[32] From the platform of the Anthropological Society meetings and the pages of its prolific publications, Hunt waged ferocious and unceasing war on the 'unnatural' notions of those liberals and radicals who suffered from the 'rights-of-man mania'. On behalf of the Anthropologicals and their new science, Hunt contested Huxley's liberal Darwinian bid to be nature's 'mouthpiece'. As Hunt saw it, the Anthropologicals rather than the Darwinians were the 'interpreters of nature's laws', and it was their duty to deliver their expert opinions on the practical applications of their science. According to Hunt, this meant that John Stuart Mill's claim for black and female suffrage was a scientific absurdity, contradicted by the 'facts of human nature' as revealed by the researches of the anthropologist.[33]

Hunt and the Anthropologicals were quickly infamous for their anthropological endorsement of slavery and the more racist manifes-

tations of British imperialism, such as Governor Eyre's bloody
suppression of black revolt in Jamaica. They prided themselves on
creating a forum for 'liberty of thought and freedom of speech'
unequalled by any other scientific society, and their members (who
included the notorious Richard Burton and the Duke of Roussilon),
went out of their way to confront middle-class morality. Nevertheless, the provocative, topical and often salacious discussions of the
meetings, which dwelt obsessively on such essential anthropological
topics as female circumcision, phallic symbolism and the anatomy of
'the Hottentot Venus', initially attracted a large and enthusiastic
membership. Within two years of their foundation, the Anthropologicals numbered over five hundred members (almost twice the size of
the Ethnological Society); they were engaged in an active publication
and translation programme; and were attempting to displace the
Ethnologicals from the meetings of the British Association.[34] The
significance of the exclusion of women for the phenomenal success of
the Anthropological Society was crystal-clear to Alfred Russel Wallace, who spelt it out for Huxley's benefit after Huxley had tried
to dissuade the backsliding Wallace from attending the Society's
meetings:

> I cannot agree with you that *"there was not the slightest reason for
> [the Society's] existence"*. It seems to me that its establishment is a
> good protest against the absurdity of making the *Ethnological* a
> *ladies' Society*. Consequently many important and interesting
> subjects cannot possibly be discussed there; – & as the Geo-
> graphical is also a *ladies' Society* the *Anthrop.* is the *only place*
> where they can be discussed.[35]

Huxley's failure to convince Wallace that his Darwinian duty lay
exclusively with the Ethnologicals undoubtedly reinforced his determination to put an end to this 'absurdity' of a 'ladies' Society' when
he was in a position to do so.

Huxley and the leading liberal Darwinians were quick to express
their outrage at the racist pronouncements and political polemics of
the all-too-successful Anthropologicals. But the Darwinians were not
so outspokenly critical of the anthropological excursions of their
competitors into that other major socio-political topic of the day, the
woman question. Here, apparently, they could find common 'scientific' ground with the Anthropologicals in the writings of Carl Vogt,
whose *Lectures on Man* was translated by Hunt and published by the
Anthropological Society in 1864.[36]

According to Vogt (currently one of the best-known European
exponents of Darwinism, which he linked with a militant materialism
and racism), the crania of men and women differed to such an extent

that they could be classified 'as if they belonged to different species'. Moreover, he claimed, 'they differ in their proportions more than many typical or race skulls'. It was the polygenist Vogt's opinion that the human races were actually different species whose separate lines of evolution might be traced back into the very remote past to a common ancestry, but whose current differences were so great as to be virtually unbridgeable. As for the sexes, comparative anatomy had demonstrated for him that the crania of adult women were more childlike than those of men, thus referring them to the inferior development of the lower races. Indeed, Vogt claimed that this anatomical difference increased with the development of the race, 'so that the European excels much more the female than the Negro the Negress'. This meant, as he interpreted it, that there could be no possibility of sexual equality among 'progressive' civilizations:

> Just as, in respect of morals, woman is the conservator of old customs and usages, of traditions, legends and religion; so in the material world she preserves primitive forms, which but slowly yield to the influence of civilization. We are justified in saying, that it is easier to overthrow a government by revolution, than alter the arrangements in the kitchen, though their absurdity be abundantly proved. In the same manner woman preserves, in the formation of the head, the earlier stage from which the race or tribe has been developed, or into which it has relapsed. Hence, then, is partly explained the fact, that the inequality of the sexes increases with the progress of civilization.[37]

Among the lower races, the occupations of the two sexes are similar – Bushmen and women share the same tasks – but among civilized nations there is a sexual division of labour, both physical and mental, which can be bridged only at the cost of a common degeneracy.

Hunt clearly found Vogt's anatomical endorsement of female inferiority and its political implications as congenial as his racism and polygenism. It was no coincidence that Hunt aired his version of the conflict over the issue of female admission to the Ethnological Society and its significance for the formation of the Anthropological Society in the preface to his translation of Vogt's *Lectures*. The 'fatal mistake' of the ethnologists lay not only in the constraints the presence of women might place on the free and open discussion of crucial anthropological topics, but in the laughable notion that women might actually engage in serious anthropological debate. Hunt and his fellow Anthropologicals were as scientifically certain of the intellectual and cultural inferiority of the female as they were of the Negro. As Vogt had shown, sexual anatomical and physiological differences were as indicative of intellectual and cultural differences as racial ones. Hunt intimated that some readers might find parts of the work offensive, but he confidently appealed to their masculine solidarity and superior scientific understanding: 'The Fellows of the Anthropological Society of London are happily neither women nor children' and should, as men of science, be quite ready to accept such of Vogt's opinions as can be 'logically deduced from well-ascertained facts'.[38]

These 'well-ascertained facts' proved equally acceptable to the Darwinians. Darwin himself in the *Descent of Man*, would reproduce Vogt's anatomical evidence in support of his evolutionary argument for the innate and continuing intellectual inferiority of women.[39] But, more immediately, Hunt's translation of Vogt's *Lectures* was clearly the inspiration of Huxley's anatomical and intellectual equation of women and blacks in his Hunterian Lectures of 1864. A year later, echoes of Vogt's claim that women were insusceptible to radical change could be discerned in Huxley's 'Emancipation – Black and White', where Huxley asserted that women were 'born conservatives'.[40] As his earlier comments to Lyell indicate, Huxley was in full agreement with Vogt on the impossibility of educating women in unconventional religious and scientific opinions.

Despite their differing political positions, the reactionary Hunt and the 'enlightened' Huxley were in fundamental agreement on the 'natural' inferiority of women and on a 'natural' hierarchy of race, and both men put their respective anthropologies to socio-political use. Nor was Hunt opposed to theories of development, providing these were located within an acceptable polygenist framework, like Vogt's.[41] Given the degree of coincidence between their anthropological systems and their shared naturalistic and anti-clerical orientation, it is not surprising that initially Hunt and his followers had tried to make common cause with the Darwinians against the more conservative and theologically oriented Ethnologicals. With Lyell and Darwin, Huxley was one of the first five Honorary Fellows to be elected to the newly formed Anthropological Society.[42] It is significant that at this stage Huxley had still not joined the Ethnological Society. It was only *after* the Anthropologicals (who had their own ideological and professional axes to grind) showed unequivocally that they were not to be recruited to the Darwinian cause by thoroughly alienating Huxley with a 'coarse attack' on his *Man's Place in Nature*, that Huxley resigned from the Anthropological Society and threw in his lot with the Ethnologicals. The leading Darwinians then rallied to take over the Ethnological Society and establish it as their institutional power-base in the human sciences. Thereafter, relations between the Anthropologicals and the Darwinians deteriorated to the point of open hostility and conflict, and Hunt's rhetoric became increasingly anti-Darwinian.[43]

The conflict was less theoretical in character than ideological and professional, and Hunt's 'anti-Darwinism' must be interpreted in this vein. His was a hegemonic struggle with Huxley and the Darwinians. The object was to define the ideological role of anthropology in Victorian society. In a period when traditional theological modes of explanation were giving way before a secular redefinition of the world, the Anthropologicals and the Darwinians offered two competing versions of a legitimating scientific naturalism. From their institutional stronghold of the flourishing Anthropological Society, Hunt and his followers were able, for some considerable time, to resist incorporation into the Darwinian anthropological model proffered by Huxley, and to offer formidable professional opposition to the takeover of London science by the Darwinians.[44] Hunt's vehemently proclaimed opposition to the 'monogenism' of the Darwinians was therefore largely strategic, and served rhetorical and political purposes. It was a convenient peg on which he could hang their ideological differences and demarcate the Anthropologicals from the competing but institutionally weaker 'Darwinian club'.

Both the Darwinians and the Anthropologicals were well aware that Darwinism was not incongruent with polygenism, and the leading Darwinians (including Huxley, Wallace and Francis Galton), were instrumental in bringing this more forcefully to the notice of the Anthropologicals. All of these Darwinians had, by the close of the 1860s, demonstrated how monogenism and polygenism might be reconciled in evolutionary biology. Thus in the process of liberating Darwinism from the charge of 'monogenism' with which Hunt and his cohort persisted in identifying it, the Darwinians made a number of significant concessions to the polygenist platform. This interpretation explains why so many prominent Darwinians came to incorporate so much specifically polygenist thinking into their interpretations of human history and racial and sexual differences. Their anthropological writings were designed not only to promote Darwinism as the key to the scientific study of humanity and society, but also to bridge the theoretical and institutional gap between the rival societies. When the two societies at last merged in 1871, a 'new' evolutionist anthropological model had been formed, shaped by the confrontations and negotiations between the Darwinians and the Anthropologicals.[45]

By 1868 the Darwinians were in the ascendancy in their power struggle with the Anthropologicals, who were in serious financial difficulties through their overly ambitious publishing activities and a

Huxley in Control, as depicted by his son-in-law John Collier in 1891. A very masculine Huxley, right hand trousered, leans casually but authoritatively on his accumulated wisdom, a symbolic cranium in hand. (Reproduced by permission of the Wellcome Institute Library, London.)

decline in membership brought about by internal dissension and their increasingly disreputable image. Huxley, grown impatient with the situation, took upon himself the task of putting an end to this 'scientific scandal'. He accepted the presidency of the Ethnological Society on condition that its Council support his efforts towards unification. With typical energy and ruthlessness, he set about re-organizing and strengthening the Ethnological Society and making the Anthropologicals more amenable to amalgamation. He pushed inept office-bearers off the Council, stepped up the Society's publications, and recruited the membership and support of those leading Darwinians, such as Joseph Hooker, who had not already rallied to the cause.[46] Finally, he tackled the pressing political problem of the Ethnological Society's derogatory image as a 'ladies' Society'.

Huxley's campaign was certainly political, and not only in the narrow institutional sense. By 1868 the issue of women's suffrage and women's rights had come to the fore with Mill's magnificent championing of the cause in the House of Commons, and the Anthropologicals had temporarily abandoned their pursuit of racial issues in order to confront this latest *reductio ad absurdum* of Mill's outmoded and unscientific political economy. They knew of 'no subject upon which [anthropology] ought to give a more authoritative decision than upon the claims of women to political power'.[47] Predictably, the consensus of the 'humble Anthropologists' on this topical and threatening issue was that sexual differences and capacities had arisen from 'the widespread action of natural laws, and are not to be annihilated by a merely human decree'. As Vogt had established, the mental and moral differences between men and women corresponded with their anatomical differences, and the latter were the 'true, irrevocable, everlasting, natural source of the practical and beneficial division of duties between men and women'. What then was the natural mission of woman? Nature (as interpreted by the Anthropologicals) answered the woman question with one resounding word – 'Maternity':

> It is woman's great function, and it should be her proud privilege, that she can bear and rear children to be men. . . . Is it possible to conceive a more contemptible and deplorable spectacle than that of the female (I will not profane the beautiful name of *woman*) who, having undertaken, and having appointed to her, by nature, those functions, in the proper fulfilment of which consists the charm and glory of the sex, deliberately neglects and abdicates the sacred duties and privileges of wife and mother, to make herself ridiculous by meddling in and muddling men's work?[48]

A certain professional self-interest may be detected in these anthropological refutations of women's claim to do the 'work' of men. As indicated above, many of the Anthropologicals were medics, and medicine was the first occupation to be assailed by women in their attempts to enter the professions. For once, Victorian prudery worked to women's advantage here, for in an age of extreme reticence about sexual matters, many considered that women would make more fitting obstetricians and gynaecologists. Women's rights advocates, such as Elizabeth Blackwell, often deployed this argument (which was based on a full acceptance of Victorian delicacy and was not without its ideological hazards), in order to promote the entry of women into medicine.[49] The extent to which the professedly anti-Darwinian Anthropologicals co-opted evolutionary arguments to ward off this feminine threat is impressive, and demonstrates yet again the compatability of Anthropological and Darwinian thought. Women, for instance, possessed less than men of that 'combativeness which is necessary not only in political life, but even in the ordinary struggles for existence'. Woman's subordination to man was 'natural and eternal' and any attempt to 'revolutionize the education and status of woman on the assumption of an imaginary sexual equality' would induce a 'perturbation in the evolution of races'.[50]

What difference there was between such arguments and Huxley's denial to women of any 'natural equality', existing or potential, lay in Huxley's extension to women of their right to legal and political emancipation on the understanding that they would not be able to overcome their biological limitations and 'compete' with men. The Anthropologicals, on the contrary, saw women's 'competition' as a real professional threat and predicted social and biological upheaval from such a violation of the laws of nature. Huxley and Hunt may not have been able to resolve all their differences in negotiating the amalgamation of the two Societies, but they could at least reach full agreement on the professional unfitness of women anthropologists. Nor had Huxley any need to expect much opposition on the issue from within the Darwinian-dominated Ethnological Society. The timely death of the aged Crawfurd had fortuitously eliminated the leading proponent of female admission, and without his patronage the 'ladies' themselves, having only the status of 'visitors' without the full entitlements of membership, could scarcely press their case. However, at least one of them tried. When Huxley gave notice of the intended 'expulsion' of the ladies from the Society, he received an impassioned petition on their behalf from Eliza Lynn Linton, who commended herself to him as a 'representative woman of the bread-winning class'.[51]

'Darwin . . . opened a new world to me'

The relation of Eliza Lynn Linton (1822–98), a minor novelist and journalist, to Victorian feminism is controversial. Her early life might have served as a blueprint for the emancipated woman of the time, for she built a reputation as an upholder of women's rights in education, family property and divorce. But in early 1868 came a *volte-face*: in a series of much-talked about essays in the *Saturday Review*, she became the great literary opponent of the nascent women's movement.[52]

As a young woman, Eliza Lynn demonstrated an interest in ethnology – it was her article on 'Aborigenes' that led to her becoming the first salaried professional woman journalist in Britain. With the failure of her marriage to the radical artisan William Linton (who was prominent in the National Chartist movement), she turned to science for consolation. All her life Eliza was torn by a contradiction she never managed to resolve, a contradiction that is reflected in all her writings on the woman question. On the one hand, she earned her own living, associated with leading radicals and intellectuals, and lived the life of an independent, strong-willed woman. Yet on the other, she clearly yearned for the more conventional Victorian role of wife and mother and its conservative social rewards. After 1865 the public humiliation of her failed marriage and her violation of her idealized womanly role precipitated one of those Victorian crises of faith so characteristic of the period. Eliza overcame her despair and found spiritual and social redemption in the certainties of science and the scientific meetings she began to frequent:

> Those Friday Evening Lectures at the Royal Institution, when Tyndall experimented or Huxley demonstrated . . . what evenings in the Court of Paradise those were! How I pitied the poor wretches who did not come to them! . . . I do not think there was one in the audience who drank in the wine of scientific thought with more avidity than I . . . [I]t strengthened, warmed, exhilarated and almost intoxicated me.[53]

Eliza's new creed was scientific naturalism: 'in science were FACTS, and these were of the kind to make a new mental era – a new departure of thought for the whole world'. In the 'substitution of the scientific method for the theological', she saw the emancipation of the human intellect from superstition, and she pinned her faith in human progress and her own moral redemption on Darwinism. 'Darwin', she later wrote, 'opened a new world to me. . . . The unity of Nature was the core of the creed to which I owe my subsequent mental progress – the Doctrine of Evolution that by which I have come to peace.'[54] In Darwinism Eliza found a substitute for William Linton's republican idealism that was more soothing to her lonely and difficult path through life, and more consistent with her growing political and social conservatism. All this found expression in the articles she published anonymously in the *Saturday Review*, which became known collectively by the title of one of them: 'The Girl of the Period'.

With attention-catching titles and vivid prose, Eliza vehemently fought university education for women, birth-control, women's suffrage and women's entry to the professions: 'The Girl of the Period' ('a creature who dyes her hair and paints her face ... who lives to please herself ... bold in bearing ... masculine in mind'); 'Modern Mothers' ('this wild revolt against nature, and specially this abhorrence of maternity ...'); 'What is Woman's Work?' ('professions are undertaken and careers invaded which were formerly held sacred to men; while things are left undone which, for all the generations that the world has lasted, have been naturally and instinctively assigned to women to do'); 'The Shrieking Sisterhood'; 'Wild Women'; 'Modern Man Haters'; etc., etc.[55]

Fundamental to Eliza's reiterated opposition to the goals of the 'shrieking sisterhood' was her insistence (which she held to the end of her life) that 'the sphere of human action is determined by the fact of sex, and that there does exist both natural limitation and natural direction'.[56] Like Huxley, she assumed naturalistic limits to women's aspirations, and she acquired these directly from the Victorian stereotype of femininity. There is a striking coincidence between Eliza's ideal woman and Huxley's 'fit companion'. Eliza's ideal was the domestically competent (like Huxley, she deplored 'dolls' who were 'hopelessly useless' and could do nothing with their brains or their hands), but inherently modest, maternally oriented girl who, 'when she married, would be her husband's friend and companion, but never his rival; one who would consider his interests as identical with her own, ... who would make his house his true home and place of rest'.[57]

Eliza's essays caused a sensation. They inspired cartoons, fashions in clothing, a satirical journal and several other 'Girl of the Period' publications. Eliza did not announce her authorship until 1883, when the essays were republished in book form, but since in certain circles it was an open secret, she recorded that her attacks on the 'Sacred Sex' caused some 'ill-blood' among her literary acquaintances. After her death, a leading advocate of female emancipation summed up what might serve as the feminist consensus on the essays: Eliza Lynn Linton found it 'more profitable' to attack rather than defend her sex.[58]

While this assessment has a good deal going for it, it would be a great mistake to dismiss Eliza Lynn Linton simply as a gifted writer, one-time radical and emancipated woman, who sold out to a reactionary anti-feminism through personal disappointment and for professional gain. Beneath the superficialities of her mawkish concern for the proprieties and her exaggerated expressions of distaste for those whom she denounced as 'shrieking' extremists and 'man-haters', Eliza consistently stated her allegiance to those three issues she regarded as the 'core of this question of woman's rights'. They were women's right to an education 'as good as, ... but not identical with, that of men' (what form this was to take remained unclear, but Eliza regarded some training in science as absolutely essential);[59] their right to property; and their right to divorce and custody of their children. These 'rights' were 'just and reasonable' and, above all, did not conflict with Eliza's insistence on the 'natural limitation of sphere ... included in the fact of sex'. Her belief in this fundamental biological determinism was paramount and absolute; it was grounded in her unshakable materialism and Darwinism. Women could not emancipate themselves from the laws of biology any more than the earth could free itself from the law of gravitation,[60] and it was this rigid scientific certainty that underpinned (and undermined) Eliza Lynn Linton's position on the woman question – and her petition to Huxley.

'It is not fair to exclude us'

Under the patronage of 'dear old Mr. Crawfurd', whom she had known as a child, Eliza had become a regular 'visitor' to the meetings of the Ethnological Society.[61] Its meetings were 'real meat and bread' to her mind, obsessed as she was with 'aborigines' and the liberating 'FACTS' of science. There she came in contact with 'clever men' like Huxley himself and into the presence of the 'living thought' instead of the 'deader reading' to which women like herself were generally restricted. She went home from the meetings feeling 'cleverer, enriched, ennobled'. In 1868, however, the prospect of her expulsion by the ruthless Huxley from this essential 'communion' with learned men provoked Eliza to step outside her paid role of deriding and attacking the 'girl of the period', to plead passionately on behalf of her right to a better education and better opportunities.

For one who subscribed so vehemently to the ideology of natural limitation and separate spheres, Eliza, when hard pressed, exhibited a very clear (and somewhat bitter) perception of the crippling effects of the social and cultural restrictions conventionally imposed upon Victorian women. 'I have been and am a newspaper writer to some extent', she told Huxley in a letter, 'but my area is limited and my powers are unequal to the best kind of work, owing to the want of those advantages and opportunities which men have.' Men could meet together in clubs and societies, talk, discuss, 'strike out new thoughts, hear a multiplicity of views', but 'we women sit at home and spin from the one poor brain unaided. Hence the comparative poverty of woman's work.' Well aware of the Victorian convention that knowledge in women was unfeminine, 'and scientific knowledge especially so', Eliza urged on Huxley the attendance of women at scientific meetings as one of the very few opportunities they had for serious discussion with men. Here they could set aside their social obligation to charm and amuse:

> You know how few opportunities we women have for getting any serious or valuable talk with men. We meet you in "Society" with crowds of friends about & in an atmosphere of finery & artificiality. Suppose I, or any woman – let her be as fascinating as possible – were to bombard you with scientific talk – would you not rather go off to the stupidest little girl who had not a thought above her pretty frock, than begin a discussion on the Origin of Species?[62]

If women were not to talk of science in society, and were excluded from scientific societies, how were they to learn about science?

In words that might have served as a paraphrase of Huxley's own 'Emancipation – Black and While', Eliza reminded him of his liberal Darwinian obligations:

> [W]hat are the facts of woman's personal condition? We are thrown into an active hand to hand struggle for existence all the same as men . . . . The battle of life is a very serious matter to some of us, and we are frequently hindered and heavily weighted. . . . [I]t is not fair to exclude us from the means of knowledge & of active thought, of extended views – such as we get from attending learned discussions – on the simple plea of our womanhood.

Eliza, being Eliza, did concede the ostensible reason, that 'in the interests of science (paramount of every other consideration)', there might be 'necessary' discussion on 'special subjects' that would render the presence of women 'hindering or indelicate'. But on such 'rare' occasions, 'we can always be got rid of' by advance warning, or a message left with the porter. 'I pray you with all my strength', Eliza wrote to Huxley, 'to keep us as attendants at the Ethnological meetings, & when you are going to discuss hazardous papers, give us warning, & we will stay away. Else let us be free still to attend –.'[63]

Although Eliza, fearing to put herself 'forward' and seem 'presumptuous', had made her case according to her own precepts, with proper regard for the proprieties and without 'shrieking', in the form of a personal letter to Huxley (even 'muddling up' her reasons – 'like a woman!'), her powerful but womanly plea did not meet with the anticipated 'fair' treatment from the relentless Huxley. The problem was that her offer of voluntary female exclusion from 'hazardous' papers did not meet the real point at issue. If women controlled their own occasional exclusion, the Ethnological would still be a 'ladies' Society', and Huxley was adamant that it must cease to be. He and the restructured Council came up with the ingenious compromise of demarcating between 'Ordinary Meetings' that would be for 'scientific' discussions to which 'ladies will not be admitted', and larger, public 'Special Meetings' where 'popular' topics could be discussed, and to which 'ladies will be admitted' by 'special invitation'.[64] With one timely stroke, this admirable (and typically Huxleyan) solution reconstituted the Ethnological as a 'gentlemen's society', paid lip service to the liberal principle of female admission, and retained the decorative drawing power of the 'fair sex'. All must have been well satisfied – except Eliza (and those she represented), who was now inexorably relegated to the more frivolous 'popular element' she deplored and exiled from the serious scientific discussions she craved. But, as a leading public advocate of the 'separate spheres' ideology, she was hardly in a position to complain.

As for those 'hazardous' but crucial 'scientific' subjects deemed unsuitable for female ears and, presumably, beyond the intellectual comprehension of such amateurs: the very first all-masculine 'Ordinary Meeting' convened by Huxley featured a series of reports on 'customs connected with childbearing amongst the natives of Australia and New Zealand'.[65] The irony of excluding women from a male discussion of such an indelicate topic as childbirth is compounded by the fact that the 'expert' author of the reports was Joseph Hooker, the leading Darwinian botanist, who had been hastily recruited to the ranks of the 'professional' ethnologists by the self-same Huxley who was so anxious to expel the 'amateur' ladies. The perversity of this would have been lost on Huxley, who, having demonstrated his concurrence with Hunt on the issue of female admission, was now busily pushing the goal of amalgamation with the Anthropologicals and Darwinian dominance of this key discipline to an ultimately successful conclusion. With Hunt's sudden death in mid-1869, the Anthropologicals lost their charismatic leader, and the remaining dissidents offered only token resistance to the forceful Huxley. By the beginning of 1871, the Ethnological and Anthropological Societies were amalgamated as the Anthropological Institute of Great Britain and Ireland, and the Darwinians were firmly in control of this politically sensitive science of 'man'.[66]

'I am at a loss to understand'

Darwin's *Descent of Man*, published in the same year, consolidated the Darwinian endorsement of many features of Hunt's polygenist platform.[67] Racial issues aside, much of the book's discussion of the mental and moral differences between men and women had previously been rehearsed by the Anthropologicals. Darwin was as insistent as any Anthropological on the biological basis of the continuing intellectual inferiority of women, and as much opposed to Mill's environmentalist interpretations. Like the Anthropologicals (and Huxley) before him, he brought Vogt's anatomical 'observations' and their built-in social implications to his aid. By asserting the instinctively maternal and inherently modest traits of the human female, and the male's innate aggressive and competitive characteristics, Darwin provided naturalistic corroboration of deeply entrenched Victorian values and proffered an evolutionary justification of woman's narrow domestic role and contemporary social inequalities. Above all, he followed Huxley's lead by arguing on evolutionary grounds that the higher education of women could have no long-term impact on their social evolution and was, strictly speaking, a waste of resources.[68]

Recent scholarship has begun to undermine the traditional historiographic distinction between Darwinism and Social Darwinism, arguing that 'Darwinism was "social" from the start', and that 'Social Darwinism' is the 'artifact of a professional discourse that increasingly pretended to divorce science from ideology'.[69] The process, it now seems clear, began with the 'young guard' Darwinians themselves, who, under Huxley's leadership, sought to advance their interconnected professional and social interests by relating science and Darwinism in particular to the social stability of the nation. It was Huxley's great achievement as a propagandist that he so skilfully distanced the scientist from the ideologue, while promoting science and the scientist as the independent and neutral arbiters of pressing social and political problems. He was the 'master of concealed debate', and while he took the heavy-handed Spencer to task, Huxley made his own subtle accommodation of ethical 'oughts' to biological 'ises'. His winning 'social stratagem' of the 1860s was to bring Darwinian anthropology and biology to the aid of a rapidly advancing liberal bourgeoisie who, with the decline of religion, lacked a compelling ideological defence against equality and democracy. It was primarily with Huxley's dexterous hands that ideologically neutral Darwinism erected the necessary barriers, by proving that the inferior could not compete in an open society.[70]

The triumph of the Darwinians over the Anthropologicals must therefore be seen as as much an ideological as an institutional victory. It was not merely a question of 'style' – that Hunt and the Anthropologicals alienated scientific and social support by their tasteless and destructive confrontation with Victorian morality and organized religion, and their total lack of 'gentlemanly' comportment in debate.[71] It was more the fact that the more 'stylist' (i.e., more socially conformist) Darwinians, and Huxley in particular, occupied the higher moral ground politically and ideologically as well. Rather than directly counterposing his anthropology to liberal bourgeois politics and ideology in the manner of Hunt, Huxley's Darwinian anthropological model indirectly accommodated them. Where Hunt biologized a range of reactionary political and social positions on the interrelated Negro and woman questions, Huxley argued the more socially appealing and insidious liberal line that these groups might achieve 'emancipation', while imposing strict biological limitations on the consequences.

It was a two-edged ideological weapon that Huxley deployed with considerable finesse and success in the face of his own demonstrated violation of the liberal principles he professed. When Sophia Jex-Blake turned to Huxley for assistance in her fight to gain medical qualification at the University of Edinburgh, she got no practical help whatever from this certified 'emancipator' who had rejected women's egalitarian claims in the name of science, but who had publicly pledged to 'work heart and soul' towards the 'practical ends' of women's emancipation.[72] Jex-Blake and a handful of pioneering women students had been reluctantly accepted into the medical faculty under new regulations for the admission of women, which stipulated that they be instructed in separate classes. Matters came to a head in 1872 when their attempts to comply with the regulations and receive their separate instruction in anatomy were blocked by the University Court, which denied the legitimacy of their instructor's qualifications but refused to permit him to prove them by examination. Jex-Blake begged Huxley to examine the women's instructor (who had already successfully filled the position of anatomical demonstrator at the Surgeon's Hall), and provide him with a certification that would be recognized by the University Court. But the eminent anatomist and dedicated 'emancipator' proved singularly reluctant to involve himself in this feminine 'storming of the citadel'. Instead,

Huxley doled out a carefully measured *soupçon* of ideological support. He 'sympathized' with Jex-Blake's cause (even though he did
not think that women were on average as intelligent as men), but
expressed his professional solidarity with those professors of anatomy, physiology and obstetrics who objected to teaching mixed
classes of young men and women. He himself had consistently refused on moral grounds to admit women to his own lectures on
comparative anatomy. Nevertheless, the same Huxley who had categorically refused to teach the 'virgins' of Bedford Square, assured Jex-
Blake that he would 'not hesitate' to teach anything he knew to a
class of women. It was, therefore, with 'great regret' (and a totally
uncharacteristic lack of resourcefulness) that Huxley was 'compelled
to refuse' her plea.[73] The women students were forced to bring
an action against the University, which had recommended that they
give up their claim to graduation (the only legal passport to practice) and accept informal (and professionally useless) certificates of
'proficiency'.

Yet again, when Jex-Blake's papers were referred to him for his
professionally disinterested scrutiny, Huxley endorsed her failure at
the final examinations. These had been taken under the combined
stress of the women's contested legal action against the University
(which they eventually lost), and continual intimidation and harass-
ment from the medical staff and male students. Huxley found that
'certain answers were not up to the standard'. Huxley and Nature
had sat in judgement, and Jex-Blake was 'plucked', if not from
'Nature's university', most definitely from the University of Edinburgh. She had failed to compete, presumably in a 'fair field', and
most assuredly with 'no favour'. Once more Huxley manipulated the
situation to his ideological advantage, on this occasion with a letter to
The Times (which had publicized Jex-Blake's failure and Huxley's
part in it). Lest Miss Jex-Blake might think that his decision was
'influenced by prejudice against her cause', for the last time Huxley
flourished his tarnished liberal credentials: without seeing any reason
to believe that women were on average as strong physically, intellectually or morally as men, he could not shut his eyes to the fact
that many women were much better endowed in these respects than
many men. To exclude such women from the profession of medicine
went against all justice and the best interests of society, and Huxley
(with the exclusion of women from the Ethnological Society and the
newly amalgamated Anthropological Institute safely behind him),
was 'at a loss to understand' it.[74]

It is notable that when the London School of Medicine for Women
was founded by Jex-Blake in 1874, Huxley was not among those

qualified lecturers who gave their time and help to the School. Nor did this famous Darwinian, who made the tag of a 'liberal education' synonymous with his name, ever use his unparalleled opportunities, professionally or publicly, to advance the cause of women's higher education, to which he was supposedly so devoted, in any practical way. With professional anthropology now a male preserve, Huxley's interest in the woman question in science was over. It had served its purpose, and he quietly dropped it when, in the more conservative climate of the 1870s, the advocacy of women's emancipation came to be seen as more of a liability than a strategic advantage to the Darwinians, by then effectively running Victorian science from the epicentre of the influential X Club (an exclusively masculine enterprise, although the select 'x's' occasionally brought their 'yv's' to the more frivolous group events).[75] In any case, a woman's successful graduation from 'Nature's university' meant that she had learned to live in harmony with Nature's laws, and, as liberally interpreted by Nature's dominant Darwinian 'mouthpiece', these all but precluded her graduation from any Victorian institution of higher education.

That Huxley continues to be historically evaluated as an 'enlightened' advocate of women's emancipation is a tribute to the manoeuvrability of his position on the woman question. The historical reality is that he suborned women's emancipation to the Darwinian control of anthropology. He not only used his considerable professional powers to exclude women from organized science, but, in conjunction with the leading Darwinians, he also subtly reinforced late-Victorian assumptions of white male supremacy and contributed to the scientific anti-feminism that characterized evolutionary biology and anthropology in this period.[76] In effect, Huxley excluded women from science in the name of science and redefined that science to ratify their exclusion. It could be argued that the impact of his two-edged ideological position on the woman question was even more damaging to nineteenth-century feminist ideology than that of Hunt's total opposition. For while Huxley appeared to proffer ideological support to the feminists in the name of liberalism, he paved the way for the scientific subversion of their liberal egalitarian roots through his rejection of their egalitarian claims in the name of Darwinism. For many feminists, themselves deeply committed to naturalistic scientific explanations and to the new Darwinism, the only recourse from the concerted Darwinian drawing of naturalistic limits to their claims, was to retreat from the egalitarian ideal, and to assert that woman was 'different but equal'. They claimed a biologically based 'complementary genius' for woman, a 'genius' rooted in her innate maternal and womanly qualities. Eliza Lynn Linton, the contradictions notwithstanding, is best located among those advanced women of the nineteenth century whose confidence in the liberating powers of science, and whose opposition of naturalistic interpretations of human nature and society to conventional theological wisdom and authority, ultimately betrayed them, when science, especially Darwinism, gave a naturalistic, scientific basis to the class and sexual divisions of Victorian society.[77]

It is only comparatively recently, in the wake of the second wave of feminism, that feminist scholars have been able to transcend Eliza's dilemma and emancipate themselves from the sociobiological laws of the latter-day 'Darwinians', by shifting the focus from the 'woman question in science', to the 'science question in feminism' – by asking how a science 'so deeply involved in distinctively masculine projects can possibly be used for emancipatory ends'.[78]

Notes

I should like to thank Jim Moore and James Secord for discussions and criticism, John Greene for his earlier encouragement of my work in this area, and the following institutions and libraries for permission to study manuscript material: the Imperial College of Science and Technology, the British Library, and the Royal Anthropological Institute.

1. E. Lynn Linton to T.H. Huxley, 11 Nov. 1868, Imperial College Archives, Huxley Papers (HP), 21.223-6.

2. T.H. Huxley, 'A Liberal Education' (1868), quoted in John C. Greene, *Science, Ideology, and World View: Essays in the History of Evolutionary Ideas* (Berkeley: University of Calfornia Press, 1981), pp. 141–3.

3. Ibid. See also Frank M. Turner, 'The Victorian Conflict Between Science and Religion: A Professional Dimension', *Isis*, 69 (1978), 356–76 and Gay Weber, 'Science and Society in Nineteenth Century Anthropology', *History of Science*, 12 (1974), 260–83.

4. See George W. Stocking, 'What's in a Name? The Origins of the Royal Anthropological Institute (1837–71)', *Man*, 6 (1971), 369–90; Ronald Rainger, 'Race, Politics, and Science: The Anthropological Society of London in the 1860s', *Victorian Studies*, 22 (1978), 51–70; and Evelleen Richards, 'The "Moral Anatomy" of Robert Knox; The Interplay Between Biological and Social Thought in Victorian Scientific Naturalism', *Journal of the History of Biology*, in press.

5. See Elizabeth Fee, 'Nineteenth-Century Craniology: The Study of the Female Skull', *Bulletin of the History of Medicine*, 53 (1979), 415–33 and *idem*, 'The Sexual Politics of Victorian Social Anthropology', in Mary S. Hartman and Lois Banner, eds., *Clio's Consciousness Raised: New Perspectives on the History of Women* (New York: Harper Torchbooks, 1974), pp. 86–102.

6. See Robert M. Young, 'Darwinism *is* Social', in David Kohn, ed., *The Darwinian Heritage* (Princeton, N.J.: Princeton University Press, 1985), pp.

609–38 (617).

7. Greene, *Science, Ideology, and World View*, p. 2.

8. Archives, Royal Anthropological Institute, 'Council Minute Book, 1844–1869', Ethnological Society of London (ESL) Minutes.

9. James Hunt, 'Dedication to Broca', in Carl Vogt, *Lectures on Man: His Place in Creation, and in the History of the Earth* (London: Anthropological Society, 1864), p. viii.

10. See Lynn Barber, *The Heyday of Natural History, 1820–1870* (London: Jonathan Cape, 1980), p. 132 and Margaret Alic, *Hypatia's Heritage: The History of Women in Science from Antiquity to the Late Nineteenth Century* (London: Women's Press, 1986), pp. 178–81.

11. C. Lyell to T.H. Huxley, 16 March 1860, HP, 6.32.

12. T.H. Huxley to Lyell, 17 March 1860, HP, 30.34; published in Leonard Huxley, *Life and Letters of Thomas Henry Huxley*, 2 vols. (London: Macmillian, 1900), I, 211-12. (hereafter *LL*).

13. T.H. Huxley to E. Haeckel, *LL*, I, 289.

14. T.H. Huxley to E. Dyster, *LL*, I, 125.

15. Barber, *Heyday of Natural History*, pp. 125–38.

16. Eveline, Countess of Portsmouth, to Henrietta Huxley, 21 Oct. 1874, HP, 28.107.

17. Barber, *Heyday of Natural History*, p. 133.

18. Quoted in ibid., p. 134.

19. Andrew C. Ramsay, 15 March 1856, Imperial College Archives, Lett's Diary no. 1, Ramsay Papers/1/24, 42V. I am indebted to James Secord for this reference.

20. *LL*, I, 181. Women, primarily as wives and daughters of British Association members, had been admitted to all sections of the meetings in 1839, but they were confined to separate galleries or railed-off areas. The first woman member of the Association was admitted in 1853, but even as late as 1876 women were not permitted to hold office. See Alic, *Hypatia's Heritage*, pp. 179–81.

21. *LL*, I, 183–4. This account is attributed to Professor Farrar, Canon of Durham.

22. See Stocking, 'What's in a Name?', pp. 380–1; John W. Burrow 'Introduction', in Charles Darwin, *The Origin of Species*, reprint edn (Harmondsworth, Middlesex: Penguin Books, 1968), p. 4; M.J.S. Hodge, 'England', in Thomas F. Glick, ed., *The Comparative Reception of Darwinism* (Austin: University of Texas Press, 1974), pp. 3–31 (11); and Michael Ruse, *The Darwinian Revolution* (Chicago: University of Chicago Press, 1979), pp. 251–2.

23. T.H. Huxley to E.P. Wright, 8 March 1862, HP, 29.115.

24. Quoted in Mario A. Di Gregorio, *T.H. Huxley's Place in Natural Science* (New Haven, Conn.: Yale University Press, 1984), p. 169.

25. T.H. Huxley, 'Emancipation – Black and White' (1865), in *idem, Collected Essays*, 9 vols.; reprint edn (New York: Greenwood Press, 1968), III, 66–75 (73–4).

26. For an assessment of Huxley's anthropological writings and their motivations, see Di Gregorio, *T.H. Huxley's Place*, pp. 129–84 and Adrian

Desmond, *Archetypes and Ancestors: Paleontology in Victorian London, 1850–1875* (London: Blond and Briggs, 1982).

27. T.H. Huxley to J. Lubbock, 18 Oct. 1867, British Library, Avebury Papers (Correspondence of Sir John Lubbock, V), 49642.63.

28. See Stocking, 'What's in a Name?', p. 376.

29. Hunt, 'Dedication to Broca', p. viii. See also *idem*, 'On the Origin of the "Anthropological Review" and Its Connection with the Anthropological Society', *Anthropological Review*, 6 (1868), 431–42 (433).

30. ESL Minutes, 27 Nov. 1860, 6 Feb. 1861, 20 Feb. 1861.

31. Royal Anthropological Institute Archives, Anthropological Society of London (ASL), 'Council Minutes', 5 Aug. 1863.

32. See Richards, '"Moral Anatomy" of Robert Knox'.

33. [James Hunt], 'Race in Legislation and Political Economy', *Anthropological Review*, 4 (1866), 113–35; *idem*, 'Anniversary Address to the Anthropological Society of London, Jan. 5, 1864', *Journal of the Anthropological Society of London*, 2 (1864), lxxx–xciii; *idem*, 'Anniversary Address, Jan. 1, 1867', *Journal of the Anthropological Society*, 5 (1867), xliv–lxx; *idem*, 'On the Negro's Place in Nature', *Memoirs of the Anthropological Society*, 1 (1863), 1–64.

34. Stocking, 'What's in a Name?', pp. 377, 380.

35. A.R. Wallace to T.H. Huxley, 26 Feb. 1864, HP, 28.91. Wallace continued sporadically to attend the meetings of the Anthropological Society until 1868, when Huxley excluded women from the 'ordinary meetings' of the Ethnological Society.

36. See Stocking, 'What's in a Name?', p. 377. Not all of the Darwinians shared Huxley's more liberal views on racial issues and the Governor Eyre incident. See Douglas A. Lorimer, *Colour, Class and the Victorians: English Attitudes to the Negro in the Mid-Nineteenth Century* (Leicester: Leicester University Press, 1978), pp. 131–200.

37. Vogt, *Lectures on Man*, p. 81.   .

38. Hunt, 'Dedication to Broca', and *idem*, 'Editor's Preface', in Vogt, *Lectures on Man*, pp. xii–xiii. Paul Broca was the acknowledged French leader of craniometry who used his researches to support his assumption of the intellectual inferiority of women and blacks. See Stephen Jay Gould, *The Mismeasure of Man* (New York: Norton, 1981), pp. 73–112.

39. Charles Darwin, *The Descent of Man, and Selection in Relation to Sex*, 2nd edn, 1874 (London: John Murray, 1889), pp. 557, 566.

40. T.H. Huxley, 'Emancipation – Black and White', p. 71.

41. Hunt consistently maintained his commitment to a vaguely defined naturalistic developmentalism. See Richards, '"Moral Anatomy" of Robert Knox'.

42. The others were the polygenist Crawfurd and the 'developmentalist' Richard Owen. See ASL Council Minutes, 18 Feb. 1863.

43. See Richards, '"Moral Anatomy" of Robert Knox'.

44. On the takeover of London science by the Darwinian 'young guard', see Turner, 'The Victorian Conflict Between Science and Religion'.

45. For characterizations of this 'new' evolutionist anthropology, see Stocking, 'What's in a Name?'; Weber, 'Science and Society in Nineteenth

Century Anthropology'; and Lorimer, *Colour, Class, and the Victorians*, pp. 131–61.

46. T.H. Huxley to J. Hooker, 24 Jan. 1868, HP, 2.140. See Stocking, 'What's in a Name?', p. 383.

47. Luke Owen Pike, 'On the Claims of Women to Political Power', *Journal of the Anthropological Society*, 7 (1869), xlvii–lxi (xlvii); G. Harris, 'On the Distinctions, Mental and Moral, occasioned by the Difference of Sex', *Journal of the Anthropological Society*, 7 (1869), clxxxix–cxcv. For historical background, see Jill Liddington and Jill Norris, *'One Hand Tied Behind Us': The Rise of the Women's Suffrage Movement* (London: Virago Press, 1984).

48. J. McGrigor Allan, 'On the Real Differences in the Minds of Men and Women', *Journal of the Anthropological Society*, 7 (1869), cxcv–ccxix (ccxii).

49. See Joan Burstyn, *Victorian Education and the Ideal of Womanhood* (London: Croom Helm, 1980), p. 85 and Regina Morantz, 'The Lady and her Physician', in Hartman and Banner, *Clio's Consciousness Raised*, pp. 38–53 (48).

50. Pike, 'On the Claims by Women', pp. liii, lix; McGrigor Allan, 'On the Real Differences', p. ccxiii. McGrigor Allan is here quoting Broca's criticism of the feminists. See Paul Broca, 'On Anthropology', *Anthropological Review*, 6 (1868), 35–52 (50).

51. Lynn Linton to T.H. Huxley, 11 Nov. 1868, HP, 21.223–6.

52. Biographical details on Eliza Lynn Linton are available in: George S. Layard, *Mrs. Lynn Linton: Her Life, Letters, and Opinions* (London: Metheun, 1901); Herbert Van Thal, *Eliza Lynn Linton: The Girl of the Period* (London: George Allen and Unwin, 1979); and F.B. Smith, *Radical Artisan: William James Linton, 1812–97* (Manchester: Manchester University Press, 1973).

53. Eliza Lynn Linton, *The Autobiography of Christopher Kirkland*, 3 vols.; reprint edn (New York: Garland, 1976), III, 96. This work is a dramatization of Lynn Linton's own life in a male persona.

54. Ibid., p. 79.

55. Eliza Lynn Linton, *The Girl of the Period and Other Social Essays* (London: Richard Bentley and Son, 1883).

56. Ibid., 'Preface'.

57. Ibid., p. 1.

58. See Van Thal, *Eliza Lynn Linton*, p. 76.

59. See Lynn Linton to Mr Benn, 1881, in Layard, *Mrs. Lynn Linton*, p. 203.

60. Lynn Linton, *Autobiography*, III, 2–4.

61. Ibid., III, 171.

62. Lynn Linton to T.H. Huxley, 11 Nov. 1868, HP, 21.223–6. This passage has also been published by Burstyn (*Victorian Education*, p. 44) to evince the Victorian convention that women were never to talk 'business' with men.

63. Lynn Linton to T.H. Huxley, 11 Nov. 1868, HP, 21.223–6.

64. 'Report of the Council', *Journal of the Ethnological Society of London*, n.s., 1 (1868–9), vii–xv (xiv).

65. 'Ordinary Meeting' (23 Feb. 1869, with Huxley in the Chair), *Journal of the Ethnological Society of London*, n.s., 1 (1868–9), 68–75.

66. On the domination of the Anthropological Institute by the Darwinian

'ethnologicals', see Stocking, 'What's in a Name?', pp. 383–6.

67. Ibid.

68. See Evelleen Richards, 'Darwin and the Descent of Woman', in David Oldroyd and Ian Langham, eds., *The Wider Domain of Evolutionary Thought* (Dordrecht, Holland: Reidel, 1983), pp. 57–111.

69. James Moore, 'Socializing Darwinism: Historiography and the Fortunes of a Phrase', in Les Levidow, ed., *Science as Politics* (London: Free Association Books, 1986), pp. 38–80 (39). See also Young, 'Darwinism *is* Social'.

70. See Desmond, *Archetypes and Ancestors*, pp. 158–64; Michael Helfand, 'T.H. Huxley's "Evolution and Ethics": The Politics of Evolution and the Evolution of Politics', *Victorian Studies*, 20 (1977), 157–77; and Eric J. Hobsbawm, *The Age of Capital, 1848–1875* (London: Wiedenfeld and Nicolson, 1975), p. 268.

71. See Stocking, 'What's in a Name?', pp. 380–1; cf. Richards, '"Moral Anatomy" of Robert Knox'.

72. T.H. Huxley, 'Emancipation – Black and White', p. 71. See also Margaret Todd, *The Life of Sophia Jex-Blake* (London: Macmillan, 1918), pp. 383–4, 415–18 and E. Moberly Bell, *Storming the Citadel: The Rise of the Woman Doctor* (London: Constable, 1953), pp. 62–109.

73. T.H. Huxley to S. Jex-Blake, 28 Oct. 1872, in *LL*, II, 387.

74. T.H. Huxley, letter to *The Times*, 8 July 1874, *LL*, I, 417.

75. *LL*, I, 258. On the influence of the X-Club, see Roy M. MacLeod, 'The X-Club: A Social Network of Science in Late-Victorian England', *Notes and Records of the Royal Society of London*, 24 (1970), 305–22. A rough correlation may be drawn between Darwinian dominance of a learned society and its resistance to the admission of women. Thus the Royal Society, the Geological and Linnean Societies, and the Royal Microscopical Society (all of which had a heavy preponderance of Darwinians), did not admit women until the twentieth century. See Alic, *Hypatia's Heritage*, p. 181.

76. See Weber, 'Science and Society in Nineteenth Century Anthropology'.

77. See Flavia Alaya, 'Victorian Science and the "Genius" of Woman', *Journal of the History of Ideas*, 38 (1977), 261–80 and Richards, 'Darwin and the Descent of Woman'.

78. Sandra Harding, *The Science Question in Feminism* (Ithaca, N.Y.: Cornell University Press, 1986), p. 29.

9

Ideology, evolution and late-Victorian agnostic popularizers

BERNARD LIGHTMAN

> The lines between science, ideology, and world view are seldom tightly drawn.
>
> John C. Greene[1]

In 1885 *The Agnostic*, a short-lived journal published in the heyday of Victorian unbelief, summed up the painful difficulty confronting agnostic writers should they wish to propagate their world view. How could agnosticism, based on an elaborate epistemology that required familiarity with complex scientific and philosophic concepts, ever appeal to the masses of English society? According to *The Agnostic*, there could only be two possible responses to this question: either agnostics could spread the good news of agnosticism among the people in a debased and simplistic form, or, if 'we be resolved that it shall not degrade itself, then, in the present condition of humanity, it must be content with an audience that is fit but few'.[2] The author of the 'article argued for the second alternative. He urged agnostic writers to resist the temptation to simplify and vulgarize the high philosophy of agnosticism in order to win more converts.

Not all agnostics agreed, despite their shared allegiance to scientific naturalism. Although both middle- and lower-class agnostics opposed Christian claims to possess divinely revealed truth by insisting on the limits of human knowledge, they fashioned their philosophy to accommodate particular audiences. Middle-class scientific naturalists such as T. H. Huxley, who first formulated agnosticism during the 1860s and 1870s, did not aim at proselytizing among the masses. They used agnostic philosophy to attack other members of the intelligentsia who represented the old social and political order dominated by the Church and the aristocracy. But during the mid-1880s a new school of agnostics, led by Charles Albert Watts, came upon the scene. They were willing to simplify the epistemological and scientific basis of agnosticism in order to reach a mass audience. Furthermore, they believed that the key to popularizing agnosticism lay in presenting it as a religious creed that had evolved out of Christianity. Through bold publishing ventures, these men became the chief popularizers of the evolutionary world view in late-Victorian culture.

The evolutionary world view of the new agnostics was inextricably connected to an ideology that reflected the delicate position of lower middle-class freethinkers committed to gradual social change. On the one hand, the new agnostics were prepared to follow the elite agnostics in reading into an evolving nature a message that sanctioned the necessity of slow progressive change and ruled out the radicalism of both Secularists and socialists. On the other hand, the new agnostics of the 1880s and 1890s were responding to urgent social problems, particularly in London, which led them at times to oppose their mentors by finding support in evolution for moderate democratic reforms. The elite agnostics reacted by rejecting the claims of the new agnostics to scientific expertise and by disputing that agnosticism could be given a religious gloss. The right to be known as a genuine agnostic therefore became the prize contested by lower middle-class agnostics and the elite scientific naturalists. With it went the status of being a legitimate interpreter of the correct ideological meaning of evolution in a time of political ferment.

C.A. Watts and the new agnosticism

The moving force behind the new agnosticism was Charles Albert Watts (1858–1946), son of Charles Watts (1836–1906), a leading English Secularist. As a teenager, C.A. Watts worked at his father's printing business at 17 Johnson's Court and thereby became immersed in the world of radical publishing. His plans for popularizing agnosticism were conceived at a time when Secularism was divided by bitter quarrels that directly involved his father. Though at first allied with the militant atheist Charles Bradlaugh (1833–91) and his National Secular Society (NSS), the elder Watts refused in 1877 to defend Bradlaugh's right to republish Knowlton's *Fruits of Philosophy*, a pamphlet on birth control. Feeling that Bradlaugh had brought disgrace upon the NSS by his prosecution in the Knowlton affair, he joined with the founder of Secularism, George Jacob Holyoake, in a bid for continued respectability by seceding from the NSS to form the British Secular Union (BSU).[3] Throughout the early 1880s Watts became more and more distressed as his father's brand of non-militant Secularism declined in popularity. In 1880 Bradlaugh was elected Liberal MP for Northampton, but his avowed atheism was used as grounds to ban him from Parliament. Sympathy for Bradlaugh grew as he was re-elected four more times. When in 1886 he finally was permitted to take the oath of allegiance and sit legally in the Commons, his notoriety had revitalized the NSS. The BSU remained a tiny dissident minority within Secularism. Watts's father, his standing in the movement drastically reduced, accepted an invitation to a free-thought 'pastorate' in Toronto. He remained in Canada from 1884 until after Bradlaugh's death.[4]

Watts's respect for his father and the tradition of non-militancy led him to consider a new strategy for placing the ideas of dissident Secularism at the forefront of organized freethought. Impressed by the success of the middle-class scientific naturalists, he decided to model his strategy on their methods. Huxley and his allies, Herbert Spencer and John Tyndall, were at the peak of their power during the 1880s. Yet as part of the new professional middle class they had originally found themselves, like the dissident Secularists, on the periphery of the English intellectual establishment. By harnessing their fortunes to the rising star of science and by developing agnosticism into a formidable body of doctrine, the middle-class scientific naturalists had forced their way into the charmed circle of power. Although the dissident Secularists were neither scientists nor members of the professional middle class, the younger Watts thought he could use elements of the successful strategy adopted by the scientific naturalists in combination with non-militant methods previously adopted by his father to offset Bradlaugh's popularity. He would broaden the appeal of the ideas embraced by dissident Secularists so that people from the more respectable classes would be attracted to the movement. He would move towards an alliance with eminent middle-class unbelievers and away from Secularism's radical working-class roots.

Agnosticism was the key to Watts's new strategy, the banner round which he hoped to rally dissident Secularists and middle-class unbelievers. Other philosophic tags, such as 'atheist', 'infidel' and 'freethinker', were all associated with forms of unbelief that were aggressive and even obnoxious. An 'agnostic', however, represented the most up-to-date phase of scientific unbelief. His doctrine was free from offensive connotations and, thanks to the middle-class scientific naturalists, it had an air of respectability that would help to distance dissident Secularists from the vulgarity of militant freethought. Science, and especially evolutionary theory, was an important component of the new agnosticism. Watts's emphasis on evolutionary theory established a new direction for Secularists, who made little use of Darwin until the early 1880s. Bradlaugh's rather crude anti-clericalism and

love of Bible-bashing, based on an outmoded form of biblical criticism, was to be given up for modern scientific arguments against
Christianity.[5]

The reorganization of dissident Secularism was the main goal
pursued by Watts. The press was his chosen means. Formerly,
Secularists had relied on the power of the press to strengthen their
movement. Bradlaugh's Freethought Publishing Company, G.W.
Foote's weekly paper *The Freethinker*, and the republication of
Knowlton's *Fruits of Philosophy* were just a few of the projects they
undertook. But Watts probably viewed these efforts as amateurish; he
believed that freethought literature should be written and packaged
with a different appeal. Now he would focus the energies of Secularists on reaching likely converts through the publication of quality
pamphlets, books and periodicals. These would be produced in
adherence, more or less, to a policy of non-political alignment.
Eschewing Bradlaugh's attempt to work through the political system
to achieve reform, Watts published material that examined religious
and scientific issues apart from the distracting, and sometimes
disunifying, controversies surrounding national party politics.

When Watts took over his father's publishing business in 1884, he
had a chance to experiment with his new strategy for dissident
Secularism. That year he brought out *The Agnostic Annual* (later *The
Rationalist Annual*), modelling it on successful up-market periodicals.
As editor he also recruited leading middle-class unbelievers to lend
the journal prestige. The first issue featured a symposium on agnosticism with pieces from Huxley and Francis Newman. The idea of a
symposium was taken from the highly popular symposia of *The Nineteenth Century*. In January 1885 Watts began publishing *The Agnostic*
with the aim of establishing 'a monthly periodical of cultured liberal
thought, which, by its moderation and ability shall commend itself to
the attention and support of advanced thinkers of every grade'. Later
that year Watts & Co. brought out Albert Simmons's *Agnostic First
Principles*, a succinct summary of Spencer's *First Principles* (1862). The
book was intended to be an authoritative guide for new agnostics
interested in evolutionary theory. Watts called it an 'Agnostic textbook' and praised Simmons for aiding 'our efforts to popularise (in so
far as it can be popularised) Philosophical Agnosticism'.[6]

In the world of publishing Watts had a huge advantage over
Bradlaugh. Watts knew the business inside out and was able to make
a profit as well as advance the cause of freethought. Bradlaugh and
other Secularist leaders, by contrast, were largely self-taught journalists; their publishing ventures were seldom self-supporting. As Watts
saw his ventures of 1884 and 1885 succeed, he began to plan an
ambitious breakthrough in freethought publishing. He realized that

certain avenues to power, used advantageously by the scientific naturalists, were not open to the new agnostics. There were no academic posts they could expect to win, no scientific organizations they could hope to infiltrate. Watts knew they would have to rely on the power of the press as an educational institution. Vast resources would be required to publish immense quantities of freethought works. Channels of distribution would have to be arranged to bypass the legitimate book trade, which remained hostile to anti-Christian literature. To overcome these obstacles *Watts's Literary Guide* was launched in 1885.[7]

Intended as a publisher's circular (now a familiar means of publicity but then a novelty), this new journal served as a bridge between Watts and the scattered freethinking readership. It advertised the publications of Watts & Co., reviewed current books, and, beginning in 1893, added a monthly supplement condensing important works on progressive thought and science. A number of Watts's publications also conducted fund-raising appeals. 'The Agnostic Press Fund' was set up to aid in issuing a series of popular leaflets on the principles of the new agnosticism. Larger works, in a form suitable for wide distribution, would follow. The 'Propaganda Press Fund' supported the establishment in 1890 of the Propaganda Press Committee, with Watts as secretary and Holyoake as chairman. In 1893 the organization became known as the Rationalist Press Committee, and in 1899 it was incorporated under the Companies Acts and renamed the Rationalist Press Association (RPA).[8]

Watts hoped to turn his headquarters at Johnson's Court into a propaganda machine for freethought and agnosticism that would outdo any of Bradlaugh's publication efforts and would rival the Society for Promoting Christian Knowledge and the Religious Tract Society. By capitalizing on the appeal of the new agnosticism and disseminating it through innovative publishing strategies, he set about transforming dissident Secularism into a respectable, middle-class organization.[9]

W.S. Ross and the crusade against Christianity

Watts had a powerful ally in William Stewart Ross (1844–1906), whose *nom de plume*, Saladin, was taken from the twelfth-century Sultan of Egypt and Syria renowned for his courage and magnanimity in battle against Christians during the Third Crusade. Although Ross was not involved directly in Watts's creation of the RPA, he belonged to the Holyoake tradition of non-militant dissident Secularism and had adopted independently a similar publishing strategy. Born at Kirkbean, Galloway, in 1844, Ross was the son of Joseph Ross, a farm servant and a devout presbyterian. At the age of twenty he went to the University of Glasgow to prepare for the ministry, but he took up writing instead. He went to London and started his own press in 1872, William Stewart & Co., which published educational works on English history and literature. Attracted at first to freethought by Bradlaugh's leadership of the NSS, Ross came to despise him for the Knowlton affair. He joined the breakaway BSU and in 1882 became joint editor with the elder Watts of its organ, *The Secular Review*. When Watts emigrated two years later, Ross held the editorship alone.[10]

Ross now joined C.A. Watts in pushing dissident Secularism towards the new agnosticism. In January 1885 *The Secular Review* was given the new subtitle, *'A Journal of Agnosticism'*. Four years later the periodical was renamed *The Agnostic Journal and Secular Review*. In the editorial column, 'At Random', Ross consistently attacked both vulgar Secularism and orthodox Christianity. His literary talents exceeded those of most Secularist writers, and he had a knack for relating satirical anecdotes. By 1886 Bradlaugh had become so sick of Ross's ridicule that he refused to speak at Secularist branch meetings where *The Secular Review* was on sale.[11]

But although Ross often wrote passionately, as if he were indeed leading a crusade, he insisted that criticism of established religion should avoid vulgarity. He agreed with C.A. Watts that the views of an advanced thinker must be stated 'like a scholar and a gentleman', that the best arguments for Secularism were drawn from philosophy and modern science, and that the less said about party politics the better. Ross had a high personal regard for Watts, the son of his old comrade on the *The Secular Review*. And after Ross's death in 1906 C.A. Watts wrote, 'We have lost a Chief whose name will live in our movement – a truly heroic soul, whose chivalry was almost incomparable.' Of the many collaborative projects undertaken by Watts and Ross, the most significant was *The Agnostic Annual*, edited by Watts, printed by Watts & Co., and published by W. Stewart & Co. until 1896. Ross needed the support of Watts in his publishing ventures. Subsidizing *The Agnostic Journal* drained Ross's resources to the point where he had to appeal to subscribers in 1891 to help him out of debt. But Ross remained a prolific writer – his works included *God and His Book* (1877) and *Why I am an Agnostic* (1889). Watts was always ready to supply an outlet for his scintillating articles in *The Agnostic Annual*.[12]

Watts's stable of agnostic propagandists

Although Ross played a significant role as printer, publisher and writer in proselytizing for the agnostic creed, Watts was the mastermind behind the new strategy of inundating the reading public with material on agnosticism and evolution. Watts himself wrote little because he had a low opinion of his writing ability, so he relied on the pens of others. Like Ross, they were popularizers – self-improving, non-professional individualists with a commitment to reputable Secularism. Richard Bithell (1821–1902) proclaimed his availability as early as 1883 in *The Creed of a Modern Agnostic*. He published *Agnostic Problems* in 1887 and, with Watts & Co., brought out *The Worship of the Unknowable* (c. 1889) and *A Handbook of Scientific Agnosticism* (1892). Born at Lewes, Sussex, Bithell was the son of a smithy. He taught mathematics and chemistry for the British and Foreign Schools Society and eventually received a Ph.D. from Göttingen University and a B.Sc. from London. In 1865 he entered the banking house of the Rothschilds where he worked until his retirement in 1898.[13]

Another important member of Watts's group of popularizers was Frederick James Gould (1855–1938), who with Watts and Bithell was a founding member of the Propaganda Press Committee. Gould's father, who mended cheap jewellery and gave violin lessons, was almost penniless when Frederick was born. To save money, the boy was sent at the age of ten to St George's Chapel, Windsor Castle, to sing in the choir, though it meant a painful separation from his parents. The pious atmosphere at Windsor, afterwards perpetuated in a school Frederick attended in the village of Chenies, led to his conversion in 1871 to zealous evangelicalism. This was followed by increasing doubts, and by the early 1880s Gould was an open participant in the Secularist movement. Subsequently, he wrote several important works, including *Stepping-Stones to Agnosticism* (1890) and *The Agnostic Island* (1891), both published by Watts & Co. He taught for the London School Board from 1877 to 1896, but resigned to work in the Ethical Movement, and later he became secretary of the Leicester Secular Society.[14]

Samuel Laing (1811–97) was perhaps Watts's greatest catch. His set of popular books – *Modern Science and Modern Thought* (1885), *A Modern Zoroastrian* (1887), *Problems of the Future* (1889) and *Human Origins* (1892) – won him an influence with the general public on a par with some of the chief thinkers of the day. Watts could always count on him to supply copy for *The Agnostic Annual*. Ross also found Laing a helpful ally. The lead article in the first issue of the *Agnostic Journal* was Laing's credal recitation of agnostic beliefs, which had been solicited by none other than the Liberal statesman W.E. Gladstone. Unlike Gould, Bithell, Ross or even Watts, Laing had a middle-class background and a Cambridge education. He associated dissident Secularism with wealth and status as chairman of the London,

Brighton and South Coast Railway for over thirty years, and as a Liberal MP almost continuously from 1852 to 1885.[15]

Gould, Bithell, Laing and Ross all pursued goals that served Watts's strategy, and with considerable success. They explicitly voiced an interest in reaching an audience composed of 'younger readers, and of the working classes who are striving after culture', as Laing put it. And they had a missionary zeal to acquaint this audience with recent scientific discoveries and their bearing on major problems of the day. These new agnostics did not claim to be original thinkers; they were only gifted popularizers, able to give simple, clear and vivid accounts of complex scientific and philosophic theories. Furthermore, they were synthesizers of the first magnitude. They desired to demonstrate that modern science could present an integrated and rational world view, encompassing every realm of thought.[16]

The grounds of belief in the existence of a unified scientific world view lay, for the new agnostics, in their certainty that all natural phenomena were governed by fixed and uniform laws. The unity of modern science only mirrored the unity of nature. One particular natural law, the 'law of evolution', was of special significance to the new agnostics. Likening Darwin's evolutionary principle to Newton's law of gravity, Laing declared it to be 'a fundamental law accepted as axiomatic by all men of science, and as the basis of modern thought, to which all religions and philosophies have to conform'. Evolution applied to the development of both the organic and the inorganic worlds; it applied to man as a physical being and to the products of man's so-called spiritual being, including religion and ethics. Only one law of evolution, however, was so all-encompassing, and it was not appropriately associated with the name of Charles Darwin. The new agnostics were in fact primarily attracted to the cosmic evolutionism of Herbert Spencer, and they often ranked him as Darwin's superior.[17]

To Watts, Spencer was the 'great master'. Simmons referred to Spencer in Agnostic First Principles as 'the greatest philosopher that the world has ever seen'. 'By the side of Spencer', he declared, 'Darwin is a dwarf.' Bithell admitted that he was indebted to Spencer more than to any other writer. And Laing's Modern Science and Modern Thought, which modestly offered summaries of the main facts in all departments of knowledge, resembled nothing so much as an abridgment of Spencer's ten-volume System of Synthetic Philosophy (1862–96). Laing asserted that Spencer's 'wide influence' came precisely from the way in which he applied evolutionary principles beyond mere nature 'to the subjects which more immediately concern the mass of thinking minds, such as history, politics, and the problems of social life'. Ross alone objected to the excessive enthusiasm of agnostics

The new agnostic propagandists of the 1880s (clockwise from top left): C.A. Watts, W.S. Ross ('Saladin'), F.J. Gould and Samuel Laing. (From Gould, *The Pioneers of Johnson's Court* [London: Watts and Co., 1929], pp. 3, 7, 45 and Gordon G. Flaws, *Sketch of the Life and Character of Saladin* [London: Watts and Co., n.d.].) (Reproduced by permission of the Rationalist Press Association.)

who hailed Spencer as a genius, but he could nevertheless recommend 'the passing of an Act of Parliament to encourage the teachings of Herbert Spencer' in place of the doctrines of the Anglican Church. In so far as Ross held that all things 'have a tendency to rise to higher levels', he showed implicit adherence to the Spencerian view of evolution as cosmic and goal-directed.[18]

Paeans of praise aside, the new agnostics were philosophically inspired by Spencer's adumbration of the agnostic position in the doctrine of 'The Unknowable', which appeared in his *First Principles*, the introductory volume of the *Synthetic Philosophy*. Just as Spencer had made the notion of the limits of knowledge central to that doctrine and the premise of his entire system, so Bithell spoke for the new agnostics when he affirmed in his *Agnostic Problems* that 'the essential and distinguishing characteristic of a true Agnostic is that he seeks to define more clearly than has hitherto been done the boundaries of his knowledge and the limits of his intellectual powers'. Yet belief in the limits of knowledge did not rule out, either for the new agnostics or for some of the elite scientific naturalists, any assertion regarding the existence of a mysterious deity. Indeed, the capitalization of 'The Unknowable' seems otherwise merely gratuitous. Agnostics, Bithell maintained, 'are Theistic in recognizing the existence and activity of a Supreme Power immanent in the process of evolution'. What distinguished the new agnostics from theistically inclined elite scientific naturalists such as Spencer was the alacrity with which the former acknowledged their god, and their readiness to draw theistic inferences from the evolution process. The new agnostics frankly celebrated evolution as a manifestation of the power of the Unknowable. The law of evolution acquired its cosmic and teleological character by courtesy of Spencer's deity. Whereas Spencer was careful to speak of his Unknowable god in scientific and detached terms, the new agnostics often wrote so as to engage the emotional and religious sensibilities of their readers. Whether it was Bithell encouraging his readers to 'worship the Unknown and the Unknowable', or Gould reminding his audience of the 'unspeakable wonder' of the 'Unknowable Reality', the new agnostics busily exaggerated theistic themes they had found in Spencer, Huxley and other elite scientific naturalists. Watts was so taken by the religious possibilities in agnosticism that he tried to establish an Agnostic Temple in southwest London.[19]

Socialism and the threat of revolution

The Spencerian version of evolution and agnosticism not only supplied the new agnostics with a god; it also authenticated their view of themselves and their place in the world. It enabled them to cultivate a self-image of responsible citizens. During the 1880s and 1890s, English society came under pressure from a number of urgent problems. A faltering economy, rising unemployment, revelations of shocking poverty and widespread labour unrest caused intellectuals grave concern. Within the Liberal party a group of important thinkers began to reformulate classical political theory, abandoning the old idea of private property as the concrete embodiment of man's worth and the association of liberalism with unrestrained competition and atomistic individualism. They saw the State as an agent of social reform and sought support for their views from evolutionary biology. This, however, only increased the anxiety of elite scientific naturalists, who feared that the 'new liberalism' would lead to 'regimental' socialism and the redistribution of wealth. Some of them, already alienated by Gladstone's Irish Home Rule policy, deserted the Liberal party and became Unionists. They joined the chorus of voices between 1885 and 1900 calling for a conservative, though not Tory, social and political philosophy that would undermine socialism. Huxley, Tyndall and others argued persistently that there were natural limits to social innovation that ruled out socialistic short-cuts to some supposed heaven on earth.[20]

Because the new agnostics had recently entered the bourgeoisie or, like Laing, were well established there, they shared many of the fears of the elite scientific naturalists. And the particular bourgeoisie to which they and the scientific naturalists belonged was based in London, where the social stresses of the 1880s were most acutely felt. Ross warned that revolution could result if the workers' demand for bread and jobs was not met. Laing, too, was anxious about the threat posed by the 'barbarians' who were 'accumulating in the slums of . . . great cities'. In his *Agnostic Problems* Bithell referred to the 'horrible' conditions disclosed a few years earlier in a sensational anonymous pamphlet, *The Bitter Cry of Outcast London*, while Gould discussed in his autobiography how 'the great Dock Strike [of 1889] agitated the district, and drew me into a closer study of the Labour problem'. Like Huxley and the other scientific naturalists, the new agnostics were also afraid that social unrest would increase support for political radicalism, including both Secularism and socialism. However, Bradlaugh was one of the bitterest enemies of socialism, and Secularism declined in part because of his opposition. By the late 1880s socialists had won many converts from Secularist societies and a number of Secularist leaders had defected, the most spectacular being Bradlaugh's protégé, Annie Besant. The growing power of socialism constituted a more immediate problem for the new agnostics, with their links to dissident Secularism, than for the elite scientific naturalists, because Huxley and his colleagues had little reason to fear that their professional ranks could be infiltrated by socialists.[21]

To discredit both the social reforms associated with the 'new liberalism' and socialism as an organized movement, the new agnostics adopted an intellectual strategy similar to that of the elite scientific naturalists. They used evolutionary theory to legitimate a conservative vision of social order. It was 'the ease with which analogies could be drawn between biology and social life' that rendered evolutionary theory ideal for Watts's purposes. Despite the fact that he held to an official policy of non-political alignment in his journals and associations, he and his cohorts had a definite political orientation: the majority of the RPA's directors and members were liberals who had 'little quarrel with the established order'.[22] But in the present state of English politics there was little benefit to be realized by pinning their hopes on a particular party. The Conservatives were too enmeshed in the coils of the Anglican Church, whereas the Liberals' alliance with Dissent and their tolerance of state reformism made a working relationship difficult. The new agnostics therefore saw the dissemination of their creed as a non-partisan endeavour.[23]

The evolutionary theodicy of the new agnosticism

Within their overall strategy for undermining radicalism and socialism, the new agnostics employed at least two main tactics. There was, first, the straightforward approach. Socialism could be classified as maladaptive, as it was contrary to nature and science. Ross declared that 'unnatural Socialism' was doomed to failure because it attempted to 'reconstruct society upon the rubbish heaps of scientific error and the quicksands of philosophic misconception'. Laing argued, more specifically, that socialist societies could not win the struggle for survival because they destroyed 'all individual initiative and enterprise in material life'. One of the bluntest attacks on socialism came in an article by Frederick Millar, 'Freethought and Individualism', published in 1889 in *The Agnostic Journal* with the editor's commendation, 'Socialism and Darwinism are at daggers drawn'. Millar, one of the members of Watts's Propaganda Press Fund, had recently debated the issue with Besant. His position was that the modern scientific freethinker 'cannot logically become a Socialist'. Whereas socialism aimed at equality by eradicating 'competition from the affairs of men', the political creed of Darwinism could only be 'Individualism'. 'If Darwinism and Evolution teach anything upon matters political and social', Millar asserted, 'it is that human society is an organism governed by natural laws, to which free scope should be given.' Without mentioning Besant by name in the article, he ridiculed a certain prominent freethinker who had become a socialist four or five years earlier. An 'Evolutionist Socialist' to him was just as incomprehensible as a 'Spencerian Salvationist'.[24]

A second tactic adopted by the new agnostics was more subtle than explicitly opposing evolutionary theory to socialist ideals. While elite scientific naturalists presented to their educated audience a secular theodicy based on evolution, in which God became virtually identified with the natural laws guaranteeing progress, the new agnostics frankly attempted to justify the ways of God to man in theological terms. They argued that God's goodness and omnipotence could be seen in the evolutionary process at work in nature and society, despite the existence of evil and suffering in the world. The evolutionary theodicy of the new agnostics went something as follows:

We should be reconciled to the existence of evil because evolutionary theory reveals to us that on the whole, and in the long run, all things work together for the good. Progressive and continuous change in nature and society, ordained by God as part of the nature of things, will eventually triumph and bring a heaven on earth. Our role is to be patient, and allow the process to unfold naturally, even though it may require that we or others suffer. Above all, we must resist the temptation to tamper with the evolutionary process by adopting schemes that try to effect radical social innovation. Should we pursue this path, we would be attempting to thwart the will of God and bring upon ourselves needless suffering.

Like their elite mentors, the new agnostics claimed a deeper insight into the nature of things than clergymen could attain, but at the same time they assumed the role of interpreters of God's inscrutable ways. Their theodicy was, ironically, less secular in its appeal. It gave the elite doctrines a distinctive popular twist: God was explicitly left in. Evolution became at once a naturalistic, a theistic and a teleological process.[25]

Bithell especially was concerned with the problem of evil. Nature perpetrates many horrors, he admitted in *Agnostic Problems*, yet God is the author of nature and, as First Cause, is responsible for everything. How is this puzzle to be solved? Bithell suggested that, as no pain or sense of evil was felt by the vegetable world, the problem of physical and moral evil was a corollary of the evolution of higher forms of life. It indicated 'nothing more than the development of higher sensibilities, and more acute powers of discrimination', which no rational man would give up 'merely for the sake of ridding himself of the alternations of Pain and Pleasure, Right and Wrong'. From this point of view, Bithell claimed, the problem of evil vanishes, for, rather than being a 'serious defect in the Divine government of the world', the existence of evil indicates 'a manifestation of superior creative energy, or a process of evolution carried to a higher stage'.[26]

Bithell's belief that the existence of evil is part and parcel of the evolution process, an inevitable by-product of the laws of nature, was echoed in various ways by his colleagues. Laing, for example, viewed the harmony of the universe as coming out of the ever-changing relationship between the opposing forces of the evolutionary process, which he referred to as the 'law of polarity'. Where there was good there must be evil. For some unknown reason, Laing declared, an inscrutable deity governs the universe through 'a fundamental law which may make the polarity of good and evil a necessary condition of existence'. Any increase in good, whether it be a rise in the scale of living or a growth in population, was accompanied by more evil, increased cost of living or aggravated unemployment. 'When "merry England" dwelt in rural hamlets and villages', Laing wrote, 'the "bitter cry" of East London could scarcely have been written. Turn it as you like, increase of population means increase of poverty.' By maintaining the impossibility of evolutionary progress 'without its counterpart of suffering', Laing could justify the existence of poverty in capitalist society.[27]

The new agnostics also pointed out that although evil inevitably exists, science decclares that evil is bound to disappear, albeit slowly, through the progressive course of evolution. Gould assured the readers of *The Agnostic Annual* that as the social organism gradually became better adapted, the result would be enhanced happiness and the lessening of evil – in short, progress. Laing was no wild-eyed optimist, but he had to admit that 'progress is certainly towards higher, and very probably towards happier conditions'. Ross gave the same doctrine an idealistic gloss. Pain and suffering were not absolutely real; they belonged to the illusion of appearances. 'As in our evolutionary ascent we approach nearer to our re-absorption in the Absolute', he rhapsodized, 'pain and suffering and sorrow shall pass away.'[28] Bithell, more sensitively, recognized that physical evil was inflicted on good and bad men alike. He drew a parallel to evils 'in social and political life' that 'shock our moral sense'. We see worthless men lapped in luxury while the honest and virtuous struggle with adversity and want, and 'members of the aristocracy (so called) squandering their resources in a gambling hell, while those who slave and toil to provide the means are left to pine in misery'. But happily, Bithell reminded his readers, there is 'another principle at work' alongside the 'Evolution of Evil', which is a 'tendency towards the *Elimination* of Evil'.[29]

Why, then, did evil tend to disappear? The new agnostics never doubted that despite the seeming amorality of the universe, a higher principle was at work, pushing the evolutionary process in an ethical direction. Evolution, as Bithell remarked, was one of 'the manifestations of the Unknowable', and this Spencerian deity, immanent throughout the universe, produced ultimately beneficent effects. Bithell praised the 'growing minority of thoughtful and religious persons' who were 'trying to understand their true relation to God the Unknowable, and to study the ways in which He manifests Himself in the universe, in society, and above all in the Soul'. Laing did not fear the outcome of clashes between radicals and conservatives because he believed that an inscrutable God lay behind the evolutionary process, which itself produced opposing social and political forces. People could always 'trust to what religious men call Providence, and scientific men Evolution, for the result' of every conflict.[30]

The new agnostics' evolutionary theodicy was addressed to a popular audience, many of whom were either religious or from religious backgrounds. It made sense of the world as they experienced it and explained why attempts at radical social change, no matter how well intentioned, were doomed to fail. Although efforts 'to bolster up the unfit and the unworthy, at the expense of the fit and the worthy, may to all appearances, succeed for a time', Millar warned, 'in the end it will be found that nature is mightier than governments, and that "the laws for nature are tougher than red tape"', for 'no legislation can reverse the order of nature existing in human societies'. A manly determination was needed to accept the deplorable cruelty that resulted from unrestricted competition between individuals, but overly emotional socialists, filled with pity for the poor, wished to 'neutralize the action of a natural law, the reign of which will ensure' the future welfare of the entire race.[31]

Ultimately, the evolutionary theodicy of the new agnostics was designed to reconcile humanity to the present state of things – to create contentment in the current stage of a dynamic, self-adjusting, divinely sanctioned process. The fate of the reverent agnostic, who strove to know the laws of nature in order to shape his conduct in conformity with those laws, could be contrasted to that of the impious radical who ignored the lessons of science and tried to meddle with evolution. Whereas the agnostic prospered from his knowledge, the radical suffered from ignorance. Here, indeed, were words of comfort for those self-improvers bent on following in the footsteps of the successful new agnostics. Gould urged his readers to 'act in harmony with the purpose of the universe'; Laing suggested that the 'true destiny of man' was to further the evolutionary process by improving himself and others through 'constant struggles upwards'. Bithell believed that 'the wise and good man is he who succeeds in making himself acquainted' with the inflexible laws through which God controls the world and who submits 'unreservedly' to them. Submission to the Unknowable, adapting to the laws of nature, 'conforming' to 'physical or social laws' – all were one and the same for Bithell. By pointing to the gradual progress in the universe, by advocating gradual reform, by counselling submission to the forces that be, and by stressing the limits to social innovation, the new agnostics pursued a strategy for containing revolutionary pressure at its critical point. Theirs was a theodicy designed to engage the religious sensibilities of a lower middle-class audience.[32]

Religious agnosticism and elite scientific naturalism

The conservatism of the new agnostics' interpretation of evolution was of a piece with the conservative – even traditional – thrust of their religious thought. 'Reverent and devout agnosticism', as Laing put it, sometimes translated into a tolerant attitude towards Christianity that must have struck Secularists like Bradlaugh as rank heresy. All of the new agnostics (with the possible exception of Watts) had been raised in families saturated with the popular evangelicalism of the day. Ross confessed that, because his mother had used the Bible to teach him to read, he retained an emotional attachment to Christianity even though his intellect rejected its doctrines. Nor was the agnostic debt to Christianity seen in such personal terms alone. For Bithell, agnosticism was not 'alien to the Christian or Jewish Churches'. In 'the bosom of these Churches it has been fostered', he declared, 'and, by a gradual process of evolution, extending through 4,000 years, has acquired consistency, clearness, and force.' Agnosticism, then, could be seen, not as the negation of Christianity, but as the next step in its orderly progressive development. Theologically, as well as socially, according to Gould, 'there need be, and there will be, no revolution'. Laing celebrated the emergence of 'Christian Agnosticism'. Referring to Christianity as 'one of the forces of Evolution', he wrote that 'the lines of Agnostic Christianity and of Agnosticism pure and simple . . . have converged so closely that the difference between them is almost reduced to a name'. Secularists should therefore be tender with the forms and creeds of Christianity, it being 'far better that the transformation requisite to bring them into accordance with the evolution of modern thought caused by the discoveries of science, should take place gradually and spontaneously from within, rather than forcibly and abruptly from without'. Secularists must 'trust with cheerful faith to evolution to bring about gradually such changes of form as may be required to embody changes of spirit'.[33]

The new agnostics also retained a place in their world view for non-secular and non-empirical religious concepts with affinities to Christian supernaturalism. Ross was interested in mysticism, spiritualism and theosophy. His belief that psychic evolution tends to the merging of all individuals into the divine 'Absolute' led him to study Eastern religions. (Huxley also looked into Eastern thought while writing on evolution and ethics, but his motivation was clearly different.) Evolution, according to Ross, could be described legitimately as 'the upward passing through Karma to Nirvana'. Nor was this hankering after the East an isolated view among the new agnostics. Watts shared Ross's interests. Laing rejected spiritualism as fraudulent but attempted to rehabilitate the old Persian religion of Zoroastrianism, the essence of which, he maintained, was the belief in a good, but not omnipotent, God in constant conflict with an evil being of almost equal power. Laing thought Zoroastrianism was fully consonant with developments in modern science, especially evolutionary theory and his own law of polarity. Indeed, Zoroastrianism was ideally suited to a theodicy that allowed for the necessary existence of evil, yet absolved God of all responsibility for it.[34]

This flirtation with Eastern thought, mysticism, spiritualism and theosophy strained the already fragile relationship between the elite scientific naturalists and the new agnostics. On matters of principle, Huxley and his colleagues were willing to support even the militant Secularists. Huxley signed a memorial in 1883 condemning the debarment of Mrs Besant and Bradlaugh's daughter Hypatia from classes at University College; Leslie Stephen criticized the exclusion of Bradlaugh from Parliament; Huxley, Stephen and Spencer all supported a petition objecting to the blasphemy laws under which G.W. Foote was imprisoned in 1883, and to the severity of his sentence. Otherwise the elite agnostics generally held aloof from members of the NSS, disclaiming any sympathy with their coarse atheistic philosophy. The followers of the Holyoake tradition were better suited to their tastes, and Tyndall reportedly subscribed to *The Secular Review* because he liked its fair-mindedness and absence of abuse. Still, however, the moderate members of the BSU were not always respectable enough for the elite agnostics.[35]

When Watts began to disassociate himself from the militant Secularists by promulgating the new agnosticism in the mid-1880s, he was able to enlist the support of some of the elite agnostics, although they

DANGERS OF DOGMATISM.

Brown (a mild Agnostic, in reply to Smith, a rabid Evolutionist, who has been asserting the doctrines of his school with unnecessary violence). "ALMOST THOU PER-SUADEST ME TO BE A CHRISTIAN!"

Conflicts between unbelievers as seen by *Punch* (5 June 1880): the manner rather than the matter of one's convictions is associated with respectability in religion.

remained suspicious of men claiming to be agnostics with former links to Secularism. In 1882 Spencer had refused to travel with Holyoake on the same boat to the United States, fearing damage to his reputation, but a few years later, according to Gould and Watts, he called at 17 Johnson's Court and was the first subscriber to *The Agnostic*. If this is true, however, Spencer certainly did not see fit to publicize his support for Watts, nor is there mention of it in his letters or autobiography. Huxley, for his part, wrote a brief article for the first issue of *The Agnostic Annual* (1884) as part of the symposium on agnosticism. But he came to feel that Watts had tricked him into contributing, and relations between the two were never good.[36]

What most distressed Huxley was the religious conservatism of the new agnostics. In 1889 he ridiculed Laing's attempt to turn agnosticism into a 'creed'. He also took a pot shot at the unscientific application of the theory of polarity outside its proper domain. Laing felt disconcerted when he read Huxley's attack, for previously their relations had been cordial. In 1866 Laing published an account of the artefacts, bones and human remains found in burial mounds and shell refuse heaps near Keiss Castle. Huxley thought enough of his *Prehistoric Remains in Caithness* to add a supplement of fifty pages, describing and illustrating the human skulls. Laing in turn referred to his colleague in *Modern Science and Modern Thought* as one of the 'first scientific authorities'. But now, four years later, Huxley's criticisms prompted him to point out a passage in another essay by his 'guide, philosopher, and friend' that seemed to apply the law of polarity to phenomena outside the domain of magnetism and electricity.[37]

What was at stake in this debate was nothing less than the legitimacy of conflicting claims to the title 'agnostic'. Focusing on the theory of polarity, Huxley purported to show that the very basis of Laing's agnostic creed was unscientific. Huxley and the other elite scientific naturalists could welcome the work of the new agnostics if it did not deviate from their views, but when there was disagreement, the elite agnostics reminded the popularizers of their non-professional status. This, however, was an affront to men who believed they were doing their utmost to advance the evolutionary world view of the elite naturalists. When Ross leapt to Laing's defense in *The Agnostic Journal*, he impeached more than Huxley's 'frivolous laugh' at the doctrine of polarity. Huxley may have coined the term 'agnosticism', but his attitude towards the extension of his philosophy by non-professional men was plainly supercilious. To the new agnostics, Huxley's insensitivity to the religious and mystical dimension of the doctrine of evolution was born of an undue proprietary concern for a theory and a philosophy that was destined to become the intellectual common coin of the realm.[38]

Reform and Laing's 'Rational Radicalism'

Strained relations between the two agnostic groups were not only the result of disagreements over the religious possibilities of the evolutionary world view; the new agnostics could, at times, alienate the elite agnostics by reading into nature an ideological message that was too reform-minded for conservative, upper middle-class tastes. The successful strategy of scientific naturalism had permitted Huxley and his friends to push their way into the intellectual elite. They had turned away from their youthful radical-liberalism and, by the late 1880s, had become part of the political establishment. Such respect, for example, had Huxley come to command that in 1892 he was made a Privy Councillor by the Tory Prime Minister Salisbury. The new petit-bourgeois agnostics, on the other hand, occupied a social and political position that entailed interests more in line with those of working-class Secularism. They held social and political views that shifted between conservative theodicy and moderate reform. Their hostility to radicalism was tempered with sympathy for attempts to engage directly with concrete social problems such as poverty and unemployment, and this affected their understanding of the evolutionary process.

The speed at which social change could be expected to take place within the process was a matter of interpretation. The new agnostics were willing to go much further than the elite scientific naturalists in emphasizing the dynamism of evolution. This led to a conflict between the two agnostic groups that is perhaps best epitomized in Laing's controversy with Huxley in 1890 over the relative merits of democracy and aristocracy. To some extent, Laing was the political spokesman for the popularizing agnostics. Even Ross, who was less of a democrat than Laing, deferred to his fellow agnostic in political matters on account of his vast experience as a Member of Parliament.[39]

Laing, anticipating his retirement from the Commons, first traced the growth of democratic conviction through his long political career in the January 1884 issue of the *Fortnightly Review*. He was astonished to find how far he had moved from original sympathy with Peelite liberalism towards the radicalism of Chamberlain. He sought to justify this 'gradual process of "political evolution"' by depicting it as a healthy experience shared by many other sincere Liberal politicians. In his own case, the conversion to 'Rational Radicalism' had come about through acquaintance with events in the United States, where democracy had proven, on the whole, wise and successful. The English aristocracy, by contrast, had been on the wrong side in all the great questions of foreign policy, in each case choosing to pursue a plan of action contrary to the well-considered and permanent interests of the country as a whole. To Laing, recent history demonstrated that the aristocracy could no longer rest their claim to superior power on the basis of greater political wisdom.[40]

Six years later, Huxley's essay 'On the Natural Inequality of Men' spurred Laing to enlarge on his political convictions in the *Contemporary Review*. No doubt still smarting from Huxley's ridicule of his theory of polarity the year before, he criticized the attempt of elite naturalists to use scientific arguments against democracy. If Huxley could draw upon his authority as a natural scientist to dismiss an essential component of Laing's evolutionary agnosticism, Laing could equally question Huxley's competence in political theory, inasmuch as here Laing considered himself the specialist and Huxley the amateur.

Laing began his article 'Aristocracy or Democracy' by noting how remarkable it was that Huxley, long looked upon as the most brilliant champion of advanced thought, now propounded the same fossilized Tory principles put forward ever since the controversies that surrounded the first Reform Bill. Huxley supported the aristocratic theory that the best government is selected by a small, hereditary, privileged class who, on account of superior wealth and education, understand political questions better than the masses. This was opposed by the democratic theory that the best government is obtained from the 'outcome of the varied opinions and conflicting views of a very large number of voters', comprising nearly the whole of the adult community.[41]

To adjudicate the two theories, Laing put the question: Under which system do we get the best leaders? To guarantee a scientific answer he proposed to apply the 'surest test of truth, whether in scientific, or in political and social evolution, "the survival of the fittest in the struggle for existence"'. Here Laing claimed to be adopting Huxley's own principle for determining a valid answer to the question, but he nevertheless insisted that the application of evolutionary theory would demonstrate the superiority of the democratic theory. According to Laing, Huxley had reached the wrong conclusion because he lacked the expertise required to apply evolutionary theory correctly in the realm of politics. Laing conceded Huxley's 'superiority in scientific attainments and in literary ability, but in this particular class of questions', he stated, 'I have the advantage over him of being a Specialist'.[42]

Laing went on to present the facts supporting his argument for the greater fitness of democracy. In the United States and all English-speaking colonies with self-government, representative institutions and a wide franchise, democratic principles had provided strength and enabled them to survive the struggle for existence. But in England an aristocratic political system had only served to weaken the country's position in the world. Recent affairs made it plain that 'the "classes" have not shown that superiority in political wisdom which is claimed for them over the "masses"'. Indeed, had the State been left to the guidance of the select few, Laing declared, all of the beneficial reforms of the last sixty years, such as the Reform Bills, the Education Act, Free Trade, and the repeal of the Corn Laws, would have been rejected, and England would have confronted revolution. To explain why the aristocracy had ceased to be a selection of the best of the nation, Laing set the problem in an evolutionary framework. He likened the 'hereditary' aristocracy to a useless organ that inhibits the social organism's struggle to survive. The 'enervating influences of luxury and idleness' had rendered the aristocracy unfit and out of step with the political views of the majority of their countrymen. Laing concluded that Huxley's aristocratic theory was bound to lose credence as the aristocracy themselves succumbed to the 'inevitable progress of democracy'.[43]

The Laing/Huxley controversy shows graphically how readily evolution could be adapted to suit the new agnostics' social aspirations. Just as Spencerian cosmic evolutionism was to give their dissident Secularism a broad appeal in a pervasively religious culture, so the flexibility of evolutionary theory as a social dynamic made it a potent weapon for attacking elite scientific naturalists who temporized about democratic reforms, as well as for criticizing unscientific socialists and radical Secularists who were too impatient to wait for the inevitable. Here the 'tensions and ambivalences' implicit in the evolutionary world view, which John Greene has analysed with such distinction, were played out and, indeed, only become fully comprehensible, against the backdrop of particular social and political events.[44]

Notes

The author wishes to express his appreciation to Gordon Stein, Sydney Eisen, Mark Francis and John Bicknell for helpful suggestions on early drafts of this essay. Material from the Huxley Papers appears with the permission of the Imperial College of Science and Technology; quotations from the Tyndall Papers are reproduced here by courtesy of the Royal Institution.

1. John C. Greene, *Science, Ideology, and World View: Essays in the History of Evolutionary Ideas* (Berkeley: University of California Press, 1981), p. 2.

2. Robert Aitken, 'Agnosticism and the Masses', *Agnostic*, 2 (Oct. 1885), 436.

3. Susan Budd, *Varieties of Unbelief: Atheists and Agnostics in English Society, 1850–1960* (London: Heinemann, 1977), pp. 35–52.

4. Edward Royle, *Radicals, Secularists, and Republicans: Popular Freethought in Britain, 1866–1915* (Manchester: Manchester University Press, 1980), p. 29;

Gordon Stein, *Freethought in the United Kingdom and the Commonwealth: A Descriptive Bibliography* (Westport, Conn.: Greenwood Press, 1981), p. 60; Warren Sylvester Smith, *The London Heretics, 1870–1914* (New York: Dodd, Mead and Co., 1968), p. 52.

5. J.M. Robertson, *A History of Freethought in the Nineteenth Century*, 2 vols. (London: Watts and Co., 1929), II, 439; Budd, *Varieties of Unbelief*, p. 31; A. Gowans Whyte, *The Story of the RPA, 1899–1949* (London: Watts and Co., 1949), p. 25; Royle, *Radicals, Secularists, and Republicans*, pp. 169, 171.

6. [Charles Albert Watts], 'Notes and Scraps', *Agnostic*, 1 (Feb. 1885), 95; and 2 (July 1885), 335.

7. Royle, *Radicals, Secularists, and Republicans*, pp. 165, 263; Whyte, *Story of the RPA*, p. 51.

8. Whyte, *Story of the RPA*, pp. 21, 27; 'The Agnostic Press Fund', *Agnostic*, 1 (May 1885), 1 (separate pagination from rest of volume); 'The Rationalist Press Committee', *Agnostic Annual* (1895), p. viii.

9. Although the development of the RPA after 1900 lies beyond the scope of this paper, it should be noted that its activities, especially the launching of the RPA Cheap Reprint series in 1902, can be seen as part of the same strategy that Watts began in 1884.

10. 'William Stewart Ross', *Dictionary of National Biography*, 2nd supp., III, 232; Joseph McCabe, 'William Stewart Ross', in *idem*, *A Biographical Dictionary of Modern Rationalists* (London: Watts and Co., 1920), pp. 683–4; Victor E. Neuburg, 'William Stewart Ross', in Gordon Stein, ed., *The Encyclopedia of Unbelief*, 2 vols. (Buffalo, N.Y.: Prometheus Books, 1985), II, 561–2.

11. David Tribe, *President Charles Bradlaugh, MP* (Hamden, Conn.: Archon Books, 1971), p. 260.

12. [William S. Ross], Saladin, 'To Correspondents', *Agnostic Journal*, no. 24 (13 April 1889), 233; 'Saladin: Extracts from Letters of Condolence', *Agnostic Journal*, no. 59 (15 Dec. 1906), 380; Royle, *Radicals, Secularists, and Republicans*, pp. 161, 183.

13. Whyte, *Story of the RPA*, p. 46; McCabe, *Biographical Dictionary*, p. 78; 'Random Jottings', *Literary Guide and Rationalist Review*, no. 79 (1 Jan. 1903), 8.

14. F.J. Gould, *The Life-Story of a Humanist* (London: Watts and Co., 1923); McCabe, *Biographical Dictionary*, pp. 300–1.

15. Robertson, *History of Freethought*, II, 422, 426; Budd, *Varieties of Unbelief*, pp. 133, 186; S. Laing, 'Articles of the Agnostic Creed, and Reasons for Them', *Agnostic Journal*, no. 24 (1889), 1–2, 17–18; 'Samuel Laing', *Dictionary of National Biography*, XXII, 948–50.

16. Samuel Laing, *Human Origins* (London: Chapman and Hall, 1892), p. 3; F.J. Gould, *Tales from the Bible* (London: Watts and Co., 1895), p. 3; Richard Bithell, *The Creed of a Modern Agnostic* (London: George Routledge and Sons, 1883), p. 17; S. Laing. *Modern Science and Modern Thought*, 1885 (London: Chapman and Hall, 1893), p. v.

17. Laing, *Modern Science*, p. 245; F.J. Gould, *Stepping-Stones to Agnosticism* (London: Watts and Co., 1890), p. 32; Richard Bithell, *Agnostic Problems*, 1887 (London: Watts and Co., 1888), pp. 48–9; S. Laing, *A Modern Zoroastrian*, 1887 (London: Chapman and Hall, 1894), p. 181.

18. [C.A. Watts], 'Notes and Scraps', *Agnostic*, 2 (1885), 432; Albert Simmons, *Agnostic First Principles* (London: Watts and Co., 1885), pp. 1, 4; R. Bithell, 'Preface', in ibid., p. v; Laing, *Modern Science*, p. 225; Saladin and Joseph Taylor, *Why I am an Agnostic* (London: W. Stewart and Co., 1889), p. 10; Saladin, 'At Random', *Agnostic Journal*, no. 24 (1889), 138; [*idem*], 'To Correspondents', *Agnostic Journal*, no. 24 (1889), 313.

19. Bithell, *Agnostic Problems* (London: Williams and Norgate, 1887), p. 16; Gould, *Stepping-Stones*, p. 88; Laing, *Modern Zoroastrian*, p. 139; Bernard Lightman, *The Origins of Agnosticism: Victorian Unbelief and the Limits of Knowledge* (Baltimore: Johns Hopkins University Press, 1987); Saladin and Taylor, *Why I am an Agnostic*, p. 37; S. Laing, *Problems of the Future* (London: Chapman and Hall, 1892), pp. 48–9; Gould, *Stepping-Stones*, p. 91; R. Bithell, *The Worship of the Unknowable* (London: Watts and Co., n.d.); Richard Bithell, *A Handbook of Scientific Agnosticism* (London: Watts and Co., 1892), p. 21; R. Bithell, 'An Agnostic View of Theism and Monism', *Agnostic Annual* (1896), p. 46; Laing, *Problems of the Future*, p. 229; Bithell, *Worship of the Unknowable*, p. 12; Gould, *Stepping-Stones*, p. 95; 'Notes and Scraps', *Agnostic*, 1 (Jan. 1885), 48.

20. L.S. Jacyna, 'Scientific Naturalism in Victorian Britiain' (Ph.D. thesis, University of Edinburgh, 1980), pp. 278–9, 293–302; Michael Freeden, *The New Liberalism: An Ideology of Social Reform* (Oxford: Clarendon Press, 1978).

21. Gareth Stedman Jones, *Outcast London: A Study in the Relationship between Classes in Victorian Society* (Oxford: Clarendon Press, 1971); Saladin, 'At Random', *Agnostic Journal*, no. 24 (1889), 90; Laing, *Problems of the Future*, p. 408; Bithell, *Agnostic Problems*, p. 61; Gould, *Life-Story*, p. 67; Royle, *Radicals, Secularists, and Republicans*, pp. 37, 200, 232, 235, 237.

22. Budd, *Varieties of Unbelief*, pp. 131–2.

23. Royle, *Radicals, Secularists, and Republicans*, p. 220; Whyte, *Story of the RPA*, pp. 41–2.

24. Saladin and Taylor, *Why I am an Agnostic*, p. 98; Laing, *Modern Zoroastrian*, pp. 253–6; [Saladin], 'To Correspondents', *Agnostic Journal*, no. 25 (1889), 57; Royle, *Radicals, Secularists, and Republicans*, p. 155; Frederick Millar, 'Freethought and Individualism', *Agnostic Journal*, no. 25 (1889), 65.

25. The efforts of scientific naturalists to formulate a new theodicy based on evolutionary theory are explored in Robert M. Young, *Darwin's Metaphor: Nature's Place in Victorian Culture* (Cambridge: Cambridge University Press, 1985), p. 199 and James R. Moore, 'Crisis without Revolution: The Ideological Watershed in Victorian Britain', *Revue de synthèse*, 4th ser. (Jan.–June 1986), 53–78.

26. Bithell, *Agnostic Problems*, p. 112; *idem*, *Creed of a Modern Agnostic*, pp. 95–7, 103–5.

27. Laing, *Modern Science*, p. 222; *idem*, *Modern Zoroastrian*, p. 175.

28. F.J. Gould, 'Is Progress an Illusion?', *Agnostic Annual* (1898), 36–7; Laing, *Modern Zoroastrian*, p. 177; Saladin, 'At Random', *Agnostic Journal*, no. 24 (1889), 393.

29. Bithell, *Creed of a Modern Agnostic*, pp. 101–2, 137.

30. Bithell, *Agnostic Problems*, pp. 125–6; *idem*, *Creed of a Modern Agnostic*,

p. 83; Laing, *Modern Zoroastrian*, p. 4; Gould, *Stepping-Stones*, pp. 66–77; Bithell, *Creed of a Modern Agnostic*, pp. 102, 121; Laing, *Problems of the Future*, p. 227; Saladin, 'At Random', *Agnostic Journal*, no. 24 (1889), 26.

31. Millar, 'Freethought and Individualism', p. 65.

32. Gould, *Stepping-Stones*, p. 65; Laing, *Modern Science*, p. 104; Bithell, 'Preface', in Simmons, *Agnostic First Principles*, p. viii; Bithell, *Creed of a Modern Agnostic*, p. 152; *idem*, *Worship of the Unknowable*, p. 10.

33. Laing, *Modern Zoroastrian*, p. 173; Saladin, 'At Random', *Agnostic Journal*, no. 59 (1906), 379; R. Bithell, 'The Evolution of the Agnostic Idea', *Agnostic Annual* (1890), p. 34; F.J. Gould, 'Agnostics yet Englishmen', *Agnostic Annual* (1899), p. 8; Laing, *Problems of the Future*, pp. 197, 212; *idem*, *Modern Science*, p. 296; *idem*, *Problems of the Future*, p. 219.

34. Saladin, 'At Random', *Agnostic Journal*, no. 24 (1889), 27; and no. 25, (1889), 26; Saladin and Taylor, *Why I am an Agnostic*, pp. 73, 100; [C.A. Watts], 'Notes and Scraps', *Agnostic*, 1 (1885), 48; Laing, *Problems of the Future*, p. 182; *idem*, *Modern Zoroastrian*, pp. 171, 202–3.

35. Leonard Huxley, *Life and Letters of Thomas Henry Huxley*, 1900; 3 vols. (London: Macmillan, 1913), II, 343 and III, 368–9; Leslie Stephen, 'Mr. Bradlaugh and His Opponents', *Fortnightly Review*, 28 (1880), 176–87; Smith, *London Heretics*, p. 66; Frank M. Turner, 'Victorian Scientific Naturalism and Thomas Carlyle', *Victorian Studies*, 18 (1975), 327; Whyte, *Story of the RPA*, p. 22.

36. Budd, *Varieties of Unbelief*, p. 92; Mark Francis, 'Herbert Spencer and the Mid-Victorian Scientists', *Metascience*, 4 (1986), 4; Frederick James Gould, *The Pioneers of Johnson's Court* (London: Watts and Co., 1929), p. 2; Charles A. Watts, 'Some Reminiscences of No. 17 Johnson's Court', *Literary Guide*, Jan. 1924, p. 19; Royal Institution of Great Britain, Tyndall Papers (Correspondence), 9.3106 (R.I. MSS T., 13/F12, 90.); Imperial College Archives, Huxley Papers, 28.199.

37. L. Huxley, *Life and Letters*, III, 105; Lightman, *Origins of Agnosticism*, pp. 140–4; Laing, *Modern Science*, p. 83; *idem*, *Modern Zoroastrian*, pp. v–vi.

38. Saladin, 'At Random', *Agnostic Journal*, no. 24 (1889), 105–6, 219; *idem*, 'Professor Huxley and Agnosticism', *Agnostic Journal*, no. 24 (1889), 262; Saladin and Taylor, *Why I am an Agnostic*, p. 105.

39. Saladin, 'At Random', *Agnostic Journal*, no. 25 (1889), 282.

40. Samuel Laing, 'Rational Radicalism', *Fortnightly Review*, 35 (1884), 74–5, 78–9, 88.

41. Samuel Laing, 'Aristocracy or Democracy: A Reply to Professor Huxley', *Contemporary Review*, 57 (1890), 525, 527.

42. Ibid., pp. 525, 527–8.

43. Ibid., pp. 529–31, 536.

44. Greene, *Science, Ideology, and World View*, pp. 8, 158.

10

Ernst Haeckel, Darwinismus and the secularization of nature

PAUL WEINDLING

Darwinism in Germany – 'Darwinismus' – was a movement promoting liberal, rational and secular values in perceptions of nature and society. Yet in-depth study of German Darwinists reveals them to be less materialistic and more committed to idealistic and pantheistic beliefs than one might otherwise expect. The expansion of the highly innovative German system of education and research produced not only outstanding advances in experimental science, and industrial exploitation of natural resources, but also new rituals, symbols and ideals. Nature was venerated as a healing and restorative force; organic doctrines were used to promote ideals of a cohesive and unified society. By the 1890s Darwinian organicism, which was originally allied with anti-clerical political liberalism, had been successfully deployed in support of quite different cultural values. These values were given expression not only in freethinking organizations such as the Monist League, founded in 1905, but in a range of religiously inspired welfare organizations that believed in the virtue of nature itself to improve the quality of human life.[1]

By viewing Darwinismus as a 'secular religion', features of German culture can be brought to light that have previously been overlooked because it has been assumed that Darwinismus represented merely an increasingly mechanistic analysis of nature. It is well known, for example, that scientists in the nineteenth century were highly polarized over the relations of Darwinism and Christianity. Among Darwinists there was particular antagonism to the Roman Catholic Church, and in Germany this hostility was complicated by the legacy of Catholic–Protestant tensions, which flared up during the anti-Catholic 'Kulturkampf' of the 1870s.[2] Darwinismus had from the first been associated with expectations for a national state that would guarantee freedom of conscience, free speech and free trade. Darwinismus thus represented a challenge to the established churches as well as to aristocratic privilege. During the 1870s, German Darwinists lacerated the Catholic Church for being superstitious, primitive and oppressive, but in general they remained silent about Protestantism. Most of these Darwinists came from Protestant families (although there were noted exceptions such as the comparative anatomist Carl Gegenbaur). For personal and psychological reasons, therefore, the tenor of Darwinismus was not categorically hostile to religion.

From the 1880s the conflict with the churches gradually subsided and another religious affinity of Darwinismus came to the fore. Now the antagonist was socialism. Between 1878 and 1890 the Sozialist- ische Partei Deutschlands was banned and persecuted in Germany. But Bismarck's autocratic and manipulative policy failed, and once the SPD had been rehabilitated, it polled so well and enjoyed such strong support among the working classes – especially among the women – that it looked set to become Germany's strongest political party. In the context of rapid industrialization, the spectre of mass democracy accompanied by revolutionary secularization of economic and domestic life, provoked accusations that another secularizing heresy, Darwinismus, was socialistic. This was a source of increasing embarrassment to many biologists. They sought, accordingly, to dif- ferentiate Darwinismus from the taint of materialist radicalism, while insisting that science could provide solutions to pressing problems of poverty and disease. Now Darwinismus was construed to support an integrating creed of social welfare, typified by the 'national social' ideals of Pastor Friedrich Naumann. The National Social Association, which he headed, flourished from 1896 until 1903, but Naumann thereafter remained influential in the endeavour to humanize soulless industrial society. He proclaimed the necessity of fusing 'Darwin with Rousseau' as a means of promoting the ideals of national duty and social efficiency. Naumann's religious liberalism appealed to many Darwinian biologists and professors of medicine as a middle ground between radical individualism and radical socialism. The professorial elite saw themselves as a modern and humane 'broad church' of experts who would dictate solutions for social problems and guide the Volk towards an orderly, sober and industrious future.

My purpose here is to shed light on how Darwinismus moved from its early alliance with political liberalism to perform this corporatist and integrative social function. I shall proceed by analysing the ca- reer of Germany's best known and most successful 'prophet of Dar- winismus', the biologist Ernst Haeckel (1834–1919). Haeckel is of interest because he established the fundamental scientific characteris- tics of Darwinismus as a fusion of the evolution doctrine with cell theory and embryology. But while he typified contemporary German excellence in such laboratory sciences (by contrast with the British field-naturalist tradition), his scientific Weltanschauung had a much wider cultural importance. As Germany's outstanding popularizer of Darwinismus, Haeckel underwent a public metamorphosis from a mechanistic-minded anti-clerical in the 1860s to a Darwinian pantheist in the 1890s. Commentators have tended to dismiss this change as a 'scientific decline', Haeckel as 'Darwin's Dachshund'.[3] Instead, I shall suggest that Haeckel's broad appeal was based on substantial scientific achievement, and provides valuable insight into German culture and public opinion at the end of the nineteenth century.

Haeckel and his times

Great scientists are usually immortalized for their discoveries rather than for their feelings about nature, their philosophies or their personalities. With Ernst Haeckel it is the reverse. He is best remembered for having inspired a love of nature in a generation of biologists, for his combative temperament and for his synthesis of mechanistic views with philosophical idealism. Indeed, Haeckel's originality as a morphologist and expert on the minute but beautifully complex *Protozoa* has often gone unappreciated, and the relationship of his specialized scientific work to his evolutionary world view is left unexplained. It is important to remember that Darwinismus meant not only Darwin's theory of natural selection. Darwinismus combined a range of exciting new scientific theories, such as the protoplasmic theory of the simplest form of life and embryological theories of the recapitulation of the history of organic forms, with an inspiring new philosophy of nature. And in Haeckel, it is claimed, Darwinismus had its greatest European representative. He was admired not only by scientists, but by other makers of world views like Freud; he was read avidly by the burgeoning industrial classes, ranging from capitalists like Alfred Krupp and scientifically educated professionals to rank-and-file workers, among whom Darwinian-influenced naturalist authors, such as Zola, were also highly popular. Haeckel has, moreover, been accused not only of abetting socialism, but of fathering Nazi ideology. His views thus repay further study in an effect to reconcile such conflicting interpretations.

It is a commonplace that Darwin's theory of natural selection replaced a harmonious view of nature with one based on chance and struggle in which man was descended from the apes. In Darwinismus, however, the theory did not entail a pessimistic philosophy of purposeless conflict. Although ethical progressivism in nineteenth-century German philosophy was countered both by the philosophical pessimism of Schopenhauer and by the anti-Christian and anti-ethical convictions of Nietzsche, in Haeckel's thought a compelling world view emerged in which even the most minute beings reveal beauty, harmonious order and the germs of intellectual and social life. Darwin was cautious in drawing philosophical conclusions from his work; he delayed publishing his views of human origins and declined to prejudge the burning issue of the origins of life. It was Haeckel's genius to address these very issues in his research, and to show how the facts of nature were relevant to the most pressing social and ethical problems of the day. In minute creatures, which he christened *Plastiden*, Haeckel found the answer to the riddles of new species and the origins of life – indeed, as the title of his best-seller of 1899 proclaimed, to 'the riddles of the universe'.

What was the secret of his success? Besides undoubted scientific gifts, Haeckel had a personality that helped to mould his biology into a secular surrogate for religion. He possessed a deep sensitivity for natural beauties. This was expressed in his scientific drawings, in his landscape paintings and in his lyrical accounts of his travels. The letters collected after his death in his remarkable home in Jena, the Villa Medusa, show his passion that the beauty and order of nature should be interpreted in progressive and optimistic terms. These letters were written not only to family members and fellow scientists, but to poets, philosophers and interested ordinary men and women. Such was the ardour of his correspondence that it once blossomed into love.[4] While it was the fiery and combative side of Haeckel's character that attracted attention to himself, it was his capacity to inspire love and devotion that maintained it. This can be seen in his relations with his students. Haeckel surrounded himself with patriotic and nature-loving cohorts. He referred to star pupils as his 'golden sons'; he wrote of the admiration and devotion of his assistants. Students petitioned him for lectures on Darwinismus during the 1860s, when the subject was still taboo among academic conservatives. A flock of students accompanied him on his regular walks in the Thuringian hills round Jena, where on the summits lay 'the spirit of freedom'. They followed the prophet of Darwinismus to Italy and the Adriatic, where he converted an abbot to the doctrine and used his monastery for marine biological research.[5]

The secret of Haeckel's persuasive powers also lay in social conditions favourable to his organicist synthesis. Between 1860 and 1920 science played a major role in the transformation of Germany from a predominantly agrarian and politically fragmented society to an industrial and imperial power. These years saw the rise of science-based industries and professions, while scientific achievement became a means of national cultural assertion. Germany was reputed to lead the world in biological and medical research, and in its system of higher education. While a mass readership snapped up fiery popularizations of secular enlightenment, such as Alexander von Humboldt's *Kosmos*, the middle classes were keen to capitalize on the knowledge produced by the expanding universities. They sent their sons and daughters there to acquire scientific training for professional careers. Everywhere an appetite was whetted for a more optimistic and relevant explanation of the world than that of traditional theology, which was promulgated by churches tied closely to archaic and repressive social forms. Theology in any event had been shaken by historical scholarship. Biology, however, promised to extend the idea of social progress to the history of life, from its earliest origins to advanced industrial society. A vast new market, in short, had emerged for Haeckel's organicist synthesis.

But to large-scale social conditions must be added local and personal factors. Haeckel was sent to university to study medicine and become a doctor. This was his father's intention for him, but his own heart lay elsewhere. As a child he had formed the ambition of becoming a naturalist. He assembled vast collections of natural objects, and by the age of seventeen he had painted a picture of the 'Parliament of Birds', showing the different species assembled. This allusion to the Frankfurt Parliament of 1848 shows an early ability to conceptualize nature in social terms.[6] But despite Haeckel's precocious audacity, which he continued to evince as an enthusiastic young student and lecturer, there remained something of the child in his pursuit of natural history. The German scientific community, like the wider culture, was organized along patriarchal and hierarchical lines. There were numerous authority figures – institute directors, public educators, *Doktorvater*. Just as every child looks towards parental authority both for guidance and for standards against which to rebel, so Haeckel defined himself as a naturalist in relation to a series of father-figures. This helps to explain the remarkable conversion he underwent to become one of the most outspoken and anti-clerical Darwinists.

Conversion

The first great scientific father-figure in Haeckel's life was the Berlin professor Johannes Müller, who was regarded as the foremost German morphologist and comparative anatomist of the time. Müller inspired Haeckel with enthusiasm for marine biology and suggested to him that he concentrate his research on the minute *Radiolaria*. Haeckel accompanied Müller on an expedition to Heligoland in 1854, and to Italy in

'Parliament of Birds, consisting of a representative from each family, designed and painted by Ernst Haeckel, 1850'. (Reproduced by permission of Ernst-Haeckel-Haus, Friedrich-Schiller-Universität Jena.)

1858. Later that year Müller died. It is a sign of their closeness that Haeckel was among the pall bearers at his funeral.

Müller was steadfastly opposed to evolution and Haeckel shared his views, including contempt for materialists such as Carl Vogt and Ludwig Büchner, who have been generally regarded as preparing the way for Darwinismus.[7] This makes Haeckel's conversion to evolution all the more remarkable. Undoubtedly, Müller's death in the year prior to the publication of the *Origin of Species* relieved Haeckel of a psychological burden that might otherwise have restrained him from supporting Darwin's theory. But the period from 1860 to 1864 was, for Haeckel, intellectually rich and emotionally full. A number of influences at the time must be balanced. Haeckel first read the *Origin of Species* in 1860. Although his initial reaction was favourable, he was then in the conservative milieu of Berlin where frank discussion of the book was impossible. Only on moving to Jena did he find colleagues who favoured evolution, with whom he could explore the revolutionary implications of natural selection for the study of comparative anatomy. The University of Jena enjoyed a liberal intellectual tradition, the legacy of Goethe (whose many scientific contributions included the concept of morphology) and the great *Naturphilosophen*, Oken and Schelling. The administration was keen to attract adventurous young scientists, and Haeckel was granted a teaching post for his monograph on *Radiolaria*. In 1864 he was promoted to a professorship of zoology. At this time his career owed much to the indulgent patronage of the Grand Duke of Weimar, and the university administrator Moritz Seebeck. They tolerated his enthusiasms and often followed his advice on appointments. Eventually, with colleagues such as the botanist Strasburger, the anatomist Gegenbaur, and the physiologist Preyer, Haeckel earned the university the reputation of being the 'citadel of Darwinism'.

After Müller's death, and during his early years at Jena, Haeckel came under the paternal influence of another gifted scientist. This was Max Schultze, whose introduction of the protoplasmic theory of the cell coincided with the dissemination of the *Origin of Species* in Germany. Darwin himself had little understanding of the theory's bearing on evolution, but it quickly became fundamental to German Darwinism. According to Schultze, the important feature of the cell was not its membrane, but its constituent granular substance, which he supposed to be common to the cells of plants and animals. This gave Haeckel the idea that the granular substance, or *Plasma*, was equivalent to the first-formed life. All organisms were plasmatic bodies, differing only in degree of organization. The protoplasmic theory thus supported Darwin's view that all species may have derived from a single common ancestor. Haeckel set himself the task of discovering what the historical progression of species had actually been. Among the thirty thousand letters to Haeckel in the Villa Medusa there is an important collection of letters from Schultze. These show how he became an intellectual mentor and father-figure in the years to 1865, when Haeckel began his phylogenetic research. Schultze's last letter to Haeckel, in which he discusses the theory of embryological re-capitulation of evolution, was adorned by Haeckel with a mysterious series of hearts and arrows. It suggests that a deep emotional attach-ment underlay their scientific exchanges.[8]

During this period Haeckel was moderate and balanced in his support for Darwin. But then came rapid change. Once again, death was a catalyst. In 1864 Haeckel's beloved first wife died. It was a traumatic shock, and Haeckel began to feel his character hardening. He wrote to his parents of his feeling that a life of sensuousness must give way to a life of the intellect devoted to science and mankind. Haeckel now set to work on a major polemic in support of Darwin-ismus, the *Generelle Morphologie*. In it he invoked Darwin's theories to support a range of progressive metaphysical and nationalist view-points. He went far beyond the arguments of the temperamentally more cautious Darwin. Immediately after the book's publication, in 1866, Haeckel went to visit Darwin at Down House. He regarded this as a momentous spiritual occasion, when his view of the unity of nature was confirmed. Darwin, however, quickly recognized that his German supporter distorted his views; he feared that Haeckel's ferocity would provoke unnecessary opposition. He warned him that 'you have in part taken what I said much stronger than what I intended'. Such chidings evoke the image of Darwin as a new father-figure, mature in criticism, generous in praise, whom Haeckel manoeuvred into a position of arbiter of fundamental values. Their relationship may be epitomized in Haeckel's response to Darwin's gift of a copy of *The Descent of Man*, where the work of 'Professor Haeckel' receives fa-vourable mention. Haeckel was ecstatic. He regarded it as the greatest compliment he had ever received.[9]

Thereafter, although Haeckel accorded natural selection only a limited role in his exposition of evolution, he regarded himself as first and foremost a Darwinist. Darwinismus for him functioned as an ideology of human progress. It represented a whole constellation of evolutionary theories, including Lamarck's principle of direct adapta-tion to the environment and the recapitulationist principle of the historical development of organic forms. Haeckel set greater store on causality than Darwin did; he believed that the discovery of the correct historical succession of organisms constituted a mechanical explanation of morphology. He sought to trace this succession by means of his 'biogenetic law': namely, that 'ontogeny' (Haeckel's term for embryological development) recapitulates 'phylogeny' (his term for the history of life). According to this law, the fundamental forms through which embryos pass, such as the early stage of 'gastrulation', actually recapitulate the historic forms of primitive animals, such as the *Gastraea*.

Further evidence of Haeckel's ability to combine in Darwinismus elements that did not derive from Darwin's theories was his insistence on distinguishing 'lower' from 'higher' forms of life. The liberal-minded Darwin hardly ever used these terms. Haeckel, however, drew on the concept of the 'physiological division of labour', employed by non-Darwinians such as Milne Edwards, Bronn and Spencer, to show that a centralized nervous system was an essential factor in the evolution of higher organisms. He argued that the more an organism was differentiated, the greater its need of centralized co-ordination and control. Such organisms he referred to as 'cell states', and he compared their nervous systems to the German imperial telegraph. Already at Messina in 1859 Haeckel had become enraptured by the beauty of the social *Radiolaria*, united into colonies by networks of protoplasm. He regarded their symmetrical forms as reminiscent of crystals, so providing an instructive parallel between physical and biological organization. At this time the Italian nationalist movement, the Risorgimento, was gathering force; many German minds were impressed with the inevitability of unification, and Haeckel himself became enthusiastic for a united Germany. Here scientific, social and intellectual developments converged: the observation of the social *Radiolaria*, the birth of nationalism and Haeckel's reading of Darwin in 1860. Haeckel admired Darwin for much more than the discovery of natural selection. As a British naturalist, Darwin not only illustrated the beneficial influence of political liberalism and *laissez-faire*; he also stood, in Haeckel's view, for nothing less than a liberating ideal of natural and social order, an emancipatory ethic and a new cosmology.[10]

Personal loss

The combination of compelling social vision and deep personal conviction in Haeckel's Darwinismus was a weakness as well as a strength. It made Haeckel especially vulnerable to scientific criticisms, and when these came, old friendships were broken, to be replaced with enmity and bitterness.

The first great break came with the cellular pathologist Rudolf

Virchow, who had introduced Haeckel as a student to cell theory. Virchow was also a leading liberal politician, and during the 1860s Haeckel shared Virchow's radical enthusiasms. But, a decade later, liberals were divided over the issue of German unification under Prussian hegemony. National Liberals supported Bismarck and gave unity priority over freedom; others remained left-wing radicals. Virchow became a leading critic of Bismarck and held egalitarian beliefs. Haeckel believed that hierarchical authority was progressive, and he admired Bismarck more and more. (Indeed, when the ageing and deposed Bismarck visited Jena, Haeckel proclaimed him 'Doctor of Phylogeny'.) This, then, was the background to the bitter dispute that erupted between Haeckel and Virchow in 1877. The issue was whether Darwinismus was a proven law, suitable for teaching in schools, or whether it was only a hypothesis. Haeckel argued that Darwinismus should be the framework for education in the newly united nation. Study of cellular organisms, he maintained, provided the basis for a natural religion of duty, teaching the division of labour and the subordination of the individual to the social whole. Unicellular organisms such as the *Radiolaria* showed that the single cell was the basic unit of mental life and, as such, could be termed a *Seelenzelle*. The sense of duty had arisen in higher organisms – the highest of all being the German Empire – through the evolution of social instinct. Virchow, however, ridiculed the attempt to construct a Darwinian cosmology based on hypotheses rather than facts. Hypotheses, he argued, were inappropriate for teaching in schools.[11]

The dispute with Virchow marked a waning of Haeckel's scientific reputation. The offers of prestigious chairs at Vienna and at the new Reich University at Strassburg, which Haeckel had previously declined, were not renewed. And his friendship with Gegenbaur, which had tied him to Jena, now deteriorated. Gegenbaur moved to Heidelberg, full of doubts about Haeckel's evolutionary synthesis. Another longstanding colleague, August Weismann, also began to question aspects of the Darwinian creed. In 1882 he rejected the inheritance of acquired characters, arguing that evolution took place by means of natural selection alone operating at the sub-microscopic level of the inherited germ plasm. Haeckel, however, insisted on the Lamarckian principle of the inheritance of acquired characters, and this led to a furious debate over the origins of sexual dimorphism in insects. Weismann pointed to hereditary determinants, whereas Haeckel insisted on environmental factors such as nutrition. Thus arose the paradox of Haeckel, the leading German Darwinist, attacking the all-sufficiency of natural selection as a mechanism of evolution. (Darwin of course also increasingly accepted Lamarckian explanations, and Spencer, like Haeckel, resisted Weismann's so-called neo-Darwinism.)[12]

But most painful for Haeckel were ruptures with former students – his 'golden sons'. Oscar Hertwig, for example, criticized Haeckel's biogenetic law and his monophyletic theory that all life was derived from a single first-formed organic particle. He substituted a law of causal ontogeny, according to which complete recapitulation was impossible because organisms were ever adapting to new environments. (Nor were Haeckel's views reinforced by the discovery that he had used pictures of the same embryo to demonstrate the common descent of different species.) Despite his scientific differences with Haeckel, Hertwig continued to share many features of his mentor's social philosophy. But just as Hertwig's organicist synthesis was scientifically more advanced, incorporating reduction division of chromosomes and Mendelism, so ideologically it was more pluralistic and adaptable than Haeckel's Darwinismus.[13]

To the very end, Haeckel's dogmatic Weltanschauung set his friends and colleagues at odds. Darwinians and Lamarckians, mechanists and vitalists, struggled to partition or annex his intellectual kingdom; and when Haeckel had established the Phyletic Museum at Jena as a public shrine to his evolutionary philosophy, his successor at the university, Ludwig Plate, denied him access. Some students did remain close to the tragic Lear-figure, even though his Darwinismus grew increasingly out of touch with the latest research into chromosomes and embryology. Among them was the zoologist Richard Semon, whose theory of inherited memory much impressed Haeckel at a time when he was endeavouring to perpetuate his own memory – the Phyletic Museum and the bequest of the Villa Medusa – in the face of numerous critics. Semon's life ended tragically: with the collapse of the monarchy in 1918, he shot himself on the old imperial flag. Haeckel was also distressed at the proclamation of a republic, which, according to his theories, was an atavistic relapse into a lower form of unhierarchical social organization, a throwback to a primitive form of cell state. He died the following year.[14]

Public prophet

Haeckel's chief compensation for his personal losses was increasing popular success. His dispute with Virchow had convinced him of the need to keep up 'Propaganda' (as he told his friend, the poet Allmers) for the evolutionary world view. From 1877, therefore, he began to publicize Darwinismus as never before, first by issuing a popular edition of his lectures, then by advertising 'Monism' as a link between science and religion.[15] Rational and empirical features in evolutionary theory now gradually gave way to mystic idealism in Haeckel's published work. The appearence of *Die Welträthsel* – 'the riddles of the universe' – in 1899 marked the end of this transition, but commentators have differed sharply as to its overall significance. Haeckel's Darwinian version of pantheistic nature philosophy was expressed in vibrant and vivid language, giving *Die Welträthsel* an immense popular appeal. But the work has very few specifically Darwinian features. It was in fact well adapted to a period in which vitalism excited sympathy and Darwinian evolutionary mechanisms were viewed with caution.[16]

Haeckel's writings were not only popular, but extremely influential as well. They contributed significantly to the developing sciences of psychology, sociology and psychoanalysis; their theories of recapitulation penetrated a wide range of disciplines, ranging from psychology to biometrics. But in his latter-day works especially, Haeckel deviated from the straight and narrow of the scientific path on to a broad social terrain. These works were read avidly across the political spectrum, among socialists and extreme nationalists alike, and they inspired new evolutionary ethics. For Haeckel maintained, with all the authority of the nation's leading Darwinian, that a correct understanding of evolutionary laws was essential for the solution of social and political problems.

Darwinismus in Haeckel's hands now became the basis of a new Social Darwinism. Organicist biology served to integrate diverse social interests by supporting corporatist ideology. Social welfare reforms and nationalism took precedence over the attack on hereditary privilege. In the 1890s the vicious hostility that liberal Social Darwinists had reserved for Catholic ultramontanism was redirected against socialism. For example, a group of biologists in Baden, including the zoologist H.E. Ziegler, deployed Weismann's new theory of an immutable germ plasm to criticize socialist Darwinists. Once liberals, they now argued for nationalist collectivism. Other Social Darwinists – a minority, it is true – opposed the preservation of the 'unfit' through the extension of social welfare and therapeutic medicine. One influential group of eugenically minded doctors urged a transition from individual-patient oriented medicine to a system in which the doctor was a state official. The health of the social organism was now considered to be the state's responsibility.

Social Darwinist organicism served to promote national unity and to create a more sympathetic attitude to welfare reforms both within the state and among landowners, industrialists and the middle classes. At the same time, however, political factions and interest groups attempted to float diverse forms of Social Darwinism as an objective basis for national unity and social progress. On the conservative side, Friedrich Alfred Krupp abetted efforts in the early 1890s to form a national party. He was a frustrated biologist, complaining that his iron-willed father had not allowed him to study botany; and he admired Haeckel's popular anthropology. In 1900 he supported a prize essay competition on evolution, from which he hoped a national political platform might emerge. Ziegler, the anti-socialist zoologist, managed the competition and arranged for the publication of the best essays in *Natur und Staat*. This experiment in a biologically based collectivist politics – or *Sammlungspolitik* – was a disaster. The prize-winning essays that supported state socialism merely exposed the bewildering varieties of Social Darwinism. There was something in them for everyone to attack, from Aryan racial theorists to sociologists like Tönnies. The reception of *Natur und Staat* exposed the basic conflicts among racist ideology, scientific monism, organicist sociology and technocratic eugenics.[17]

The first decade of the twentieth century saw many rival initiatives for biologistic social programmes. Each had a distinctive name, presenting the intellectual trademark of its organizers: *Volkshygiene*, *Sozialhygiene*, *Rassenhygiene*, *Politische Anthropologie* and *Eugenetik*. Feminists and sexual reformers advocated *Mutterschutz*, the creation of favourable conditions for the survival of the offspring of unmarried mothers, showing that they, too, were profoundly inspired by biology. Adherents of these movements for biomedical reform argued that their new scientific creeds amounted to modern rational religions. And the sectarian diversity of these creeds does suggest that there was widespread public support for biologistic solutions to social questions. Indeed, the concept of a popular national opposition can be extended into the realm of the biomedical reformers. It is a rewarding exercise to disentangle these reformers from Aryan ideologues such as Houston Stewart Chamberlain. Aryan racism as an organized phenomenon originated as late as 1894 in the small Gobineau Society; it was more criticized than supported by Social Darwinists. Such groupings were in fact quite marginal in comparison to the movements for biomedical reform, and Chamberlain's attack on Haeckel as a superficial popularizer is indicative of this. The racism of the reformers and Aryan racism had distinct and incommensurate appeals.[18]

Whether the scientific roots of Nazi racism stem from Haeckel has been controversial. He used racist concepts of a hierarchy of distinct human types and spoke of 'lower' and 'higher' races. His *Anthropogenie* of 1874 included a tirade against the 'black international of the Vatican', which should be obliterated by the 'heavy guns' of science. He predicted that whole libraries stuffed full of 'clerical wisdom and backward philosophy [*Afterphilosophie*]' would 'melt into nothing' when illuminated with 'the sun of embryology'. Haeckel's savage anti-Catholicism had yet to be reformulated into an anti-semitic racial theory vested in a party political organization, but the foundation of the Monist League in 1905, which was accompanied by a ceremonial renunciation of the church, did not meet these terms of reference. The League's effective leadership was taken over by Wilhelm Ostwald, the Nobel prize-winning chemist, before the 1914–18 war. Ostwald was a relentless materialist and reductionist; his creed of 'Energetics' differed sharply from Haeckel's doctrine of pantheistic monism, which Ostwald referred to scornfully as 'Klimbim'. Haeckel, now over seventy, was chiefly a father-figure for the League, and its diverse membership reflected the breadth of his influence. It included distinguished naturalist authors such as Boelsche (best known for his lyrical accounts of love-life in nature), leading nationalists and socialists, and even dissident Protestant clerics. Ostwald staunchly upheld pacifist beliefs during the war, while Haeckel joined in the chauvinistic and militaristic propaganda.

Clearly, the Monist League was a very 'broad church', and internally divided as well. It was too complex and ambiguous to be seen as the scientific standard-bearer for national socialism. Haeckel's own position, furthermore, was deeply ambivalent. He did make the occasional anti-semitic remark, but he never supported an organized attack on the Jews as an inferior race. He held pacifist convictions for most of his life, but renounced these to sign the sabre-rattling Pan-Germanist manifestos. His language, his ideals and his practices thus place him within the distinctive context of Wilhelmine intellectual imperialism, as a critic of the repressive old regime who was nevertheless concerned to develop revitalized forms of national power. Haeckel used biology to shore up a form of corporatist social thought that differed fundamentally from the hereditarian social pathologies current under the Nazis.[19]

Darwinian social organicism in its monistic form corresponded to the integrating needs of Wilhelmine imperialism. Yet after the demise of the Kaiserreich, it continued to be a source of inspiration for budding scientists and social reformers. Among Haeckel's last assistants was the radical Julius Schaxel, who during the 1920s advocated a socialist biology and supported the Urania movement for public lectures. While Schaxel became more a disciple of Trotsky than of Haeckel, the latter's influence was indelible. Haeckel's monistic evolutionary creed, indeed, promoted rational attitudes to all aspects of life and could be a liberating social force. After inspiring Freud, it opened new scientific perspectives on the interaction of psyche and reproductive physiology for a new generation of sexual reformers. Magnus Hirschfeld's renowned Institute for Sexual Science, founded at Berlin in 1919, had a Haeckel Room for public meetings. There speakers were flanked by busts of Haeckel and Darwin, the patron saints of a scientific ethics that was to replace Christianity. Hirschfeld, like Schaxel, shared Haeckel's conviction that knowledge must be disseminated to the public. He campaigned for reform of the laws on homosexuality, justifying his demands on the basis of a theory of intermediate sexual types. Evolutionary biology could be turned – within limits – to emancipatory social ends.[20]

A number of modern scientists have confessed to the lasting impact of Haeckel's Darwinismus on their work. Leading geneticists and eugenicists of the 1920s and 1930s continued to venerate Haeckel, and many of his neologisms were enshrined in their scientific vocabulary. Among these devotees were the embryologist Hans Spemann and the geneticist Hans Nachtsheim.[21] Although the Nazis disbanded the Monist League for its masonic and freethinking connotations, the Haeckel cult would not disappear. The Haeckel Haus in Jena became a centre for the history of racial hygiene in the 1930s and the papers of eugenicists were deposited there. The SS historians were interested in the support of socialists for the early racial hygiene movement. In 1942 an Ernst Haeckel Society was established under the protection of Fritz Sauckel, the Thuringian Gauleiter and Labour Minister. But today, despite attempts to Nazify the image of Haeckel, historians in the two Germanies agree that he is best understood in the imperial context, which gave birth to his ideas.

For the historian of science and medicine, Haeckel provides much insight into how the nineteenth-century German mind was transformed during the social upheavals of industrialization. Haeckel, who sought out a series of father-figures – Müller, Schultze, then Darwin – was finally able to mature as a Darwinist and himself father a distinctive vision of a progressive and creative universe. He himself became a cult figure, symbolizing the nation's scientific genius and special affinity for nature. This metamorphosis entailed that Haeckel loose himself from strict Darwinian views, but in doing so he made his own developmental and organicist vision of nature a resounding public success. While anti-clerical and anti-Catholic, Haeckel's naturalistic principles served to underwrite new ethics and a humanitarian creed that could be assimilated with modernized forms of Christianity in which the churches concerned themselves less with doctrine than with issues of social welfare. Indeed, a substantial body of churchmen were prepared to concede the truth of Darwinismus, despite its monistic cosmos. For Haeckel's formulation of a new secular religion of nature was based on an understanding of the immense creative potential of the evolutionary process. And diversity in nature was refracted through his Darwinismus into a plurality of ideologies.

Notes

1. On German Darwinism, see G. Altner, ed., *Der Darwinismus: Die Geschichte einer Theorie* (Darmstadt: Wissenschaftliche Buchgesellschaft, 1981); Kurt Bayertz, 'Darwinism and Scientific Freedom: Political Aspects of the Reception of Darwinism in Germany, 1863–1878', *Scientia*, 118 (1983) 297–307; Donald C. Bellomy, '"Social Darwinism" Revisited', *Perspectives in American History*, n.s., 1 (1984), 1–129; Alfred Kelly, *The Descent of Darwin: The Popularization of Darwinism in Germany, 1860–1914* (Chapel Hill: University of North Carolina Press, 1981); Gunter Mann, ed., *Biologismus im 19. Jahrhundert* (Stuttgart: Ferdinand Enke, 1973); idem, 'Ernst Haeckel und der Darwinismus: Popularisierung, Propaganda und Ideologisierung', *Medizinhistorisches Journal*, 15 (1980), 269–83; Hans Querner, 'Darwins Descendenz- und Selektionslehre auf den deutschen Naturforscher-Versammlungen', *Acta historica Leopoldina*, 9 (1975), 439–56; Paul Weindling, *Darwinism and Social Darwinism in Imperial Germany: The Contribution of Oscar Hertwig (1849–1922)* (Stuttgart: Gustav Fischer, in press); and idem and Pietro Corsi, 'Darwinism in Germany, France and Italy', in David Kohn, ed., *The Darwinian Heritage* (Princeton, N.J.: Princeton University Press, 1985), pp. 683–729. For Haeckel, see the fundamental studies by Georg Uschmann, *Geschichte der Zoologie und der zoologischen Anstalten in Jena, 1779–1919* (Jena: Urania, 1959) and idem, *Ernst Haeckel: Biographie in Briefen*, 1954 (Gütersloh: Prisma Verlag, 1984). For an English summary, see idem, 'Haeckel's Biological Materialism', *History and Philosophy of the Life Sciences*, 1 (1979), 101–18. For biographical portraits, see Wilhelm Boelsche, *Haeckel: His Life and Work* (London: T. Fisher Unwin, 1906) and Erika Krausse, *Ernst Haeckel* (Leipzig: B.G. Teubner, 1984).

2. H.J. Dörpinghaus, 'Darwins Theorie und der deutsche Vulgärmaterialismus im Urteil deutscher Katholischer Zeitschriften zwischen 1854 und 1914' (Diss., University of Freiburg im Breisgau, 1969).

3. Bellomy, '"Social Darwinism" Revisited'.

4. J. Werner, ed., *The Love Letters of Ernst Haeckel written between 1898 and 1903* (London: Methuen, 1930); M. Jodl, ed., *Bartholomäus von Carneri's Briefwechsel mit Ernst Haeckel und Friedrich Jodl* (Leipzig: K.F. Koehler, 1922); R. Koop, *Haeckel und Allmers* (Bremen: Arthur Geist, 1941).

5. Uschmann, *Geschichte der Zoologie*.

6. Krausse, *Ernst Haeckel*, p. 15.

7. Frederick Gregory, *Scientific Materialism in Nineteenth Century Germany* (Dordrecht, Holland: Reidel, 1977); Timothy Lenoir, *The Strategy of Life: Teleology and Mathematics in Nineteenth-Century German Biology* (Dordrecht, Holland: Reidel, 1982).

8. Corsi and Weindling, 'Darwinism in Germany, France and Italy', pp.

694–7. See also the Spencer – Haeckel correspondence in Senate House Library, University of London, and at the Haeckel–Haus, Jena.

9. Francis Darwin, *The Life and Letters of Charles Darwin* ..., 1887; 3 vols. (London: John Murray, 1888), I, 86–9 and III, 68; Georg Uschmann and Ilse Jahn, 'Der Briefwechsel zwischen Thomas Henry Huxley und Ernst Haeckel', *Wissenschaftliche Zeitschrift der Friedrich-Schiller Universität Jena*, Mathematische-naturwissenschaftliche Reihe, 9 (1959–60), 13.

10. Paul Weindling, 'Theories of the Cell State in Imperial Germany', in Charles Webster, ed., *Biology, Medicine and Society, 1840–1940* (Cambridge: Cambridge University Press, 1981), pp. 99–155.

11. Ernst Haeckel, *Freie Wissenschaft und freie Lehre* (Stuttgart: Schweizerbart, 1878); Rudolf Virchow, 'Die Freiheit der Wissenschaft im modernen Staatsleben', *Amtlicher Bericht 50. Versammlung deutscher Naturforscher und Aerzte* (Munich: F. Straub, 1877), pp. 65–77.

12. Frederick B. Churchill, 'Weismann, Hydromedusae, and the Biogenetic Imperative: A Reconsideration', in T. Horder, J.A. Witkowski and C.C. Wylie, eds., *A History of Embryology* (Cambridge: Cambridge University Press, 1986), pp. 7–34. On Gegenbaur, see William Coleman, 'Morphology between Type Concept and Descent Theory', *Journal of the History of Medicine and Allied Sciences*, 31 (1976), 149–75.

13. Weindling, *Darwinism and Social Darwinism*.

14. Paul Weindling, *Health, Race and German Politics, 1870–1945* (Cambridge: Cambridge University Press, 1989).

15. Roger Chickering, *Imperial Germany and a World without War* (Princeton, N.J.: Princeton University Press, 1975).

16. Niles R. Holt, 'Ernst Haeckel's Monistic Religion', *Journal of the History of Ideas*, 32 (1971), 265–80; H. Dorber and W. Plesse, 'Zur philosophischen und politischen Position des von Ernst Haeckel begründeten Monismus', *Deutsche Zeitschrift für Philosophie*, 16 (1968), 1325–39.

17. Sheila Faith Weiss, 'Race Hygiene and the Rational Management of National Efficiency: Wilhelm Schallmayer and the Origins of German Eugenics, 1890–1920' (Ph. D. dissertation, Johns Hopkins University, 1983).

18. Patrik von zur Muhlen, *Rassenideologien: Geschichte und Hintergründe* (Berlin: J.H.W. Dietz Nachf., 1977); Rolf Winau, 'Ernst Haeckels Vorstellungen von Wert und Werden menschlicher Rassen und Kulturen', *Medizinhistorisches Journal*, 16 (1981), 270–9.

19. Daniel Gasman, *The Scientific Origins of National Socialism: Social Darwinism in Ernst Haeckel and the German Monist League* (London: Macdonald, 1971). On the Monist League, see also F. Bolle, 'Monistische Mauerei', *Medizinhistorisches Journal*, 16 (1981), 280–301 and W. Breitenbach, *Die Gründing und erste Entwicklung des deutschen Monistenbundes* (Brackwede: W. Breitenbach, 1913). On the Nazi image, see Viktor Franz, *Ernst Haeckel* (Jena: W. Agricola, 1941/44).

20. Frank Sulloway, *Freud, Biologist of the Mind* (New York: Basic Books, 1979); Charlotte Wolff, *Magnus Hirschfeld: A Portrait of a Pioneer in Sexology* (London: Quartet Books, 1986).

21. Tim Horder and Paul Weindling, 'Hans Spemann and the Organiser', in Horder *et al.*, *History of Embryology*, pp. 183–242.

11

Holding your head up high: Degeneration and orthogenesis in theories of human evolution

PETER J. BOWLER

John Greene's title, *The Death of Adam*, suggests the potential crisis provoked in Western thought by the emergence of the theory of evolution. If the human race had evolved from an animal ancestry, it might have to be seen not as God's direct creation, but as the by-product of a meaningless natural process. Yet the crisis was resolved for a time, at least to the satisfaction of many thinkers. As John Greene himself has shown, particularly in the essays collected in *Science, Ideology, and World View*, the idea of progress was used to redefine mankind's crucial role within a changing universe. Evolutionism need not have been threatening, so long as it was portrayed as a process leading inexorably towards moral and intellectual improvement, with the human race at the forefront of the advance. The real crisis of faith came in the early twentieth century, when the idea of progress itself came under fire. There had always been some thinkers willing to express doubts about the future of Western society, but these doubts rose to a crescendo in the 1890s' *fin de siècle* obsession with the threat of cultural degeneration. Sensing the crucial role played by this loss of faith, historians have recently begun to take a more active interest in the origin and development of nineteenth-century degenerationism.[1] In the debate between progressionists and degenerationists, the role of evolutionism as a guarantee of progress was explicitly challenged. The purpose of this essay is to move beyond the debates sparked off by the publication of the *Origin of Species* to see how evolution theory participated in this later crisis of confidence.

The historian of evolution theory who seeks to deal with the debates of the decades around 1900 enters a world of bewildering complexity. The rise of degenerationism itself suggests that perceptions of the evolutionary link between nature and humankind had begun to change. No one could be sure what the implications of evolutionism might be. This difficulty was compounded by the grow-

329

ing dissension among biologists over the nature of the evolutionary process. Darwin had never been the undisputed leader of the evolutionists, and his theory of natural selection was now being challenged by a number of alternatives. These included complex mixtures of the Lamarckian theory of the inheritance of acquired characters and the concept of internally directed evolution, or orthogenesis. The rival theories were inevitably caught up in the debate over the validity of the belief that progress was inevitable. The relationship between the scientific and ideological positions was, however, a subtle one – far too subtle to support the claim that each theory had an innate moral or political character of its own. Generations of writers from Samuel Butler through George Bernard Shaw to Arthur Koestler have tried to convince us that Lamarckism, for instance, is an inherently more humane and optimistic theory than Darwinism.[2] Yet we shall see that Darwinism, Lamarckism and orthogenesis were *all* exploited by *both* progressionists and degenerationists. Indeed, the theory of orthogenesis, which may in principle convey a message more pessimistic than anything attributed to Darwin's natural selection, was used as support for progressionism by biologists seeking to explain the origin of humankind. As the modern sociology of science has shown, an individual's reaction to a new theory depends on a blend of previously established political and professional interests, not on a recognition of the theory's obvious moral value.[3]

Degenerationism and evolution

In the 1860s evolutionists such as John Lubbock still had to battle against the last vestige of a degenerationist view of human history based on scriptural authority. Archbishop Whately of Dublin and the Duke of Argyll insisted that the human race had degenerated from an original state of moral perfection.[4] Lubbock's evolutionary anthropology emerged victorious, establishing the progress of civilization as a continuation of the progress inherent in biological evolution. But by the end of the century doubts were being raised about the inevitability of further progress. Without denying that most of human history represented an advance from uncivilized origins, some writers now began to wonder whether the triumphal development of Western culture could be maintained. This later degenerationism was not a throwback to Whately and Argyll; nor was it a simple alternative to the prevailing faith in progress. To some extent, degeneration and progress went hand in hand.[5] Even the most optimistic progressionist knew that there had been some episodes in the past when the advance of civilization had faltered. It was thus necessary to identify the circumstances that might lead to a temporary reversal of the upward trend. Degenerationists, however, believed that their *own* era was undergoing such a process of decline. The majority of late nineteenth-century artists, for instance, seem to have believed that artistic standards were declining.[6] Many actually blamed the progress of technology for stifling artistic sensitivity. The prospect of a decline in the arts was one of the most obvious threats leading to the increased concern over cultural degeneration that surfaced in the last decade of the century. The German sociologist Max Nordau's widely read *Degeneration*, which appeared in English in 1895, stressed the degeneracy of modern art; and the notoriety achieved by artists such as Aubrey Beardsley and Oscar Wilde convinced many that Nordau was right.

Sociologists in the later nineteenth century tended to assume that social changes were accompanied by biological modifications of the human type. Progressionists among them, such as Herbert Spencer, believed that social progress could only be achieved if the mental character of the race was improved. Similarly, those who were degenerationists seldom limited their warnings to the sphere of culture: they usually attributed declining standards, at least in part, to the biological degeneration of an increasing segment of the population. Nordau himself sought biological explanations of the artists' decline, based on the susceptibility of the human nervous system to the pressures of civilized life.

The possibility that degenerative factors might be introduced into the human constitution had been popularized by Bénédict-Augustin Morel's *Traité des dégénérescences physiques, intellectuelles et morales de l'espèce humain* as early as 1857.[7] Morel and other physicians thought that environmental factors such as the taking of drugs or alcohol could produce a degenerative trend in individuals that would be inherited by their offspring. Perhaps the most obvious expression of the growing concern over human degeneracy was the criminal anthropology of Cesare Lombroso, whose *L'uomo delinquente* of 1875 introduced the notion of the criminal as a degenerate throwback to an earlier stage of evolution. Nordau, who dedicated his *Degeneration* to Lombroso, presented the artist and the criminal as equivalent cases of arrested development:

> Degenerates are not always criminals, prostitutes, anarchists, and pronounced lunatics; they are often authors and artists. These, however, manifest the same mental characteristics, and for the most part the same somatic features as the members of the above-mentioned anthropological family, who satisfy their

unhealthy impulses with the knife of the assassin or the bomb of the dynamiter, instead of with pen and pencil.[8]

Lombroso believed that the environment caused the arrest of development that produced the subhuman criminal type. Individuals reared in poor conditions would suffer thus, and in areas where the overall environment was unsuitable the whole population might be to some extent degenerate. Lombroso also identified certain races as more inclined to degeneracy than others,[9] a view supported by many racial theorists who assumed that each race had its natural home territory and would degenerate if moved elsewhere.[10] By the end of the century, fears were growing that the slums of the industrial world were becoming a breeding ground for degenerates. The eugenics movement, which advocated the imposition of restrictions on the breeding of 'unfit' individuals, was inspired largely by the fear that the white race itself was degenerating. In 1900 Karl Pearson expressed the fear that British defeats in the early stages of the Boer War were a sign of racial degeneration; he called for legislation to prevent the spread of harmful characters.[11] The growing strength of the eugenics movement in the early twentieth century indicates that many social thinkers had begun to doubt the inevitability of progress.

The fact that social degeneration was so often linked to a biological weakening of the race meant that evolution theory inevitably entered the debate. Even if evolution was progressive on the whole, there were obvious examples of degeneration in the animal kingdom, and these episodes offered a clue to the future of the human race. But there was no scientific consensus on the mechanism of biological evolution to which the degenerationists could turn for inspiration. In a world where Darwin's theory of natural selection was challenged by a host of alternatives, social thinkers of all persuasions had to explore the whole range of evolutionary mechanisms to see which best suited their purposes.

Lombroso's theory can be used to illustrate what was perhaps the most important source of tension within late nineteenth-century evolutionism. At one level, Lombroso attributed the appearance of degenerate individuals to the environment, the poor quality of which had led to stunted growth. But his claim that the criminal type was a throwback to an earlier, more bestial stage in human evolution suggests that the poor environment was only the trigger – it did not control the direction in which the degeneration took place. The true source of the degenerate character was internal, for growth had to follow a predetermined course and could only be checked at an incomplete stage. The view that incomplete growth produced a character equivalent to an earlier stage of evolution linked Lombroso's theory of the prevalent nineteenth-century belief that individual growth recapitulated the evolutionary history of the species.[12] As we shall see, this was by no means the only way in which the variation of growth was supposed to be predetermined, but the central issue was always the relative strength of internal versus external or environmental factors in determining the course of evolution.

Darwin's theory had stressed the role of environmental factors in determining the evolutionary path a species would take. Variation among the individuals of the species was completely random, and hence there could be no control by forces acting from within the organism, such as a predetermined pattern of growth. Selection of those characters best adapted to the environment offered the only directing factor in evolution. Darwin knew that his theory was incompatible with the idea of a goal towards which evolution must progress. Any notion of progress had to be applicable to a wide range of different types; it could only be assessed on the basis of increased biological efficiency.[13] Nor was progress a universal phenomenon in Darwin's view: evolution required only increased levels of adaptation, and some adaptations did not require progress toward a higher state. For example, Darwin realized that parasites were adapted to a passive life-style that invariably led to degeneration. On this basis, then, a lessening of the stimulus provided by the environment would be the most likely explanation of human degeneration, but the decline would not be inevitable if the external challenge could be renewed.

If adaptation was accepted as the critical requirement for evolutionary change to take place, the most obvious alternative to natural selection was J.-B. Lamarck's theory of the inheritance of acquired characteristics. First published early in the century, Lamarck's theory was widely accepted in the post-Darwinian era as a more humane model of adaptive evolution than natural selection.[14] Instead of relying on the elimination of the unfit, the Lamarckians supposed that in a new environment *all* members of a species would strive to adapt. They would adopt new habits, and the use of their organs in a new way would lead them to acquire new bodily characters adapting them to their way of life. If these acquired characters were inherited by the next generation, the process would be cumulative and the new adaptive characters would transform the species. In the classic example, the modern giraffe has acquired its long neck through generations of its ancestors stretching their necks to reach the leaves of trees. It was the implication that evolution follows the animals' voluntary choice of a new behaviour pattern that led many Lamarckians to claim that their theory offered a less mechanistic interpretation than the 'trial and error' of natural selection. But the theory also implied that organs no longer required would diminish in size through the cumulative effects of disuse. Degeneration in an unstimulating environment was thus still a possibility – indeed many Lamarckians claimed that the inherited effect of disuse offered a more plausible explanation of degenerative evolution than did natural selection.

Some late nineteenth-century Lamarckians were not just concerned with adaptation. The American neo-Lamarckians, including palaeontologists such as Edward Drinker Cope and Alpheus Hyatt, adopted the inheritance of acquired characters merely to flesh our their belief that the evolution of any group can be understood through a parallel with the growth of the individual organism. Progressive evolution consisted of the successive addition of stages to the growth process, produced by the inheritance of acquired characters as each generation became more specialized for the species' chosen way of life. Because new characters were thought to appear always as additions to growth, the development of the modern embryo was supposed to recapitulate the evolutionary history of its species. The analogy with growth allowed the Americans to treat evolution as a highly directed process, moving inexorably toward a predetermined goal. Hyatt went on to stress a further parallel with the life-history of the individual: he argued that each species must inevitably enter a phase of 'racial senility' in which it degenerated toward simpler characters and ultimate extinction. Here the element of directed evolution generated a belief that all species must at last complete the progressive phase of their evolution and enter a period of degeneration. The decline was not merely a response to unstimulating conditions, but an inevitable fate awaiting all species. If the human race had begun to enter its period of senility, degeneration would occur regardless of efforts to maintain the external stimulus.

The concept of racial senility remained popular among palaeontologists well into the twentieth century, but the theoretical background to the idea changed dramatically in the 1890s. Hyatt used the analogy with the individual's life-cycle to promote an almost vitalist interpretation in which the species degenerated towards extinction after exhausting its supply of evolutionary energy. Towards the end of the century, though, the analogy with growth became less fashionable, and palaeontologists sought a more mechanistic explanation of directed evolution. Racial senility was now interpreted as a consequence of orthogenesis, or evolution driven in a fixed direction by variation-trends built into the germ plasm of the species. The supposedly non-adaptive trends revealed by the fossil record were seen as evidence that many species were affected by an inbuilt tendency to evolve in a fixed direction. Racial senility remained almost inevitable, because the majority of variation-trends were thought to result in the over-development of once-useful structures up to the point where they became positively harmful to the species. Some palaeoanthropologists were willing to explain the evolution of the human species in terms of such orthogenetic trends, but were unwilling to admit that in this case the trend was leading towards racial senility.

Evolution theory thus provided a wide range of models upon which to base predictions about human progress or degeneration. In the case of natural selection or simple Lamarckism, degeneration became a contingent process arising from a lack of environmental stimulus. Even if the human species was now experiencing such enervating conditions, it would be possible to reawaken the sense of challenge and return to the path of progress. The threat of racial stagnation could thus be used as a warning rather than a firm prediction. The theories of racial senility offered a more dismal prospect in which the future was predetermined. The fact that very few scientists were prepared to apply this model to the human race suggests that degenerationism was indeed no more than an attempt to reassess the conceptual foundations of progressionism. We shall see that Hyatt himself made one of the few systematic efforts to explore the possibility of human racial senility, although he seems to have done so largely as a means of claiming that the emancipation of women would lead to social decay.

The perils of progress

Among Darwin's followers it was E. Ray Lankester who most clearly explored the possibility that evolution by natural selection might in some cases lead to degeneration rather than progress. In an address to the British Association in 1879, published a year later as *Degeneration: A Chapter in Darwinism*, Lankester pointed out that evolutionists seeking to reconstruct the history of life on the earth could not assume that simple forms were relics of earlier stages in the development of their type. In many cases these forms were the degenerate descendants of originally much higher forms. Most naturalists recognized degeneration in the case of parasites, but drawing on the work of Anton Dohrn, Lankester argued that the phenomenon was much more widespread. Any adaptation to 'less varied and *less* complex conditions of life' could result in degeneration, 'just as an active healthy man sometimes degenerates when he becomes suddenly possessed of a fortune; or as Rome degenerated when possessed of the riches of the ancient world'.[15] Some major groups of animals were the degenerate descendants of higher types. If it were not for a barnacle's embryonic stages, no one would realize that it was a degenerate crustacean. The Ascidians or sea-squirts should be regarded as degenerate members of the vertebrate type, 'standing in the same relation to fishes, frogs and man, as do the barnacles to shrimps, crabs and lobsters'.[16]

Having used analogies with human affairs to make the point in his biological argument, Lankester concluded with a discussion of the possibility of widespread cultural degeneration. Anthropologists such as Lubbock tended to assume that 'primitive' societies were relics of earlier stages in human progress, preserved unchanged in isolated areas, but Lankester argued that 'savages' such as the bushmen and the Australian aborigines might be the descendants of once-civilized peoples. Could a similar fate threaten the white race?

> In accordance with a tacit assumption of universal progress – an unreasoning optimism – we are accustomed to regard ourselves as necessarily progressive, as necessarily having arrived at a higher and more elaborated condition than that which our ancestors reached, and as destined to progress still further. On the other hand, it is well to remember that we are subject to the general laws of evolution, and are as likely to degenerate as to progress. As compared with the immediate forefathers of our civilisation – the ancient Greeks – we do not appear to have improved so far as our bodily structure is concerned, nor assuredly so far as some of our mental capacities are concerned. . . . Possibly we are all drifting, tending to the condition of intellectual Barnacles or Ascidians. It is possible for us – just as the Ascidian throws away its tail and its eye and sinks into a quiescent state of inferiority – to reject the good gift of reason with which every child is born, and to degenerate into a contented life of material enjoyment accompanied by ignorance and superstition.[17]

Thus the cultivation of science was vital to prevent any possible future degeneration, although it did not occur to Lankester that science might itself solve all humanity's material problems and remove the stimulus to progress.

At the time of publication Lankester's remarks were intended to counter the facile optimism of the progressionists. By the 1890s, however, the prospect of degeneration had become a popular concern. One of the most vivid literary explorations of the theme, H.G. Wells' *The Time Machine* of 1895, paralleled some of Lankester's ideas but added the possibility that science might be the cause of future decay. Wells' awareness of the fragility of civilization is usually linked to the 'cosmic pessimism' expressed two years earlier by Thomas Henry Huxley in his Romanes Lecture, 'Evolution and Ethics'.[18] And the various threats to civilization postulated in Wells' science fiction stories certainly reflect Huxley's vision of a universe that is essentially hostile to the aspirations of humankind. But *The Time Machine* concentrates on the possibility of degeneration without external threat. We know that Wells was interested in the idea of biological degeneration because he wrote a popular article on the topic in 1891.[19] Although Lankester is not mentioned by name, the theme of the Ascidians as degenerate vertebrates is stressed in the article, suggesting that Wells knew of Lankester's views on the subject.

The Time Machine was developed from a story entitled 'The Chronic Argonauts' that Wells began as early as 1887.[20] The projected future into which the time-traveller ventures brings out the theme of degeneration caused by a civilization so advanced that it has protected the human race from all hardships. The time-traveller meets the frail Eloi, childlike in both physical appearance and mentality. He speculates on the cause of their degeneration: 'What, unless biological science is a mass of errors, is the cause of human intelligence and vigour? Hardship and freedom: conditions under which the active, strong and subtle survive and the weaker go to the wall. . . .' But mankind had used science to conquer nature, and 'after the battle comes Quiet. Humanity had been strong, energetic and intelligent, and had used all its abundant vitality to alter the conditions under which it lived. And now came the reaction to the altered conditions.' The weak were better fitted than the strong for this life of perfect ease and security.

> No doubt the exquisite beauty of the buildings I saw was the outcome of the last surgings of the now purposeless energy of mankind before it settled down into perfect harmony with the conditions under which it lived – the flourish of that triumph which began the last great peace. This has ever been the fate of energy in security; it takes to art and to eroticism, and then comes languor and decay.[21]

The asexuality of the Eloi strikes a chord reminiscent of the recapitulation theory – the notion of degeneration as return to a juvenile state through arrested development. But there is a twist in Wells' story that avoids the recapitulationist view of degeneration and confirms that the path of decline is not marked out in advance. The time-traveller soon discovers that the fragile Eloi are the descendants only of the leisured classes. The pale, half-blind Morlocks are the end-product of a separate line of human evolution by which the industrial workers have adapted to their underground homes. Although retaining more intelligence than their open-air cousins, upon whom they now prey, the Morlocks are all the more hideous as a symbol of the degeneration produced by adaptation to a stable and stultifying environment.[22]

Wells' vision of degeneration through the removal of external stimuli may seem to reflect Lankester's Darwinism, in which natural selection is the agent adapting the species to its less demanding environment. But there has been some debate over the possibility of an additional Lamarckian influence. One scholar presents Wells as a pure Darwinist, pointing out that his *Textbook of Biology* of 1893 makes no reference to the inheritance of acquired characters.[23] Others, however, note that Wells at first resisted August Weismann's theory of the germ plasm, apparently because he had some sympathy with the Lamarckian mechanism.[24] They suggest that the degeneration of the Eloi and Morlocks follows a Lamarckian pattern, by assuming the inherited effects of disuse. Against this interpretation is the fact that in an earlier version of the story Wells commends Weismann's concept of 'panmixia', which was designed to explain the loss of disused organs through the relaxation of selection pressure.[25] By the time the final version of *The Time Machine* was published in 1895, Wells had begun to accept Weismann's critique of Lamarckism, and must have intended his readers to interpret the message of degeneration in a Darwinian light.[26]

The confusion among modern authorities over the possibility of a Lamarckian element in Wells' degenerationism shows that the importance of struggling *against the environment* could be recognized as easily by a Lamarckian as by a Darwinian. Indeed, Wells' account of the Morlocks recalls the contemporary zoological fascination with blind cave-animals, a study particularly favoured by Lamarckians such as Alpheus Packard because they felt that the inherited effect of disuse offered the most convincing explanation of the phenomenon.[27] The theme of degeneration through adaptation to less stimulating conditions was also explored by one of the last major exponents of the recapitulation theory, the Lamarckian embryologist Ernest William MacBride. MacBride developed a whole philosophy of evolution based on the claim that the various invertebrate types were the degenerate remnants of branches from the main stem of vertebrate evolution. The progressive stem 'leads straight on to the victorious Vertebrata, whose leading members have never deserted the free life of the ocean rover, and have been and are now dominant in the seas'. Weaker branches have given up the struggle to seek refuge on the ocean floor, where their inactive life-style has caused them to degenerate into the various invertebrate types. 'It is, therefore, broadly speaking true that the Invertebrates collectively represent those branches of the Vertebrate stock which, at various times, have deserted their high vocation and fallen into lowlier habits of life.'[38] MacBride also saw those human races that had adapted to a 'soft' environment as less advanced than the Anglo-Saxons. In the 1920s and 1930s he became a vigorous supporter of the eugenics movement, calling for the sterilization of the unfit to prevent further degeneration of the British people.[29]

Both theories of adaptive evolution, Darwinian and Lamarckian, could thus be used to stress the possibility of degeneration brought on by the adoption of a passive life-style. But if adaptation was the only guiding force in evolution, there could be no preordained course along which the degenerative process must run. Many late nineteenth-century evolutionists, however, were reluctant to concede that adaptation was the only guiding force. There was a persistent tendency to believe that evolution is somehow a directed process, following a more consistent course than could be explained by the hazards of exposure to an ever-changing environment. The link between such theories of directed evolution and the debate over the future of humankind must now be explored.

Racial senility and the status of women

The earliest interpretation of directed evolution arose from the claim that the growth of the individual offered the best model for understanding the evolution of the species. Many Lamarckians believed that the development of the modern embryo recapitulates the evolutionary history of its species, because, in their view, acquired characters were inherited as successive additions to the growth process. This process had been accelerated to make room for new additions, so that the old adult forms of the species became stages through which the individual had to grow to reach the new state of maturity. Some Lamarckians assumed that the addition of new stages to growth followed a regular pattern, a continuation of the growth process in the same direction. This model offered an alternative to MacBride's position as a means of explaining degeneration. Instead of being merely a response to less stimulating conditions, degeneration might be produced by a complete reversal of the normal process of evolution. If progressive evolution worked by the addition of stages to growth, degeneration might occur through the loss of more advanced stages and a retreat back to an earlier and more primitive phase. Thus Lombroso's criminal anthropology rested on the assumption that the

criminal was a case of arrested development, a throwback to a brutal
and prehuman stage of evolution.

The most complete expression of the link between growth and evolution came in the work of the American school of Neo-Lamarckism.
As we have seen, palaeontologists such as Cope and Hyatt began
from the assumption that the growth of the individual must serve as a
clue to the evolution of its species.[30] In Hyatt's case this insight came
from his training under the Harvard zoologist Louis Agassiz. But,
whereas the German-educated Agassiz saw the analogy between
individual growth and the history of life as rational proof of a fixed
plan of creation, his American students adopted an evolutionary
interpretation of the analogy during the 1860s. Equally in defiance of
Darwin, however, they retained the view that variation must consist
of additions to growth directed along a path already marked out by
the existing growth process.

Cope and Hyatt soon realized that the inheritance of acquired characters would explain how new steps were added on to growth, but
they retained the belief that many aspects of evolution were directed
along a fixed path and were not open to modification by the hazards
of adaptation. The pattern of individual growth still represented a
valid model for evolution, and Hyatt in particular was anxious to
stress that the pattern was normally completed by a phase of senile
degeneration and death. In evolution the decline began when a group
lost its original vigour and the individual members could no longer
cope with the challenges posed by the environment. Unable to make
a positive response, they reacted to the now-unfavourable conditions
by simplifying themselves, acquiring non-adaptive characters, and
passing them to their offspring, so that the process became cumulative. Hyatt also insisted that racial senility regressed the group as a
whole along the path of growth back to a simple form very similar to
the juvenile phase from which it began. As in the case of the individual, the end result of the decline was death or extinction. Hyatt
thus turned Lamarckism into an extremely pessimistic model of evolution in which each group possessed a mysterious and limited store of
vital energy. Once this was exhausted, the degeneration and extinction of the group was as inevitable as the death of the individual.

Hyatt's detailed research was confined to the fossil invertebrates,
but he certainly thought that his scheme was valid for other groups.
Could it, then, be applied to the human race? If so, the implications
would be disturbing: however progressive the earlier phases of human
evolution, our species must sooner or later run out of evolutionary
energy and begin to degenerate. Hyatt made only one attempt to
apply his theory of racial senility directly to human affairs, but when
he did so, he linked the prospect of degeneration to another con-

troversial issue of the time, female emancipation. In effect, he argued that to give women equal political rights would diminish the psychological differences between the sexes and would thus encourage a degenerative trend in the species. These claims were expounded in a lecture given in Boston in 1897, subsequently published in the British journal *Natural Science* under the title 'The Influence of Woman in the Evolution of the Human Race'.

Hyatt remarked that it was appropriate for his ideas to be published first in Britain, since

> some leading Englishmen are so sadly deficient in knowledge of the subject and its importance that they consider the question of whether the suffrage shall or shall not be granted to women as a huge political joke, rather than as a question dealing with matters of importance to the future evolution of civilised races. People do not yet recognise that the tendency of evolution is quite as often toward retrogression and extinction as in the direction of progression; the former being indeed the final result both in the life history of the individual and of his family, and finally of the race to which he belongs.[31]

Would the encouragement of similar habits in men and women be biologically harmful? Hyatt insisted that it would, but to do this he had to develop a series of indirect arguments. He suggested that a number of human characters are already 'phylogerontic', indicating that our species is already beginning to move into its senile phase. But since he made no effort to survey the course of primate evolution, he was unable to specify these senile characters in detail. He merely concentrated on the difference between the sexes, advancing two quite different arguments to show that increasing sexual differentiation was a sign of evolutionary progress. This in turn would allow him to claim that giving votes to women would constitute a move in the opposite direction and would hasten the degeneration of the race. *degeneration = threat to status quo?*

Hyatt's first point was that many phyla seem to have evolved from an original hermaphrodite form. The appearance of two distinct sexes was thus a feature of progressive evolution, and any tendency for the sexes to merge back again would be degenerative. This trend was not a major feature of Hyatt's own interpretation of invertebrate evolution, and it is not clear how it could be applied to the higher vertebrates. At best, Hyatt was offering an analogy to the earliest phases of the evolution of life on the earth. Because this analogy might have appeared somewhat remote, he moved on to consider the most recent phase of human evolution, arguing that significant differences between men and women had only appeared since the development of a more civilized life-style. No hard evidence was provided to back up the claim; Hyatt simply asserted that our remote ancestors would have exhibited little sexual differentiation because the habits of both sexes would have been very similar. This allowed him to argue that 'the divergence of the sexes is a marked characteristic of progression among highly civilized races'.[32] To his own satisfaction, at least, he had established that any decrease in the differences between the sexes would run counter to the progressive trend in human evolution.

Hyatt was now in a position from which he could evaluate the demand for the emancipation of women. It was obviously 'of the highest importance' for the human race to 'avoid all movements tending to the increase of the natural and possibly inherent tendencies toward retrogression'.[33] Any move to diminish the difference between the sexes by encouraging similar habits in men and women would run against the pattern of progressive evolution in the past and would accelerate the move into the degenerative phase. There was some risk of men becoming effeminate in the modern world, but far more dangerous was the tendency for women to become 'virified' or mannish. The burden of maintaining the sexual differential was thus thrust upon women, who had to be impressed with the fact that their demands for equality constituted a biological danger to the race. Hyatt hoped women themselves would accept the need to retain their distinctive character within the existing social arrangements, but if their demands continued they would have to be resisted. Significantly, Hyatt modified the pessimistic message of his general evolutionary theory to hold out the prospect that, for humankind at least, racial degeneration might be avoided:

> It seems obvious that the time has come when thoughtful men and women should be warned, if this be possible, that their organisations are not of such a kind that they can rely upon continuous and certain progress. The laws of evolution point distinctly to a future in which retrogression and extinction is perhaps certain; but man's past history and the same laws also hold out hopes for the maintenance of progress through an indefinite time, if he is capable of controlling his own destiny through the right use of experience and the wonderful control over nature that his capacities have enabled him to attain.[34]

Here Hyatt joined the majority of Lamarckians who believed that human intelligence had gained the power to take control of the race's evolution. Degeneration was not inevitable.

Hyatt was not alone in his fear that the emancipation of women might be dangerous. Others made use of the prevailing assumption that civilization was an essentially masculine creation to justify the subordination of women.[35] Any attempt by women to acquire equal status was seen as threatening masculine domination and was resisted by the claim that it would undermine the structure of the family and hence of society itself. Biology was also readily invoked to support the claim that women were inferior. Cope argued openly against emancipation on the grounds that women were less developed (in the evolutionary sense) than men.[36] Another American, Joseph LeConte, adopted a more tolerant view, but still held that evolution had given men and women essentially different characters.[37] Hyatt appealed to the same idea, that progressive evolution enhanced sexual differentiation, but used his own theory of racial senility to argue that equal rights for both sexes would be *biologically* dangerous for the race. The argument could have been applied equally to men – especially artists and writers – who were thought to be in danger of becoming effeminate. Hyatt, however, chose to concentrate on resisting the emancipation of women, and the highly contrived use of his theory of racial senility in this context suggests that in so doing he was merely giving expression to a longstanding prejudice.

Although Hyatt hoped that humankind might be able to resist the trend, his theory in principle implied that degeneration was practically inevitable once a species had exhausted its supply of evolutionary energy. The possibility that there might be such a biological limit to human progress was also noted by Nordau in *Degeneration*, but he, too, evaded the full implications of the idea. He seemed to accept that racial exhaustion was inevitable in the long run, but then insisted that the current problems of civilization were not a sign that the critical point in human evolution had now arrived. The problems of modern culture were only temporary, 'because humanity has not yet reached the term of its evolution; because the over-exertion of two or three generations cannot yet have exhausted all its vital powers. Humanity is not senile. It is still young, and a moment of over-exertion is not fatal for youth; it can still recover itself.'[38] Here Nordau accepts the view that the evolution of the race can be compared to the life-cycle of an individual; yet he evades its darker implications by simply asserting that racial senility is not yet upon us. Irreversible degeneration may be inevitable in the distant future, but Nordau's underlying optimism shines through in his assumption that the current problems have a much less serious cause. The pressures of civilization have indeed placed heavy demands on human biology, but the race still has enough energy to adapt, and only a few 'degenerates, hysterics and neurasthenics . . . are fated to disappear'. In the end, even Nordau held out the hope of further progress once humankind had adjusted to its new way of life.

The fact that Hyatt and Nordau did not face up to the prospect of an inevitable decline suggests that even in the 1890s there were limits beyond which the pessimistic implications of degenerationism could not be explored. In the scheme outlined by Lankester and Wells, degeneration is self-induced and could be prevented by constant exposure to new challenges. In another science-fiction novel, *War of the Worlds*, Wells supposed that the Martians, although physically degenerate, had continued to advance intellectually because they had been confronted by the threat of their planet drying up. *The Time Machine* does create the impression that degeneration is ultimately inevitable, but this is because of physical, not biological, limitations. The gradual extinction of the Sun will mean the end of life on the earth, whatever the achievements of humankind.[39] As a realistic prediction, this was far less depressing than Hyatt's vision of a biological limit to the energy of the human race. It was one thing to warn against degeneration through a collective failure of effort and initiative, quite another to suppose that the slide into racial senility was biologically inevitable. Indeed, of all the available theories of evolution, Hyatt's analogy between the history of the individual and the species was in principle the most depressing. The air of inevitability it gave to the degenerationist position was unacceptable in a world where most thinkers still hoped that progress would be resumed in some future environment. The theory of racial senility was useful in special cases, such as Hyatt's opposition to female emancipation, but it was too fatalistic to be accepted as the main explanation of the *fin de siècle* cultural malaise.

Orthogenesis and human evolution

Reluctance to apply the theory of racial senility to humankind was paralleled by the way in which the related concept of orthogenesis featured in debates over human evolution. Hyatt and other supporters of the recapitulation theory saw degeneration as a return to the group's youthful or primitive state, but during the 1890s theoreticians began to disagree. One turn-of-the-century account of degeneration in biology and sociology devoted much of its space to arguments demonstrating that 'Degenerative evolution ... can in no way be regarded as constituting a return to the primitive condition.'[40] Instead of adopting Hyatt's theory, many invertebrate palaeontologists regarded the apparent degeneration of the fossil cephalopods studied by Hyatt as an example of orthogenesis, or evolution directed by internally programmed trends that would force variation inexorably in a certain direction, even when the results were non-adaptive.[41] Vertebrate palaeontologists handled the concept of racial senility in a rather different way. In many vertebrate families the fossil record seemed to reveal consistent trends in the development of certain structures, such as a steady increase in the size of horns throughout the group's history. It was as though variation had gained a sort of 'momentum': a structure originally developed for a useful purpose continued to enlarge even when it became a positive encumbrance. In cases such as the 'Irish elk', the size of the antlers produced by the orthogenetic trend was supposed to have led to the extinction of the species. Racial senility still ended in extinction, but the cause was the over-development of once-useful structures, not the exhaustion of the species' evolutionary energy.

This version of racial senility still gave a highly pessimistic view of the future. If all species were subject to orthogenetic trends, then all would sooner or later be driven to extinction. In 1920, the palaeontologist Francis A. Bather attacked the theory of orthogenesis in an address to the British Association by pointing out the implications of the belief that evolution was governed by internal rather than external factors:

> If we are to accept the principle of predetermination, or of blind growth-force, we must accept also a check on our efforts to improve breeds, including those of man, by any other means than crossing, and elimination of unfit strains. In spite of all that we may do in this way, there remain those decadent races, whether ostriches or human beings, which "await alike the inevitable hour". If, on the other hand, we adopt the view that the life-history of races is a response to their past environment, then it follows, no doubt, that the past history of living creatures will have been determined by conditions outside their control, it follows that the idea of human progress as a biological law ceases to be tenable; but since man has the power of altering his environment and of adapting racial characters through conscious selection, it also follows that progress will not of necessity be followed by decadence.... [42]

In other words, evolution controlled by external factors may not guarantee progress, but at least it does not make degeneration inevitable. There were, in fact, very few attempts to predict the future demise of the human race as the result of orthogenetic trends. Some years after Bather's attack, an American physician, George W. Crile, invoked orthogenesis to explain certain human disabilities, such as the tendency towards peptic ulcers.[43] But this was an exception. Despite the widespread belief among palaeontologists that over-development was a major cause of extinction among vertebrates, the prospect of an inevitable decline in the biological character of the human species was too pessimistic to be entertained.

Yet the basic idea of orthogenesis *was* applied to human evolution by the simple expedient of assuming that the dangerous consequences of over-development would not apply in the case of a trend towards increasing brain size. To understand how this was possible, we must look at the assumptions shared by almost all of the early twentieth-century palaeoanthropologists who sought to explain the origins of humankind. As I have shown elsewhere,[44] there was widespread reluctance to believe that the human race might have originated through an unlikely combination of circumstances. Darwin had to some extent anticipated the modern view of human evolution by suggesting that the change to an upright posture may have been the decisive step paving the way for the enlargement of the brain, by freeing the hands for tool-making and other activities that would encourage the growth of intelligence.[45] But few took up this suggestion: in general it was assumed that intelligence had increased steadily throughout vertebrate and especially primate evolution, so that the human mind could be seen as the inevitable product of a longstanding evolutionary trend. As Grafton Elliot Smith argued, our early human ancestors stood upright because they had at last developed enough intelligence to appreciate the advantages of doing so. They did not get their intelligence *as a result* of standing upright to adapt to a change in their environment.[46] The popular assumption that the extinct Neanderthal race constituted a distinct species or even genus of humanity encouraged the belief that there had been evolutionary trends at work capable of driving several different forms in parallel towards human status.

Given the widespread assumption that the human race originated not from a unique adaptive change, but from a consistent trend in primate evolution, it was inevitable that comparisons would be made with evolutionary trends found elsewhere in the fossil record. Palaeoanthropologists simply borrowed the palaeontologists' technique of describing the evolution of each group as being governed by its own characteristic set of trends. Of course the primate trend towards brain development could be ascribed to a functional cause. Elliot Smith championed the view that the interaction of eye, hand and brain had increased the intelligence of the primates as an adaptation to their original arboreal habitat. But if many palaeontologists were convinced that non-functional trends were at work throughout vertebrate evolution, it was hardly surprising that some, at least, should consider the possibility that human evolution might also be the result of internally programmed orthogenesis. The trick would be to do this in a way that evaded the widespread belief that such trends normally ended with over-development and extinction.

Orthogenetic trends in human evolution were postulated by a

number of early twentieth-century palaeoanthropologists, including Earnest A. Hooton in America and the young Wilfrid Le Gros Clark in Britain.[47] One of the most consistent supporters of the idea was Sir Arthur Smith Woodward, best known for his role in the notorious Piltdown fraud of 1912.[48] Although he is now remembered only as the original champion of 'Piltdown man', Woodward was a vertebrate palaeontologist rather than an anthropologist, and he had always supported the view that evolution was driven by forces somehow built into the germ plasm of the species.[49] In his 1915 British Museum *Guide to the Fossil Remains of Man* (which of course included the Piltdown remains), Woodward explicitly linked his theory of orthogenesis with the evolution of the human brain:

> Now, the study of many kinds of fossils has shown that when, in successive generations, one part of the body begins to increase in size or complication much more rapidly than other parts, this increase rarely stops until it becomes excessive. As a rule it passes the limit of utility, becomes a hindrance, and even contributes to the extermination of the races of animals in which it occurs. In the case of the brain, however, a tendency to overgrowth might become an advantage, and it seems reasonable to imagine that such an overgrowth in the early apelike animals eventually led to the complete domination of the brain, which is the special characteristic of man.[50]

Here Woodward openly admits that orthogenetic trends normally end in extinction through over-development. But he puts aside any pessimistic implications by claiming, in effect, that we cannot have too much intelligence. This may appear naive in view of latter-day technological threats to human existence, let alone the mechanized slaughter of the trenches at the time that it was published. Woodward, however, seems to have been quite happy to turn orthogenesis into a progressionist explanation of human origins. He, at least, did not see the brain as a danger to the species.

The link between orthogenesis and progress suggests once again that the degenerationism of the 1890s was only skin deep. We have seen that Hyatt's theory of racial senility was rarely applied to the human race, with even Nordau backing away from the possibility that the species had exhausted its evolutionary energy. Degeneration was a problem of contemporary society, and although this could be linked to biological stresses in the human race, it was not necessarily a sign of racial senility. The scientists studying the origins of the human race automatically made progressionist assumptions. Their self-imposed task, to explain the ascent from the apes, left no room for a major degenerative trend. In palaeoanthropology, the progressionism we associate with the Victorian era survived well into the twentieth century.

Of all the available theories of evolution, orthogenesis and the concept of racial senility were in principle the most pessimistic, as they implied that each group was driven inexorably towards extinction. Darwinism and Lamarckism explained evolution through an interaction with the external environment, and were thus able to admit the possibility of degeneration without making it inevitable. In a climate of opinion where the threat of degeneration was seen more as a warning than as a definite prediction, there was an obvious preference for biological analogies that allowed some degree of flexibility in the future. The theory of racial senility was simply too pessimistic for many thinkers to take its implications seriously. It was used occasionally in the degeneration debate of the 1890s, but even its strongest supporter, Hyatt, allowed for the possibility that the human race could resist the trend. In view of the infrequent use of racial senility as an explanation of human degeneration, it would seem more relevant to ask why the idea of racial senility was so popular among vertebrate palaeontologists. For if the implications of orthogenesis were evaded in the case of the human race, why were the palaeontologists so anxious to see rigid trends elsewhere in evolution?

The answer to this question must surely lie in the strength of the anti-Darwinian feeling that still existed in many areas of biology. The one thing that the palaeontologists did *not* want was evolution governed by 'chance', and – however naive it may now seem – Darwin's reliance on 'random' variation as the raw material of natural selection was seen as the basis for a theory of totally undirected evolution. The simplest ways of guaranteeing that evolution worked in an orderly, predictable manner, were to compare it with the growth of the embryo, as in Hyatt's theory, or to postulate rigid variation trends, as in the theory of orthogenesis. Biologists did not set out to create theories in which extinction became inevitable, but they accepted this pessimistic implication as a necessary by-product of their deeper concern for the regularity of the evolutionary process. Precisely because racial senility was only a by-product, it was relatively easy to ignore or evade it when dealing with human evolution.

What, in conclusion, can be said about the general relationship between the various theories and the debate over the progressive character of social and biological evolution in the late nineteenth and early twentieth centuries? It seems evident that there were no clearcut moral or political lessons to be drawn from particular theories. The most intrinsically pessimistic theories were seldom used by the

[Margin notes, handwritten:] Degenerationism has a moral dimension. Compensating for the doubts evolution cast on religion. Evolution left humans with an insecure place in the world – one governed by chance mutations + adaptations. Degeneration & orthogenesis tries to put some order back in.

degenerationists, while orthogenesis was transformed into a progressionist account of human origins. Darwinism and Lamarckism were both adapted to support either the progressionist or the degenerationist positions. Far from having in-built implications for the issue of progress, each theory was capable of being exploited by either side of the debate. This suggests that we must be careful in assessing how moral or political labels are pinned on to scientific theories. Each professional or political group, sometimes each individual, makes its own decision on whether or not to support a theory and on how that theory will be linked to broader issues. Hyatt's opposition to female emancipation encouraged him to apply his theory of racial senility to humankind in a way that was unacceptable to those writers who were more concerned about cultural degeneration. The palaeoanthropologists, who were dealing with a topic that required an explanation of how the human mind had developed, found the progressionist version of orthogenesis a convenient way of linking the general 'laws' of palaeontology with their belief that the human race was an inevitable product of evolution. Bather's claim that the theories of racial senility offered a deeply pessimistic view of the future was thus evaded by most of the biologists who actively supported the theories.

The fact that Darwinism and Lamarckism were also used by both sides in the degeneration debate seems even more paradoxical, given the considerable number of eminent thinkers who have believed that there is an intrinsic moral difference between the two theories. I suspect that the moral or ideological characters ascribed to the theories are illusions built up by the more articulate participants in the debates. Depending upon which of its aspects is stressed, any complex theory can be turned into a panacea or a nightmare. The label that sticks in our minds will be the one that receives the best publicity. This does not mean that we should accept the scientists' frequent protestations that their theories are free from any subjective bias. All labels are suspect, including the claim that there are no labels at all. The historian sees the complex relationship between each theory and the outside world and suspects that all theories are accepted or rejected at least in part because of the labels attached to them. Participants in a moral or ideological debate will exploit any interpretation of any scientific theory that seems to suit their purpose. The scientists themselves are constrained by professional as well as political interests, and if they make their decision first on professional grounds, they will always be able to find a way of adapting the theory of their choice to their wider beliefs. Their own individual and collective perception of a theory's moral implications may also play a role in the initial decision, but this does not mean that the theory will necessarily retain the character sketched out for it by its original proponents.

Notes

1. See J. Edward Chamberlin and Sander L. Gilman, eds., *Degeneration: The Dark Side of Progress* (New York: Columbia University Press, 1985). A further indication of the growing interest was the symposium 'Degenerationism in Late Nineteenth-Century Thought' held at the Wellcome Institute for the History of Medicine, London, 24 May 1985. This article is a greatly expanded version of my own contribution to that symposium, with a new title suggested by Simon Schaffer.

2. See Samuel Butler, *Evolution, Old and New* (London: Harwick and Bogue, 1879); the preface to Bernard Shaw's *Back to Methuselah*; Arthur Koestler, *The Ghost in the Machine* (New York: Macmillan, 1967); and *idem, The Case of the Midwife Toad* (London: Hutchinson, 1971).

3. For a survey of the sociological approach to the history of science, see Steven Shapin, 'History of Science and Its Sociological Reconstructions', *History of Science*, 20 (1982), 157–211. For an application of this approach to evolutionism, see Adrian Desmond, *Archetypes and Ancestors: Palaeontology in Victorian London, 1850–1875* (London: Blond and Briggs, 1982).

4. See Neal C. Gillespie, 'The Duke of Argyll, Evolutionary Anthropology, and the Art of Scientific Controversy', *Isis*, 68 (1977), 40–54.

5. Robert A. Nye, 'Sociology and Degeneration: The Irony of Progress', in Chamberlin and Gilman, *Degeneration*, pp. 49–71.

6. Patrick Bade, 'Art and Degeneration: Visual Icons of Corruption', in Chamberlin and Gilman, *Degeneration*, pp. 220–40.

7. Eric T. Carlson, 'Medicine and Degeneration: Theory and Praxis', in Chamberlin and Gilman, *Degeneration*, pp. 121–44.

8. Max Nordau, *Degeneration*, 2nd edn (London: Heinemann, 1895), p. vii.

9. Cesare Lombroso, *Crime: Its Causes and Remedies*, translated by Henry P. Horton, 1911 (Montclair, N.J.: Patterson Smith, 1968), ch. 3.

10. Nancy Stepan, 'Biological Degeneration: Races and Proper Places', in Chamberlin and Gilman, *Degeneration*, pp. 97–120.

11. Karl Pearson, *National Life from the Standpoint of Science: An Address delivered at Newcastle, November 19, 1900* (London: A. and C. Black, 1901). See G.R. Searle, *Eugenics and Politics in Britain, 1900–1914* (Leiden: Noordhoff, 1976).

12. On Lombroso's system as a version of the recapitulation theory, see Stephen Jay Gould, *Ontogeny and Phylogeny* (Cambridge, Mass.: Harvard University Press, 1977), pp. 120–5 and J.E. Chamberlin, 'An Anatomy of Cultural Melancholy', *Journal of the History of Ideas*, 42 (1981), 691–705 (698–9).

13. For an account of how Darwin tried to define a concept of progress consistent with his theory, see John C. Greene, *The Death of Adam: Evolution*

and Its Impact on Western Thought (Ames: Iowa State University Press, 1959), ch. 9.

14. For further details of the anti-Darwinian theories, see Peter J. Bowler, *The Eclipse of Darwinism: Anti-Darwinian Evolution Theories in the Decades around 1900* (Baltimore, Md.: Johns Hopkins University Press, 1983).

15. E. Ray Lankester, *Degeneration: A Chapter in Darwinism* (London: Macmillan, 1880), p. 33. This essay is reprinted in William Coleman, ed., *The Interpretation of Animal Form: Essays by Jeffries Wyman, Carl Gegenbaur, E. Ray Lankester, Henri Lucaze Duthiers, Wilhelm His and H. Newell Martin* (New York: Johnson Reprint Corp., 1967). For a modern commentary on the same theme, see Gunther S. Stent, *Paradoxes of Progress* (San Francisco: Freeman, 1978).

16. Lankester, *Degeneration*, p. 41.

17. Ibid., pp. 59–61.

18. For example, see Norman and Jeanne Mackenzie, *The Time Traveller: The Life of H.G. Wells* (London: Weidenfeld and Nicholson, 1973), pp. 57, 122, 128 and John Batchelor, *H.G. Wells* (Cambridge: Cambridge University Press, 1984), pp. 12–13. For Huxley's original lecture, see his *Collected Essays*, 9 vols. (London: Macmillan, 1893–4), IX, 46–116. Wells studied for a time under Huxley at the Normal School of Science in London. Other recent biographies of Wells include Anthony West, *H.G. Wells: Aspects of a Life* (London: Hutchinson, 1984) and David C. Smith, *H.G. Wells: Desperately Moral* (New Haven, Conn.: Yale University Press, 1986).

19. H.G. Wells, 'Zoological Retrogression', *Gentleman's Magazine*, 271 (1891), 246–53; reprinted in Robert M. Philmus and David Y. Hughes, eds., *H.G. Wells: Early Writings in Science and Science Fiction* (Berkeley: University of California Press, 1975), pp. 158–68.

20. On *The Time Machine*, see Mackenzie, *The Time Traveller*, ch. 8 and Peter Morton, *The Vital Science: Biology and the Literary Imagination, 1860–1900* (London: Allen and Unwin, 1984), ch. 4. More generally on the theme of predicting the future, see I.T. Clarke, *The Pattern of Expectation, 1644–2001* (London: Jonathan Cape, 1979).

21. H.G. Wells, *The Time Machine: An Invention*, 1895 (London: Heinemann, 1952), pp. 39–41. See also *The Short Stories of H.G. Wells*, 1927 (London: Ernest Benn, 1960), pp. 36–7.

22. Wells, *The Time Machine*, pp. 61–3; *idem, Short Stories*, pp. 52–3.

23. Morton, *Vital Science*, p. 102. On the controversy between Lamarckians and Darwinians, see Bowler, *Eclipse of Darwinism*.

24. Philmus and Hughes, *H.G. Wells*, pp. 9–10. Unlike Morton, Philmus and Hughes (p. 10n) do find Lamarckian passages in Wells' *Textbook of Biology*, although I find their reading unconvincing. Wells' critique of Weismann is clear enough, however, in 'The Biological Problems of Today', *Saturday Review*, 78 (1894), 703–4, reprinted in Philmus and Hughes, pp. 123–7.

25. See Wells, 'The Sunset of Mankind' (from the sequence of 'Time Traveller' stories in *The National Observer*, 1894), reprinted in Philmus and Hughes, *H.G. Wells*, p. 76.

26. See Philmus and Hughes, *H.G. Wells*, p. 10.

27. See Alpheus Packard, 'The Cave Fauna of North America', *Memoirs of*

the National Academy of Sciences, 4/1 (1888), 1–156 and *idem*, 'On the Origins of the Subterranean Fauna of North America', *American Naturalist*, 28 (1894), 727–51. On Packard, see Bowler, *Eclipse of Darwinism*, pp. 134–5.

28. E.W. MacBride, *Textbook of Embryology*, vol. 1, *Invertebrates* (London: Macmillan, 1914), p. 662.

29. On MacBride's social views, see Peter J. Bowler, 'E.W. MacBride's Lamarckian Eugenics and Its Implications for the Social Construction of Scientific Knowledge', *Annals of Science*, 41 (1984), 245–60.

30. On the American school, see Bowler, *Eclipse of Darwinism*, ch. 6 and Gould, *Ontogeny and Phylogeny*, pp. 85–100. For a concise statement of Hyatt's theory, see his 'Cycle of Life in the Individual (Ontogeny) and in the Evolution of the Group (Phylogeny)', *Proceedings of the American Academy of Arts and Sciences*, 32 (1897), 209–24.

31. Alpheus Hyatt, 'The Influence of Woman in the Evolution of the Human Race', *Natural Science*, 11 (1897), 89–93 (89).

32. Ibid., p. 90.

33. Ibid., p. 92.

34. Ibid.

35. Sandra Siegal, 'Literature and Degeneration: The Representation of "Decadence"', in Chamberlin and Gilman, *Degeneration*, pp. 199–219. For an example of this trend, see Frederic Harrison, 'The Emancipation of Women', *Fortnightly Review* n.s., 50 (1891), 437–52.

36. See Susan Sleeth Mosedale, 'Science Corrupted: Victorian Biologists Consider "The Woman Question"', *Journal of the History of Biology*, 11 (1978), 1–56 (on Cope, pp. 24–32). Other studies of this theme include Lorna Duffin, 'Prisoners of Progress: Women and Evolution', in Sara Delamont and Lorna Duffin, eds., *The Nineteenth-Century Woman: Her Cultural and Physical World* (London: Croom Helm, 1978), pp. 57–91; Evelleen Richards, 'Darwin and the Descent of Woman', in David Oldroyd and Ian Langham, eds., *The Wider Domain of Evolutionary Thought* (Dordrecht: D. Reidel, 1983), pp. 57–111; and *idem*, this volume.

37. On LeConte, see Lester D. Stephens, 'Evolution and Women's Rights in the 1890s: The Views of Joseph LeConte', *Historian*, 38 (1976), 239–52 and *idem, Joseph LeConte: Gentle Prophet of Evolution* (Baton Rouge: Louisiana State University Press, 1982), pp. 243–50.

38. Nordau, *Degeneration*, p. 540.

39. See Stephen G. Brush, *The Temperature of History: Phases of Science and Culture in the Nineteenth Century* (New York: Burt Franklin, 1978).

40. Jean Demoor, Jean Massart and Emile Vandervelde, *Evolution by Atrophy in Biology and Sociology* (New York: D. Appleton, 1899), p. 320; see also bk 2, pp. 175–250.

41. On orthogenesis, see Bowler, *Eclipse of Darwinism*, ch. 7. The term 'orthogenesis' was coined by Wilhelm Haacke in 1893 and popularized by Theodor Eimer.

42. Francis A. Bather, 'Fossils and Life', *Report of the British Association for the Advancement of Science*, 1920, pp. 61–86 (86).

43. George W. Crile, 'Orthogenesis and the Powers and Infirmities of Man', *Proceedings of the American Philosophical Society*, 72 (1933), 245–54.

44. Peter J. Bowler, *Theories of Human Evolution: A Century of Debate, 1844–1944* (Baltimore, Md.: Johns Hopkins University Press, 1986; Oxford: Basil Blackwell, 1987).

45. Charles Darwin, *The Descent of Man and Selection in Relation to Sex*, 2 vols. (London: John Murray, 1871), I, 138–45; 2nd edn, 1874 (London: John Murray, 1885), pp. 49–53.

46. Grafton Elliot Smith, 'President's Address, Anthropology Section', *Report of the British Association for the Advancement of Science*, 1912, pp. 575–98, reprinted in *idem, The Evolution of Man: Essays* (London: Humphrey Milford/ Oxford University Press, 1924), ch. 1.

47. For details, see Bowler, *Theories of Human Evolution*, ch. 8.

48. There is a vast literature on the Piltdown affair. For example, see J.S. Weiner, *The Piltdown Forgery* (Oxford: Oxford University Press, 1955) and Ronald Millar, *The Piltdown Men: A Case of Archaeological Fraud* (London: Victor Gollancz, 1972).

49. For example, see Arthur Smith Woodward, 'President's Address, Geology Section', *Report of the British Association for the Advancement of Science*, 1909, pp. 462–71.

50. Arthur Smith Woodward, *Guide to the Fossil Remains of Man in the Departments of Geology and Palaeontology in the British Museum (Natural History)* (London: British Museum, 1915), p. 3.

12

Evolution, ideology and world view: Darwinian religion in the twentieth century

JOHN R. DURANT

As a student of the history of ideas, I am convinced that science, ideology and world view will forever be intertwined and inter-acting. As a citizen concerned for the welfare of science and of mankind generally, however, I cannot but hope that scientists will recognize where science ends and other things begin.

John C. Greene[1]

These concluding words from John Greene's fine volume, *Science, Ideology, and World View*, encapsulate the common theme of the essays it contains: namely, the close interpenetration of what might be termed purely technical considerations with much broader metaphysical and moral issues in the development of scientific theories of organic origins. This theme was, of course, present in Greene's earlier and now classic work, *The Death of Adam*. There, nineteenth-century evolutionary biology was portrayed as a product of the application to the living world of the seventeenth-century mechanical philosophy of nature as a law-bound system of matter in motion. At a philosophical level, the Darwinian synthesis was seen to represent 'The Triumph of Chance and Change'; but at an ideological level it was seen also to represent a particular mid-Victorian vision, a vision compounded of belief in lawful competition and faith in the power of such competition to generate unending organic and social progress. Greene suggested (or at least implied) that the Darwinian synthesis was not really up to the job of accounting for the flowering of life and mind in the universe; in connection with the ideas of progress and purpose, in particular, Darwin and his contemporaries faced what Greene termed 'the difficulty inherent in attempting to rid biology of normative concepts incapable of definition in purely biological terms'.[2]

After the publication of *The Death of Adam*, Greene went on to explore in another and more explicitly philosophical work, *Darwin and the Modern World View*, the metaphysical and moral inadequacies of Darwinian evolutionary theory.[3] Since then, he has continued to probe the conceptual basis of evolutionary biology and to find it inadequate as a foundation for the larger ambitions of many prominent evolutionary biologists. For example, in addition to an important analysis of Darwin as a social evolutionist, which does much to undermine the distinction between Darwin and 'social Darwinism', *Science, Ideology, and World View* contains a provocative essay entitled 'From Huxley to Huxley: Transformations in the Darwinian Credo'. Here Greene deals with the continuing attempts of twentieth-century evolutionary biologists to explain the significance of their subject for human duty and human destiny, dubbing the works of these authors 'the Bridgewater Treatises of the twentieth century, in that they seek to find in science indications and proofs concerning ultimate questions of meaning and value'.[4]

Over the years, Greene has engaged in a lively debate with numbers of Darwinian evolutionary biologists who take him to be maligning their subject for essentially ideological reasons. (Michael Ghiselin, for example, once dismissed *Darwin and the Modern World View* as 'a religious tract'.[5]) The dynamics of the debate are displayed in a recent issue of the journal *Revue de synthèse*. Greene begins with a useful summary of his view of Darwinism as science, ideology and world view. The extension of the mechanical view of nature into biology and history, he suggests, introduced four ideas fundamentally incompatible with that view: the idea of progress; the idea that mind is a part of nature; the idea that competitive struggle is the source of natural order, harmony and progress; and, finally, the idea that in nature chance plays an important part. The tensions caused by these anti-mechanistic ideas are revealed in Darwin's ambivalence about evolutionary progress, the relationship (if any) between the evolutionary laws of nature and the Creator, and the role of chance in evolution. 'Clearly', Greene concludes, 'the Darwinian world-view ... was not a seamless fabric fashioned for all time but an unstable compound of old and new ideas whose incompatibility must eventually become apparent with the progress of science and speculation.'[6]

Greene's position paper is followed by a series of strongly critical responses. In the first and most strident of these, the evolutionary biologist Ernst Mayr suggests that Greene's arguments 'reveal quite conclusively how little he understands Darwin's thought'. Rejecting every one of Greene's major claims, Mayr asserts that a century of biological research has served overwhelmingly to vindicate the theory of evolution by natural selection: 'the news of "the death of Darwin"', he concludes, 'is greatly exaggerated'.[7] In a characteristically insightful contribution, the biologist and historian Stephen Jay Gould goes some way towards explaining (if not healing) the rift between Greene and Mayr. Greene, he suggests, is concerned principally with the philosophical underpinnings of Darwin's world view, whereas Mayr is preoccupied with the scientific validity of Darwin's evolutionary theory. Yet Gould, too, is unhappy with Greene's claim that in Darwinism a mechanical philosophy of nature has finally overreached itself. While 'old-fashioned mechanism' is certainly transcended in the Darwinian emphasis on the statistical and historical character of evolutionary processes, he suggests that this emphasis is a source, not of weakness, but rather of philosophical strength; for Darwinism is consistent with but (it is strongly implied) irreducible to the terms of the mechanical philosophy.[8]

The critique of Darwinian ideology

I have begun this essay with a brief review of John Greene's work because I believe that it raises major issues, not only about the nature, origins and limitations of Darwinian evolutionary theory, but also about the more general question of the relationship between historical scholarship and the critical evaluation of evolutionary biology (or, for that matter, any other science). Greene is a historian whose work lies at the interface between the history and the philosophy of science – in short, his work has to do with both the genealogy and the justification of scientific ideas. Historically, Greene's work makes a number of bold claims about the intellectual pedigree and the ideological influence of Darwinism; philosophically, it makes a number of equally bold claims concerning the conceptual inadequacies of Darwinism and the inconsistency with which many of its major representatives have set about applying it to nature and society. As field naturalists, Greene suggests, the founders of the modern synthesis

> were keenly aware of the harmony of living things, of the perfection of adaptation, of the qualitative differences between the life of an amoeba and the life of a human being. They had to make room for these intuitions, and if the conception of nature and natural science in which they were reared left no such room, simile and metaphor must be called upon to give evolutionary biology a meaning and value that could not be supplied by the positivistic, mechanistic conceptual framework of neo-Darwinism.[9]

My point here is that while Gould is right to describe Greene as being concerned chiefly with the philosophical underpinnings of evolutionary theory, Mayr is also right to feel that against him some sort of scientific response is in order; for Greene's historico-philosophical analysis is potentially very damaging to Darwinism's scientific credibility. After all, a theory of organic origins that leaves no room for the harmony of living things, the perfection of adaptation, or the qualitative differences between the life of an amoeba and that of a human being scarcely deserves to be taken seriously.

Greene's work on the history of Darwinism returns repeatedly to this central theme. It dwells on numerous alleged contradictions between what may be termed the official theory of Darwinism and the unofficial practices of many leading Darwinians – from Darwin himself, through the founders of the modern synthesis, and on up to prominent contemporary evolutionary biologists such as the American entomologist and sociobiologist Edward O. Wilson. In theory, Greene asserts, Darwinian evolution is mechanistic, but in practice its leading exponents commonly describe it as 'creative' and 'opportunistic'; in theory Darwinian evolution is blind and purposeless, but in practice it is widely seen as 'progressive'; in theory Darwinian evolution is value-free, but in practice it is continually being made the basis for all kinds of political and religious ideologies. If Greene is correct, Darwinian biologists experience the greatest difficulty in living up (perhaps one should rather say, down) to the canons of their creed; and this, not primarily because they are particularly weak-willed or woolly-minded, but rather because their creed is somehow inadequate to the task of accounting for the development of life and mind in the universe.[10]

I believe that there is an important insight contained within this general thesis, but that there is also a rather less satisfactory aspect to it as well. The important insight has to do with the nature of Darwinian (and, arguably, all evolutionary) ideologies. Partly as a result of Greene's own careful scholarship, we have come to recognize that modern evolutionary theory arose in response to challenges that were at once scientific and ideological. Just as many of the early opponents of evolution rejected the idea because of the threat that it posed to traditional beliefs based on the notions of providence and purpose, so many of the early supporters of evolution accepted it because of the comfort that it could give to a variety of more secular beliefs; beliefs in which, for the most part, the notions of providence and purpose were transformed into the idea of orderly and law-governed progress in nature and society. The search for a natural law of social progress was the basis for a great many of the evolutionary ideologies that abounded in the late nineteenth and early twentieth centuries; and in every case of which I am aware, Greene's argument holds: 'the lessons to be drawn ... [from evolution] were prescribed in advance by preconceived ideas concerning nature, science, man, and reality in general'.[11]

This is a fundamental point, and the bulk of this essay is devoted to exemplifying its continuing relevance up to the present day by means of an analysis of one particular form of Darwinian religion in the twentieth century. Before turning to this task, however, I shall say something about the second and rather less satisfactory aspect of Greene's general thesis. This has to do, not with the relationship between Darwinism and the hope of human progress (or any other ideological commitment), but rather with the relationship between the formal philosophical content of Darwinism and the informal and frequently highly anthropomorphic terms in which many Darwinian biologists are prone to discuss the evolutionary process.

Do Darwinian metaphors matter?

If I understand him correctly, Greene believes that the colourful metaphorical language employed by many Darwinian authors is symptomatic of fundamental philosophical weaknesses at the heart of the Darwinian enterprise. On the one hand, Darwinian writings are rich in purposive terminology; but, on the other hand, Darwinian theories are entirely mechanistic and purposeless. It is this discrepancy between form and content that Greene offers as evidence of philosophical weakness in the Darwinian world view. The question we must ask is: can the use of purposive metaphors and other similarly anthropomorphic linguistic devices be justified in scientific accounts of supposedly purposeless Darwinian mechanisms?

Before addressing this question, it is important to concede Greene's major premise concerning the metaphorical character of much Darwinian literature. Typically, anthropomorphic analogies and metaphors abound in even the most dry-bone descriptions of evolutionary processes. One of the best places to find them is in textbook accounts of what is sometimes termed 'the pageant of evolution'. Fragments from this pageant have become extremely well-known in our culture. Indeed, they are often to be found in our children's books and on their nursery walls. Life appears and begins to 'exploit' the 'accumulated capital' of organic molecules in the earth's primaeval oceans; animals, long confined to water, eventually emerge to 'conquer' the dry land; dinosaurs arise to 'rule' the earth, and then – mysteriously – they disappear; mammals are swift to 'take advantage' of this mass extinction; and, finally, against all the odds, humans emerge from their forest home and soon come to 'dominate' over the rest of life on earth. The striking feature of this kind of evolutionary narrative

(margin handwritten note: Darwin's theories don't set out to provide philosophical insight so people bend evolution to their own beliefs.)

is its close similarity to conventional historical narrative: stages are set, actors are placed in them, and dramas unfold before our eyes.[12]

So far, so good; but what are we to make of such manifestly anthropomorphic and metaphorical evolutionary narratives? Several different possibilities present themselves: first, we may see evolutionary narratives as a form of 'poetic licence'; secondly, we may see them as a convenient way of attaching meaning or significance to events; and, thirdly, we may regard them as having some altogether deeper metaphysical significance. The first option amounts to saying that we recount several billion years of evolutionary change in the form of heroic folk-epics, or whatever, in order to keep ourselves and our readers interested in what is going on. The second option amounts to saying that evolutionary narrative is a way of picking out particular events that are judged on theoretical grounds to be especially noteworthy or important. The third option amounts to saying that the fact that we resort to evolutionary narratives at all is revealing of the nature of the evolutionary process itself; for example, it may be telling us that evolution really does have an underlying plot or purpose.

I must confess that I am not entirely clear as to how Greene himself would be inclined to choose among these options. Certainly, his writings give the impression that he believes the presence in evolutionary narrative of anthropomorphic analogies and metaphors has at least some sort of deeper metaphysical significance; but given the choices outlined above (and obviously there may be more) the question is, why should it? Unless the anthropomorphisms within evolutionary narrative can be shown to be absolutely indispensable, it is always open to the Darwinian to suggest that they are purely heuristic. So long as 'natural selection' can be translated without loss into purely naturalistic terms, the mere fact that we go on using Darwin's metaphor instead of some more rigorous – and probably more long-winded – formulation is of absolutely no metaphysical (as opposed to historical or psychological) significance. Of course, it may be that in using Darwin's metaphor we deceive ourselves and come to ascribe to nature powers that it does not possess. For example, we may be tempted to credit 'natural selection' with the foresight of the wise stockbreeder. This error is revealed, however, as soon as we attempt translation into naturalistic terms; for now, no such translation is available.

My point is that those who wish to reveal structural weaknesses in the Darwinian philosophy of nature need to proceed beyond the cataloguing of anthropomorphic analogies and metaphors to the identification of genuine conceptual or theoretical difficulties. At several points in his writings (and especially in relation to evolutionary ideology, where – significantly, I think – his arguments are most persuasive), Greene proceeds in precisely this fashion, and he succeeds in identifying genuine contradictions; but elsewhere he does not pursue the argument in this way, and as a result his examples often fail to bear the philosophical weight they appear to be intended to carry. Thus, for example, Greene states that Julian Huxley's dichotomy between living matter and its environment, in which life is seen as 'exploiting' the environment and even as gaining 'independence' and 'control' of it, constitutes the abandonment of the mechanistic view of nature in favour of a 'cryptic vitalism'.[13] But how so? An animal that can shiver may be able to maintain a viable body temperature in the cold, whereas another animal that cannot shiver may die. It is a perfectly reasonable (though admittedly anthropomorphic) shorthand to say that a process of evolution which generates animals that can shiver has endowed them with a measure of independence from their environment; and if a similar process is found to have generated in these same animals a tendency to huddle together in burrows during the worst of the winter, then it is another reasonable (though equally anthropomorphic) shorthand to say that it has also endowed them with a measure of control over their environment. There is no 'cryptic vitalism' here; there is merely the philosophically innocent attempt to render as clearly as possible some of the remarkable effects of natural selection.

Now compare this very weak criticism with Greene's much stronger indictment of Huxley's evolutionary ideology. As Greene shows, Huxley consistently sought to derive from the evolutionary process a basis for human hopes and aspirations. Repeatedly, this quest led him to introduce into his evolutionary writings notions that had no foundation in (that is, were literally untranslatable into) his explicitly Darwinian precepts. To be blunt about it, the reader of Huxley's corpus is faced with the repeated and at times almost embarrassingly obvious intrusion into the text of extraneous and incoherent ideological notions. The presence of these extraneous notions is all the more remarkable because of the general clear-headedness and consistency of Huxley's Darwinian biology. What makes Huxley so intriguing as a natural philosopher is precisely the fact, noted by Gould, that he never succumbed to the easy solutions of teleological or vitalistic evolutionism. Rather, for well over half a century he held his Darwinian science and his progressivist ideology together in tense and uncomfortable juxtaposition as he strove (always unsuccessfully) for a satisfactory synthesis.[14]

Huxley's life-long dilemma illustrates perfectly the point made by

Greene in the epigraph to this essay. His philosophical difficulties (which were many) were almost entirely the product of his single-minded determination to make evolutionary theory yield conclusions far beyond its legitimate domain. Huxley has rightly been characterized as 'a Victorian thinker fated to live in an unsympathetic modern age'.[15] Like so many of Darwin's contemporaries, he looked to Darwinian principles for the foundations of a scientific creed to replace what he took to be the discredited dogmas of conventional religion. This was always a sadly unpromising philosophical adventure, and Huxley fared no better or worse in it than anyone else. The theory of evolution by natural selection is a rich source of insights into organic origins, but it is a poor guide to the living of an individual human life, let alone the conduct of social affairs. Huxley never accepted this, Greene's most telling point; but at least he may be admired for having lived with the difficulties that flowed from this basic refusal without doing violence either to his science or to his most deeply held values.

Although Huxley was in many ways a nineteenth-century thinker, it would be wrong to isolate him altogether from twentieth-century intellectual culture. As Greene has shown, numbers of other prominent biologists have continued to look to evolution as a source of moral and spiritual values, and at the same time they have been joined by philosophers and theologians keen to revitalize a tradition of natural theology that had come perilously close to extinction in the early decades of this century. As a tribute to Greene's lead in this area, the remainder of this chapter is devoted to a brief critical analysis of one particular form of twentieth-century evolutionary ideology. The aim of this analysis is two-fold: first it is intended to reveal some of the sources of the continuing appeal of evolutionary ideology in the twentieth century; and, secondly, it is intended to expose some of the fundamental philosophical weaknesses of this continuing appeal, however it may be re-framed and re-formulated in the light of scientific progress and changing social concerns.

Religion in an age of science

For several decades, the American philosopher and theologian Ralph Wendell Burhoe has been the central figure within a principally Chicago-based circle of scientists, philosophers and theologians who have been associated with *Zygon: Journal of Religion and Science*. *Zygon* was founded in 1966 through the combined efforts of two organizations: the Institute on Religion in an Age of Science (IRAS) – which in turn had been founded in 1954 by the coming together of the American Academy of Arts and Sciences (AAAS) Committee on Science and Values and a group of (mostly Unitarian) theologians involved with an annual conference on 'The Coming Great Church' – and the Center for Advanced Study in Theology and the Sciences (CASTS) at Meadville Theological School, Chicago, which was established in 1965 as a direct result of the work of IRAS. For the past two decades, *Zygon* has acted both as a mouthpiece for these organizations and as a test-bed for new, post-Darwinian natural theologies. It thus provides a useful case-study of the continuing relationship between evolution, ideology and world view.

Obviously, there is much that could and should be said about the intellectual circle around *Zygon*: about biologists such as Charles Birch, Theodosius Dobzhansky and Alfred Emerson (in the early years), and Donald Campbell, Bernard Davis and Edward O. Wilson (more recently); and about an even larger number of attendant philosophers, sociologists and theologians. Time and space, however, do not permit this sort of extended analysis here. Instead, Burhoe will be taken as a key representative figure. The justification for this decision is Burhoe's pivotal place in the *Zygon* intellectual circle. In addition to having been the Executive Officer of the AAAS Committee on Science and Values, a founder member of IRAS and the first Director of CASTS, he was also the founding editor and, for over a decade, the chief intellectual guide of *Zygon* itself. Throughout his career, Burhoe has been a tireless supporter of the unification of evolutionary biology and ideology. At the risk of a certain amount of over-simplification, I shall treat his views as indicative of wider movements of thought within the *Zygon* circle.

It will be helpful to begin by setting Burhoe's scientific world view in its ideological context. Burhoe has never made a secret of the fact that his attempt to construct an evolutionary natural religion constitutes a theological response to a number of perceived tensions within post-war American society. Over and again in the early years of *Zygon*, Burhoe placed himself and his journal in the wider context of contemporary American moral, social and political problems. 'We find ourselves', he wrote in an editorial in 1968, 'in the midst of perhaps the most violent and disruptive period in all human history thus far.' This was a period, he told his readers, in which opinion polls indicated that fully two-thirds of the American population believed that religion was losing its influence, and in which there was growing doubt about 'the sacrality of the American Way, or of Western Civilization. . . . Barometers here include the rise of the Beatniks, hippies, draft resisters, and student rebellions.'[16] In case this statement should appear merely rhetorical, it is worth pointing out that the social disturbances of the late 1960s had a direct influence on the activities of *Zygon*. For example, in 1967 the geneticist Michael Lerner wrote to Burhoe in the midst of intense student unrest on the campus of the University of California at Berkeley. Lerner, Dobzhansky and Birch were concerned at the growing alienation of American young people from traditional values, and they suggested that a symposium be convened to address this issue scientifically. Burhoe duly organized such a symposium in 1972 under the auspices of IRAS; he called it 'Science and Human Purpose', and many of the papers presented there appeared subsequently in *Zygon*.[17]

Burhoe's response to these social tensions has been clear and un-equivocal. In his view, personal alienation and social unrest alike are the direct result of a steep decline of traditional religious values, and this decline has in turn been caused by the rise of science. For-merly, traditional religious values were rooted in a traditional theo-logical world view, but this has been undermined by the spectacular successes of science. Unfortunately, however, the scientific world view has not spoken directly to people's moral and spiritual needs. Consequently, a kind of ideological vacuum has come to exist within which personal values and social norms are withering away. Burhoe's prescription maps neatly on to his diagnosis. What is needed, he sug-gests, is a revitalization of traditional religious values; and in an age of science, this can only be achieved by rooting these values directly in science itself. Morality, in short, must be given unimpeachable scientific credentials:

> For us what is true and what is right and what will prevail are
> not determined by military force or by any other arbitrary human
> wishes or pressures but essentially by those forces presented in
> the scientific picture of the historical flow of events in history.[18]

This could almost be Herbert Spencer, seeking to find a scientific basis for conduct in the laws of nature in the mid-nineteenth century; in fact, however, it is Burhoe himself, setting out the aims of his new journal *Zygon* in the very first issue.

Towards a scientific theology?

It is in his conception of a 'scientific theology' that Burhoe attempts to provide traditional religious values with their much-needed scientific credentials. Scientific theology is a particular product of the union of scientific realism with theological relativism. It proceeds on two broad fronts: first, it seeks to naturalize theology by showing that the claims of conventional religion (in principle, all conventional religions; in practice, the Judaeo-Christian religions) are the first dim and highly culture-bound foreshadowings of what are today the clear findings of science; and, secondly, it seeks to naturalize ethics by showing that the claims of conventional morality (in principle, all conventional morality; in practice, the conventional morality of post-war middle-class America) are the unambiguous verdicts of modern science. The simplest way of summarizing Burhoe's efforts on each of these fronts is by setting out a table of correspondences, as follows:

Theology	Science
God/'Lord of History'	Nature/natural selection
soul/immortality	gene/culture continuity
predestination	scientific determinism
'original sin'	nature–culture conflict
salvation	adaptation
good	life/survival
evil	death/extinction

This table summarizes a series of formal correspondences between elements within traditional Christian theology and homologous elements within the post-Darwinian world view. Burhoe takes these correspondences extremely seriously. For him, the concept of God as 'Lord of History' is nothing more nor less than a poetic and pre-scientific representation of natural selection: 'It makes little difference whether we name it natural selection or God, so long as we recognize it as that to which we must bow our heads or adapt.'[19] Similarly, the idea of soul, or personal immortality, is a symbolic expression of the genetic and cultural continuity which are nature's 'rewards' to those who successfully adapt to 'her' demands; predestination is the universal chain of natural causation that unerringly assigns to each person his or her place in the world; original sin (at least in Burhoe's later writings) is the inevitable conflict between a selfish, biologically based human nature and a selfless, culturally based human society; and salvation is the biological and cultural success won by individuals and societies that find themselves in harmony with the overall scheme of things. In short, the table of correspondences portrays traditional Christian theology as a metaphorical version of literal scientific truths. If Burhoe is correct, Christianity finds its ultimate justification in population genetics.

Evolving evolutionary ethics

For Burhoe, science is not merely a revelation of truth; it is also a revelation of the growth of goodness. This growth is constituted by the evolution of the religious impulse and, in particular, by the development of the moral sense or conscience. Quoting the anthropologist Clyde Kluckhohn, Burhoe argues that 'There must be codes which unite individuals in adherence to shared goals that transcend immediate and egoistic interest.'[20] But how have such codes evolved, and how are they now to be maintained? Burhoe's answers to these questions, which lie at the heart of his scientific theology, have changed over the years in direct response to developments in the evolutionary study of social behaviour. In the 1960s, Burhoe was influenced by the many biologists, from W.C. Allee and A.E. Emerson in the United States, to Konrad Lorenz and V.C. Wynne-Edwards in Europe, who believed that self-sacrifice for the good of the community might evolve genetically, by natural selection operating between competing social groups ('group selection').[21] At that time, many group-selectionists were keen to offer their evolutonary models as explanations of the evolution of ethical conduct; and for a while Burhoe appears to have followed their lead. Around 1970, however, he came under the influence of biologists such as G.C. Williams and W.D. Hamilton, whose steadfast opposition to group selection was to inspire the so-called sociobiology revolution.[22] Following a renewed insistence upon the individual as the unit of selection, Burhoe now abandoned the notion that ethical conduct could evolve by genetic evolution; instead, he came to see ethics as the result of a purely cultural evolutionary process – a process working in direct opposition to the unethical course of nature. Encouraged by the support of others, including the American psychologist Donald Campbell, Burhoe now presented religiously based moral sanctions as key elements in the evolution of human culture.[23] Religion was, he suggested, 'the missing link between ape-man's selfish genes and civilized altruism'.[24]

The metaphor of genetic selfishness has caused a great deal of confusion and misunderstanding between sociobiologists and their critics. While sociobiologists have defended a technical, population-genetic definition of 'selfish' behaviour as conduct promoting the reproductive success (more strictly, the 'inclusive fitness') of individual actors, their critics have accused them of illegitimately importing into the evolutionary process a colloquial, ethical definition of selfishness as regard for self as a principle of conduct.[25] As with the more general question of the legitimacy of anthropomorphic metaphors in evolutionary theory, which was discussed earlier, the philosophical key to this issue is translation. In this case, it is possible to translate the metaphor of genetic selfishness into strictly Darwinian terms. (Briefly, a gene may be said to be selfish if its phenotypic effects are such as to promote its own as opposed to any other entity's reproductive success.) This means that the metaphor is at least potentially philo-

Metaphor of 'selfish gene' which is a dominant gene
not a genetic trait of
selfishness

sophically innocent. However, actual innocence depends entirely upon the maintenance of a sharp distinction between the technical, population-genetic and the colloquial, ethical meanings of selfishness; any collapse of this distinction immediately exposes the sociobiologist to the just objection that the metaphor constitutes a form of moral pollution of evolutionary theory.

It is relevant to inquire where Burhoe stands on this issue. To begin with, the very subtitle of his paper on this subject – 'The missing link between ape-man's selfish genes and civilized altruism' – gives immediate cause for concern; but we may be charitable and presume that both 'selfishness' and 'altruism' are being used here in their technical, population-genetic senses. Such charity becomes hard to sustain, however, when once we enter the text of Burhoe's paper. For in setting out the issues for the benefit of fellow symposiasts concerned with 'Sociobiology and Religion', Burhoe states that 'we are engaged here in examining his [Campbell's] thesis that sociobiology helps us understand the positive and natural role religion has played in generating altruistic behavior in admittedly selfish humans'.[26] It is extremely difficult to place any sensible interpretation on this passage that does not violate the injunctions set out in the previous paragraph. Quite simply, Burhoe is now actively exploiting the ambiguity of the term selfishness in order to make large claims for the ability of evolutionary theory to account for moral conduct.

Now of course to provide an evolutionary explanation of something is not to provide an evolutionary justification of that thing. In other words, even if we were to accept Burhoe's attempted reconstruction of the evolution of human 'selfishness' and 'altruism' (and for the reasons just given, among others, I do not think for a moment that we should), this would not of itself amount to a rational defense of morality. On the contrary, to the extent that evolutionary theory provides a causal explanation of morality in terms that are themselves amoral (for example, to the extent that morality is seen as a device promoting the survival and reproduction of individuals or groups of individuals), it may be supposed that it actually undermines morality. If this appears paradoxical, then consider the analogy with Freud's explanation of religious belief in the existence of God. Why is Freud's attempt to explain theism in psychoanalytic terms not generally regarded as a form of apologetic theology? The reason is, of course, that psychoanalysis is a purely secular (i.e. non-theistic) form of explanation. To the extent that a believer accepts a Freudian explanation of his belief, he must presumably reject any alternative religious explanation (such as, for example, a direct experience of God); and to this extent his belief itself must presumably be undermined. By the same token, to the extent that a moral agent accepts a Darwinian explanation of her actions, she must presumably reject any alternative ethical explanation (such as, for example, a utilitarian calculus of pleasures and pains); and to this extent her adherence to ethical principles is undermined.

Burhoe appears totally to ignore this philosophical problem in his published writings on evolution and ethics. At first sight, this may seem rather strange; after all, the conflict between the scientific explanation of human conduct on the one hand, and the moral justification of it on the other, is potentially fatal to his entire ideological enterprise. However, on closer inspection it becomes apparent that Burhoe is able to side-step this problem completely by transferring from religion to science not only the theological concepts of God, soul, predestination and all the rest, but also the moral concepts of good and evil. As may be seen from the table of correspondences above, Burhoe equates life with ultimate values: what survives is good, and what perishes is evil. This means, however, that so far as he is concerned Darwinian evolutionary theory is at one and the same time a scientific and an ethical framework; hence, it may legitimately be used for purposes both of explanation and of justification.

While this investment of the evolutionary process with value enables Burhoe to escape one philosophical problem, however, it does so only by burdening him with another. For the central difficulty with his, as with most evolutionary ethics, is that in the absence of precisely the sort of non-naturalistic assumptions that such ethics are designed to avoid there can be no rational grounds for accepting the equation of the evolutionary with the ethical process. Why should what survives in the Darwinian struggle for existence be not only genetically but also morally 'fitter' than what does not? Any satisfactory answer to this question must resort to ethical principles that transcend the domain of evolutionary naturalism; and such principles Burhoe is determined at all costs to abjure. At this point, Burhoe's scientific theology is in a dilemma rather similar to that of Julian Huxley's evolutionary humanism. Both philosophies would stand a better chance of working if only they had more resources to call upon than the Darwinian philosophy of nature; but, of course, if they had these resources then they would not merit the epithets *scientific* and *evolutionary* at all.

Scientific theology and the scientific priesthood

There is one last and perhaps rather obvious point to be made about Burhoe's scientific theology, and this is that it constitutes an attempt to invest the post-Darwinian, scientific world view – and its scientist-custodians – with all of the authority traditionally invested in religion and the priesthood. Burhoe's solution to the ideological crisis that he perceives in contemporary American society is to offer up to Darwinian biologists the awesome responsibility of laying down the law, not only for nature, but for human nature and human society as well. Not surprisingly, perhaps, this is an offer that one or two evolutionary biologists have found extremely attractive. Edward O. Wilson, for example, has specifically endorsed Burhoe's scientific theology in his book *On Human Nature*[27] – one of the works, it may be observed in passing, that Greene dubbed 'the Bridgewater Treatises of the twentieth century'. Here Wilson offers a sociobiological explanation of religion – as he puts it, 'When the gods are served, the Darwinian fitness of the members of the tribe is the ultimate if unrecognized beneficiary' – together with an attempt to invest the 'evolutionary epic' itself with moral and religious significance. At the same time, however, he recognizes the extremely limited success of such scientific theology as an alternative ideology for a secular society:

> Today, scientists and other scholars, organized into learned groups such as the American Humanist Society and the Institute on Religion in an Age of Science, support little magazines distributed by subscription and organize campaigns to discredit Christian fundamentalism, astrology, and Immanuel Velikovsky. Their crisply logical salvos, endorsed by whole arrogances of Nobel Laureates, pass like steel-jacketed bullets through fog. The humanists are vastly outnumbered by true believers, by the people who follow Jeane Dixon but have never heard of Ralph Wendell Burhoe.[28]

There are great ironies here. Certainly, scientific theology has had nothing like the mass ideological appeal of Christian fundamentalism in American culture over the past two decades; in this sense, Burhoe's ideological crisis has been resolved, at least in part, in ways that he could not possibly have been expected either to anticipate or to admire. But before writing off scientific theology as an ideological failure, we would do well to remember that Wilson's *On Human Nature* won the Pulitzer Prize in 1979, and that, in its sociobiological incarnation, evolutionary ideology has made great strides in America over the past decade in precisely those well-educated, liberal-minded circles in which so-called 'scientific creationism' is an object of ridicule and contempt. There may well be something in Wilson's claim, made at an IRAS symposium on 'Sociobiology, Values, and Religion' in 1979, and subsequently published in *Zygon*, that 'liberal theology can serve as a buffer' between scientific materialism and 'one of the unmitigated evils of the world', namely fundamentalist religion.[29] For

but Darwin used to rationalize + justify v. extreme regimes + ideologies eg Nazis. Darwin's theories caused religious scholars to reassess relationship of religion + science to coexist more harmoniously?

Burhoe's scientific theology demonstrates that the gulf separating many evolutionary naturalists from their fundamentalist critics is not necessarily as great as many people, including Wilson himself, appear to imagine.

Darwinian religion in the twentieth century

Burhoe's project for a scientific theology illustrates three general points about the continuing relationship between evolutionary theory and religious belief in the twentieth century. First, despite the still-powerful myth of the inevitable conflict between religion and modern science, it is clear that natural theology has continued to flourish in the post-Darwinian world. Secondly, much twentieth-century natural theology has adopted an extremely uncritical or even adulatory posture towards cosmology and evolutionary theory, embracing them not merely as state-of-the-art theories of origins but rather as revelations of ultimate truths about nature and the human condition. Thirdly, in its attempt to root human values in the evolutionary process, post-Darwinian natural theology has tended to function rather like a small but powerful megaphone, handily placed to amplify (and, more often than not, painfully to distort) technical and scientific propositions within evolutionary theory into ideological and religious pronouncements.

For a period of around twenty years, *Zygon* has served as just such an ideological megaphone. In the late 1960s, the pages of *Zygon* were filled with articles looking to the burgeoning field of evolutionary ethology for diagnoses and cures of a variety of human ills; in the late 1970s, they were filled with attempts to obtain from the newer sociobiology an equally comprehensive vision of the human condition. In both cases, the general effect was the same: theologians gave every encouragement to those evolutionary biologists who saw in their discipline far more than 'mere science'; and thus, moral, religious and political values deserving of serious consideration in their own right came instead to be adjudicated in the terms of natural science. It is particularly poignant that this wholesale naturalization of values should have been undertaken in the name of theology. For if, with John Greene, we are entitled to hope that scientists will recognize 'where science ends and other things begin', by the same token we are entitled to expect that theologians and humanists will at the very least do nothing to hinder them in the performance of so vital a task.

Notes

1. John C. Greene, *Science, Ideology, and World View: Essays in the History of Evolutionary Ideas* (Berkeley: University of California Press, 1981), p. 197.

2. John C. Greene, *The Death of Adam: Evolution and Its Impact on Western Thought*, 1959 (New York: Mentor Books, 1961), p. 295.

3. John C. Greene, *Darwin and the Modern World View* (Baton Rouge: Louisiana State University Press, 1961).

4. Greene, *Science, Ideology, and World View*, pp. 162–3.

5. Michael Ghiselin, *The Triumph of the Darwinian Method*, 1969 (Berkeley: University of California Press, 1972), p. 241.

6. John Greene, 'The History of Ideas Revisited', *Revue de synthèse*, 107 (1986), 201–27 (210).

7. Ernst Mayr, 'The Death of Darwin?' *Revue de synthèse*, 107 (1986), 229–35 (235).

8. Stephen Jay Gould, 'Commentary on Greene and Mayr', *Revue de synthèse*, 107 (1986), 239–42.

9. Greene, 'History of Ideas Revisited', p. 222.

10. Robert M. Young (this volume) makes essentially this same point the mainstay of the argument for 'a radical metaphysical reconstruction' in which purposes and values are reintegrated with material explanations of the universe. I cannot say, of course, exactly how far Greene and Young would agree on the sort of scientific revolution that is needed; but that some sort of revolution is required appears to be the common conclusion of their work.

11. Greene, 'History of Ideas Revisited', p. 219. In addition to Greene's writings, the following works provide good starting points for the study of the history of Darwinian ideology: Robert C. Bannister, *Social Darwinism: Science and Myth in Anglo-American Social Thought* (Philadelphia: Temple University Press, 1979); Richard Hofstadter, *Social Darwinism in American Thought*, 1944; rev. edn (Boston: Beacon Press, 1955); Greta Jones, *Social Darwinism and English Thought: The Interaction between Biological and Social Theory* (Brighton, Sussex: Harvester Press, 1980); and Robert M. Young, *Darwin's Metaphor: Nature's Place in Victorian Culture* (Cambridge: Cambridge University Press, 1985).

12. The narrative structure of scientific accounts of human origins has been analysed in considerable detail by Misia Landau in her doctoral disseration, 'The Anthropogenic: Paleoanthropological Writing as a Genre of Literature' (Yale University, 1981). See also Landau, 'Human Evolution as Narrative', *American Scientist*, 72 (1984), 262–8.

13. Greene, 'History of Ideas Revisited', p. 217.

14. The tension in Huxley's thought between Darwinian purposelessness and humanistic progression is discussed in more detail in my essay, 'Julian Huxley and the Development of Evolutionary Studies', in M. Keynes and G. Harrison, eds., *Evolutionary Studies: A Centenary Celebration of Julian Huxley* (London: Macmillan, in press).

15. Colin Divall, 'From a Victorian to a Modern: Huxley and the English

Intellectual Climate', in A. Van Helden, ed., *Julian Huxley: Statesman of Science* (Houston, Tex.: Rice University Press, in press).

16. R.W. Burhoe, 'Editorial', *Zygon*, 3 (1968), 113. For similar remarks, see Burhoe's next editorial, ibid., pp. 238–41.

17. See *Zygon*, 8 (1973).

18. R.W. Burhoe, *Zygon*, 1 (1966), 8.

19. R.W. Burhoe, *Toward A Scientific Theology* (Belfast: Christian Journals, 1981), p. 21.

20. Ibid., p. 33.

21. The most obvious influence on Burhoe in this direction was the American zoologist Alfred E. Emerson. Emerson was a Cornell-educated biologist with a special interest in the evolution of insect societies. Since 1962, he had been emeritus professor of zoology at the University of Chicago; and from the outset he was a key figure in the *Zygon* circle. For many years, Emerson had worked to establish a unified theory of biological and social evolution based on a hierarchical application of the principle of natural selection to cells, organisms, populations and even whole ecosystems. See his articles, 'Social Coordination and the Superorganism', *American Midlands Naturalist*, 21 (1939), 182–209; 'Ecology, Evolution and Society', *American Naturalist*, 77 (1946), 97–118; and 'The Supraorganismic Aspects of Society', *CNRS* (Paris), 34 (1952), 333–53. Burhoe thought so highly of Emerson's essay, 'Dynamic Homeostasis: A Unifying Principle in Organic, Social and Ethical Evolution' (*Scientific Monthly*, 78 [1954], 67–85), that he had it reprinted in his own journal (*Zygon*, 3 [1968], 129–68). The general theme of this essay was that natural selection operating at the higher levels of the family, the group, the species and beyond was responsible for the growth of ever more functionally integrated and harmonious social structures. According to Emerson, by adopting the 'ethical' course of acting in conformity with this evolutionary trend, individuals could and should contribute to the emergence of what he termed 'a world order of mutually cooperative social relations among nations'. For his part, Burhoe recommended Emerson's essay in glowing terms: 'I suggest that *Zygon* readers will find it a master key for the development of their own thinking on how the sciences may provide a common theology for elucidating and evaluating man's convictions about right and wrong, about his destiny and proper hopes and fears therein' (*Zygon*, 3 [1968], 130). The twentieth-century appeal to group selection as a basis for social ethics may be traced back to Petr Kropotkin's famous *Mutual Aid* (1902); this appeal undoubtedly deserves detailed historical treatment in its own right.

22. See George C. Williams, *Adaptation and Natural Selection: A Critique of Some Current Evolutionary Thought* (Princeton, N.J.: Princeton University Press, 1966) and W.D. Hamilton, 'The Genetical Evolution of Social Behavior', *Journal of Theoretical Biology*, 7 (1964), 1–16, 17–51.

23. See, in particular, Donald T, Campbell, 'On the Conflicts between Biological and Social Evolution and between Psychology and Moral Tradition', *American Psychologist*, 30 (1975), 1103–26 (reprinted in *Zygon*, 11 [1976], 167–208).

24. R.W. Burhoe, 'Religion's Role in Human Evolution: The Missing Link Between Ape-Man's Selfish Genes and Civilized Altruism', *Zygon*, 14 (1979), 135–62 (reprinted in Burhoe, *Toward a Scientific Theology*, ch. 7). This paper was given at a Symposium on 'Sociobiology and Religion' at the annual meeting of the American Psychological Association in 1978, the purpose of which was to explore the issues raised by Campbell in his presidential address to the Association in 1975.

25. For a classic confrontation on this issue, see Mary Midgley, 'Gene Juggling', *Philosophy*, 54 (1979), 439–58 and Richard Dawkins, 'In Defence of Selfish Genes', *Philosophy*, 56 (1981), 556–73. The influence of the metaphor of genetic selfishness in both the technical and the popular literatures on sociobiology is discussed in more detail in a recent pseudonymous work by John Klama, of which I am joint author and editor, *Aggression: Conflict in Animals and Humans Reconsidered* (Burnt Mill, Harlow: Longman Scientific and Technical, 1988), chs. 1–2.

26. Burhoe, *Toward a Scientific Theology*, p. 203.

27. E.O. Wilson, *On Human Nature* (Cambridge, Mass.: Harvard University Press, 1978).

28. Ibid., p. 197.

29. E.O. Wilson, 'The Relation of Science to Theology', *Zygon*, 15 (1980), 425–34 (433).

13

Persons, organisms and . . . primary qualities

ROBERT M. YOUNG

John Greene's *The Death of Adam* was one of the first books I read in the history of biology. He treated the subject as part of the history of ideas. This was a boon. (Indeed, he was kind enough to sign my copy of the book when we first met.)[1] Before that, the way he thought about science and social theory in his article on Auguste Comte and Herbert Spencer was important in my doctoral research.[2] And in recent years, his integration of history of science with broader historical issues, especially in his essay 'Darwin as a Social Evolutionist', has provided both inspiration for and reassurance about my own work.[3] His observations on that work and its influence have also been very supportive. For all these things, and for his gentlemanliness, I am grateful. Hence this essay in his honour, where I first spelled out my own view of what the history of the biological and human sciences has to say to the philosophy of science and the philosophy of nature.

Since I first drafted the essay, my own views have broadened and deepened in the direction of humanistic marxism. It is the last piece I wrote before that process was entered into. Although I have mentioned it from time to time and mined it for various purposes, it has remained unannotated and unpublished, and I offer it here more or less as it was first written. (I am aware that some of my generalizations might be cast in different terms if I were now writing it for the first time.) The process of recontextualizing its ideas was so daunting and so vertiginous that I have taken a long while to see that its basic position on the philosophy of nature still underlies the later developments in my thinking. Unlike Rip van Winkle, I keep waking up twenty years on and discovering that the fundamental issues are the same and wishing I had a greater sense that historians of science were engaged with them. Surely the reason we do history of science is to try to shed light on the meaning of life – of life itself, of humanity, and the husbanding and enhancement of generous values?

I would say that reductionism is facing in the opposite direction and
that there must be another way.

The foundations of defensiveness

The concept of mind has made it difficult for persons to be seen as
organisms. Similarly, the phenomena of biology have repeatedly
been explained in terms of secondary qualities and even less quantifi-
able concepts, while the paradigm of physical explanation requires
that appeals should only be made to primary qualities. The title of this
paper was chosen to draw attention to these hiatuses in the conceptual framework of modern science – gaps which have not disappeared
with the development of the theory of evolution or with micro-
techniques in neurophysiology and molecular biology. The question,
of course, is whether they are empirical, conceptual or philosophical
gaps.

If one looks at the history and philosophy of science from the point
of view of biology, psychology and the social sciences, it looks very
odd indeed. It is possible to write about these disciplines in terms of
the traditional historiography of Thomas Kuhn's 'scientific revolution' and Charles Gillispie's advancing 'edge of objectivity'. One can
also write about the philosophy of these disciplines as special, albeit
refractory, cases within the paradigm of explanation of the physico-
chemical sciences. Indeed, if one attempts to apply the Kuhnian
analysis of paradigms to these disciplines, it turns out that the further
one moves away from micro-processes, the more difficult it is to
apply the concept of paradigm at all. There seems to be a sort of
continuum that extends from mathematics and the physico-chemical
sciences to biology, psychology and the social sciences, and as one
moves along it, one encounters increasing difficulty in applying the
Kuhnian analysis. Kuhn himself points out that the first universally
received paradigms in parts of biology are very recent. And there are
still 'schools' of psychology – a sure sign of 'immaturity' in science.
Finally, it remains an open question what parts of social science
have yet acquired paradigms at all. 'History', Kuhn concludes, 'suggests that the road to a firm research consensus is extraordinarily
arduous.'[4]

The attractiveness of the conception of scientific progress that is
exemplified by Gillispie's approach[5] and has to some extent been
given a formal expression in the Kuhnian analysis, highlights the
historical and philosophical difficulties of the student of the biological
and human sciences. They can be relegated to the Kuhnian limbo of

pre-paradigm' sciences and await the advancing edge of objectivity, but this 'solution' has resulted in a great deal of unsatisfactory writing.

If one looks at the relevant secondary literatures, it appears that many writers have taken an implicit version of this position. The biological, psychological and social sciences are seen as laggard; the standard histories are least illuminating and their authors least keen with respect to these problems. Attempts are made to force important figures in the history of biology into an empiricist, positivist, mechanist mould. Vesalius is treated in terms of the method of observation and the alleged overthrow of Galen, while apologies are made for his physiological views. Harvey is distorted out of all recognition and becomes a positivist mechanist. John Ray is seen as a taxonomist, while the pervasiveness of his anti-mechanist natural theology is played down. Lamarck is dismissed as a romantic counter-offensive: 'the last, though one of the most explicit, of a whole series of attempts, some sad, some moving, some angry, to escape the consequences for naturalistic humanism, of Newtonian theoretical physics'.[6] With Charles Darwin, 'Biology Comes of Age'.[7] 'In the concept of natural selection, Darwin put an end to the opposition between mechanism and organism through which the humane view of nature, ultimately the Greek view, had found refuge from Newton in biology.'[8]

> Darwin did better than solve the problem of adaptation. He abolished it. He turned it from a cause, in the sense of a final cause or evidence of a designing purpose, into an effect, in the Newtonian or physical sense of effect, which is to say that adaptation became a fact or phenomenon to be analyzed, rather than a mystery to be plumbed.[9]

In the light of these renderings of important figures in the history of biology, it is not surprising to find that the eminent historian who gives them has difficulties with Darwin. He writes:

> What fundamental generalization ever came into the world in so unassuming guise as Darwin's theory of evolution? Is there any "great book" about which one secretly feels so guilty as *On the Origin of Species*? None in the history of science gives me, at any rate, such uphill work with students.[10]

This confession, repeated a year later,[11] illustrates the difficulty many scholars find in making the biological sciences fit into the official historiography of modern science. It may be that a more fruitful approach would be to abandon the attempt to force biology into this mould, to pay greater attention to the philosophies of nature of figures such as Harvey, Ray and Darwin, and to refrain from distorting and obscuring the theoretical contexts within which they saw their work.

If one turns to the philosophical literature, there is an analogous lack of enthusiasm. The philosophy of science journals publish very few articles on biology, psychology and the social sciences, and most of these are bad. Having worked as an assistant editor of one such journal, I can attest to the fact that those submitted in these fields are of a far lower standard than the run of articles on other topics in the philosophy of science. The good articles, furthermore, reflect the difficulties in this field: either they are attempts to transform teleological into mechanistic explanation, or they are maverick pieces, difficult to classify.

In the remainder of this essay, I should like to attempt a tentative diagnosis of how the history and philosophy of the biological and behavioural sciences got into such a mess and then to suggest that we might pay closer attention to a number of concepts which appear to me to be basic to these disciplines. The lack of enthusiasm of historians and philosophers with respect to these topics, coupled with the defensiveness of scientists in the primary disciplines, may have a more fundamental explanation than the refractoriness of their subject matters: they may reflect problems in the deep structure of scientific explanation. In what follows, then, I hope to point to the metaphysical foundation of methodological defensiveness in the biological and human sciences. The discussion falls into three parts: an exposition of the 'official' paradigm of explanation of modern science, a review of its symptomatic problems by means of examples of the continuing refractoriness of biology and psychology to physicalist reduction since the seventeenth century, and finally a very tentative look at the oddity of a hierarchy of concepts to which we might fruitfully direct our attention. In this last section I want to draw attention to the record of question-begging, laggard behaviour and shoddiness of much that has passed for the philosophy of the best biologists, if one judges their work by the standards of the physico-chemical paradigm. In doing so, my aim is not to indulge in historical pornography but to suggest that a patient who goes on complaining really does have a pain, although he may well be mistaken about its cause.

My argument falls somewhere between the history and philosophy of science. It would be safest to present it as straightforward history, but one of my aims is to ask if such persistent historical themes may not be of philosophical interest. This approach raises difficulties. The first is that my philosophical colleagues insist that it is simply a logical mistake to suggest that philosophical conclusions might be drawn from the history of science. Necessary conclusions cannot be drawn from contingent matters, and one is doing just that in pointing to the persistence of efforts to avoid the injunctions of the paradigm of explanation of modern science, then using this evidence to argue that there may be important problems in the assumptions of the paradigm. I am afraid that I do not feel the force of this criticism, because, to compound the putative fallacy, this is what people have persisted in doing in the history and philosophy of science. In biology, for example, the ideas and discoveries of Harvey, Descartes, Wöhler, Darwin and the molecular biologists have, in their respective periods, been used as a basis for arguing that there was no place in biology for vitalism and teleology.

My aim, however, is a more modest one. I only want to point out that the explanatory paradigm of modern science was elaborated to serve certain purposes. If it has served those purposes well, others less well, and still others very badly, it would seem open to us to look for a more useful one. If this argument leads to the well-known difficulties of utilitarian and pragmatic epistemologies, then so be it. I find that history is opportunistic. People make what they need of others' writings and of nature. I have tried to show this with respect to the admirers of Thomas Malthus.[12] I would also say that nature is manifold, and our priorities and the resolution of historical forces lead humanity to notice and shape the features of nature that resonate with the values and vision of the epoch. There are, of course, a number of overlapping and partially contradictory voices and forces at work in any period.

A second difficulty is more worrying. I shall argue that in certain key episodes in the history and philosophy of biology since the seventeenth century, purposive explanations were persistently offered by means of covert or overt appeals to concepts drawn from the idea of human 'intention'. *Ad hoc*, question-begging terms were self-consciously used in biological explanations, which disobeyed the injunction to explain all phenomena in terms of matter and motion. When one reviews this record and attempts to outline its current manifestations, it sounds very much as though one is making a straightforward appeal for the reintroduction of final causes or teleological explanation in the biological and behavioural sciences. Thus, when one draws attention to attempted explanations in terms of functions, adaptations, biological properties (to say nothing of explanation in social science in terms of the intentions of human actors), and when one alludes to the persistene of teleological, emergentist, intentional, holist, gestalt or organic theories, one seems to be making an implicit case for these points of view. In appearing to do this, one invites the traditional question of the seventeenth century and later: how do final causes push and pull? How do gestalts organize wholes, how do emergents get new properties, and so on?

I only want to make three points about this in the hope that they add up to a sort of defensive *scholium*. First, my intentions are diagnostic. I want to gather symptoms and direct others' attention to them. I am in no position to offer any other prescription than 'Dig here.' Secondly, I believe that the use of putative explanations in terms of faculties, functions, adaptations, emergents, gestalts, organic wholes, and so on, does not solve problems, but hypostatizes them and offers them in the guise of solutions. Thirdly, in pointing to the persistence of such explanations I am drawing a historical conclusion, but I believe that it has prescriptive force in the following sense: if we have such difficulty in obeying a paradigm of explanation in investigating certain aspects of nature, it may be worth while to take another look at the paradigm itself.

Now, what is that paradigm and how have people disobeyed its injunction?

The official paradigm of explanation

The paradigm of explanation of modern science is a set of interrelated ontological, epistemological and methodological decisions to which biologists and, *a fortiori*, students of psychology and the social sciences have found it very difficult to conform. The *ontological* aspect is best seen in the work of Descartes, whose ontology codified a rupture that was becoming increasingly likely because of strains apparent in the explanatory scheme of the Aristotelian tradition. In the Aristotelian scheme, formal, final, material and efficient causes had no independent status apart from the particular phenomena that could be analysed according to these four aspects of 'coming to be'. However, it became increasingly difficult to avoid anthropomorphic expressions of final causes, and for certain purposes, final causes seemed irrelevant. Material and efficient causes, on the other hand, were relatively easy to handle in numerical terms, and formal causes could be reduced to the organization of particles of matter in motion and expressed in terms of mathematical formulae. Descartes replaced the organic analysis of phenomena with a dualist ontology: *matter* was extended, divisible, passive and subject to determinist natural laws; *mind* was defined negatively as having the attributes that could not be referred to matter. It was unextended, indivisible, active and free. Its essence was thought or will. Final causes were banished from scientific explanation and had status only in the intentions of God (about which it was sometimes thought impious to speculate) and in human will (which was not subject to scientific investigation). This ontology led to well-known difficulties in explaining the interaction of mind and body in experience and behaviour: namely, how do sensations cause ideas and how do intentions cause muscular motions? As a persisting framework for thinking about nature, it made theories of learning and evolution metaphysically absurd and codified a dichotomy between dualism and the principle of continuity, which continues to plague evolutionary theory, comparative psychology and the social sciences.[13]

The *epistemological* aspect of the paradigm found expression in Descartes' *Principles of Philosophy* (1644), but various versions of the same doctrine also appeared in the writings of Democritus, Galileo, Gassendi, Boyle, Newton, and lesser figures such as Charlton, Hartley and Priestley. According to this doctrine, the material world is characterized by the *primary* qualities of extension (or size), figure (or shape), motion or rest, number, and solidity or impenetrability (some would substitute mass here). These qualities appeared to be inseparable from objects and invariant under different conditions of observation. The senses can rely on them, while all other qualities are subject to variation and illusion: for example, colours, odours, tastes, sounds and tactile impressions. These were relegated to the mental realm as *secondary* qualities, along with all of the rest of subjective experience – pleasure, pain, love, hope, fear, status and (latterly) upward social mobility.

The problem of the relationship between primary and secondary qualities forms the basis of the history of modern epistemiolgy. Although the realm of mind is supposed to have an independent existence, the status of secondary qualities is ambiguous at best. They are not properties of matter and do not persist in the absence of an observer. Rather, they are the consequences of an interaction between the attributes of matter and a perceiving organism. Secondary qualities are caused by the effects on our organs of the motion of bodies. As Edwin Burtt says, 'We cannot conceive how such motions could give rise to secondary qualities *in the bodies*; we can only attribute to the bodies themselves a disposition of motions, such that, brought into relation with the senses, the secondary qualities are produced.'[14] Thus, the distinction treats primary qualities as objective and independent of the perceiver, while the secondary ones are subjective and exist only in the consciousness of perceiving persons.[15]

This picture of reality came under severe criticism in the writings of

Foucher and other contemporaries of Descartes, and aspects of the criticism were reiterated by Bayle, Berkeley and Hume. It was quickly pointed out that primary qualities also vary and that they, too, are represented through the fallible medium of sense perception. These objections imply that the distinction is difficult to maintain on philosophical grounds. However, in the context of the seventeenth century it is clear that the uses to which the distinction was to be put determined the emphasis on primary qualities. In the cases of Galileo, Descartes and Newton, the amenability of these qualities to mathematical and geometrical treatment, and their interpretation in increasingly mechanical and corpuscular terms, were the fundamental determinants. The primarily astronomical and physical interests of seventeenth-century scientists led to a particular *definition* of external reality.

These same interests led Galileo to reject explanations in terms of final causes as irrelevant to his purposes, just as Boyle made the same move in his application of the mechanical philosophy to chemistry. According to E.J. Dijksterhuis:

> The new conception rapidly gained ground, and in the second
> half of the century the distinction between the primary, geome-
> trico-mechanical qualities, which were considered to be really
> inherent in a physical body as such, and the secondary qualities,
> which were the names for the perceptive sensations and the
> feelings of pleasure or pain experienced in consequence of, or in
> connection with, physical processes in the external world, was
> (almost) universally accepted, and in fact considered to be
> almost self-evident.[16]

A close student of the early critics of Descartes' formulations of ontological dualism and the primary–secondary quality distinction once said to me that much of Western philosophy today is in a broad sense Cartesian. However, twentieth-century critics of the modern scientific world picture have been eloquent in their attempts to draw our attention to the price that science has paid for the convenience of handling nature mathematically. A.N. Whitehead points out that the spatio-temporal relationships of material substances were seen to *constitute* nature, while their orderliness constitutes the order of nature:

> The occurrences of nature are in some way apprehended by
> minds, which are [somehow] associated with living bodies.
> Primarily, the mental apprehension is aroused by the occur-
> rences in certain parts of the correlated body, the occurrences
> in the brain, for instance. But the mind in apprehending also
> experiences sensations which, properly speaking, are qualities

of the mind alone. These sensations are projected by the mind so as to clothe the appropriate bodies in external nature. Thus the bodies are perceived as with qualities which in reality do not belong to them, qualities which in fact are purely [I think White-head is slightly lost here] the offspring of the mind. Thus nature gets credit which should in truth be reserved for ourselves: the rose for its scent; the nightingale for its song; and the sun for its radiance. The poets are entirely mistaken. They should address their lyrics to themselves, and should turn them into odes of self-congratulation on the excellency of the human mind. Nature is a dull affair, soundless, scentless, colourless; merely the hurrying of material, endlessly, meaninglessly.[17]

On the other hand, Whitehead grants that these abstractions have been enormously successful. The problem lies in accepting them as reality itself. 'Thereby', he concludes, 'modern philosophy has been ruined.' It has oscillated in a complex manner among three extremes: dualists who accept both mind and matter, monists who put matter inside mind and monists who put mind inside matter.[18]

Burtt draws out the consequences of the paradigm in nearly identical terms:

The world that people had thought themselves living in – a world rich with colour and sound, redolent with fragrance, filled with gladness, love and beauty, speaking everywhere of purposive harmony and creative ideals – was crowded now into minute corners in the brains of scattered organic beings. The really important world outside was a world hard, cold, colorless, silent, and dead; a world of quantity, a world of mathematically computable motions in mechanical regularity. The world of qualities as immediately perceived by man became just a curious and minor effect of that infinite machine beyond. In Newton the Cartesian metaphysics, ambiguously interpreted and stripped of its distinctive claim for serious philosophical consideration, finally overthrew Aristotelianism and became the predominant world-view of modern times.[19]

Lest it be thought that the force of the paradigm has weakened, it may be worthwhile to allude to some more recent expositions of it. It has been forcefully defended by Jonathan Bennett in his article, 'Substance, Reality, and Primary Qualities',[20] and in a philosophical compendium, R.J. Hirst writes:

Science can adequately explain and describe the nature of the physical world solely in terms of primary qualities; hence, while primary qualities must characterize objects, there is no need to suppose that secondary qualities must also. The latter would

be otiose, and on principle of economy, or Occam's razor, . . .
it would be unscientific to suppose that they exist as intrinsic
properties of objects. . . . Investigation of the causal processes
on which perception depends shows that the only variables
capable of transmitting information about the properties of
external objects are spatiotemporal ones, which are associated
with primary qualities.[21]

It might be thought that this doctrine is held only by philosophers or
that it has no practical effect. As I was annotating this essay for
publication, I came across the following in a book review in *The New
Statesman*, written by a distinguished mathematical physicist, Felix
Pirani:

Much of modern science is rooted in the method of reduction,
which entails, in the first place, studying the parts to understand the whole. For societies, you study individuals; for individuals, their organs; for organs, their cells; for cells, their
molecules; for molecules, their atoms; for atoms, their protons
and electrons. . . .

Nobody would deny that this is one successful way of working, but many scientists insist that the parts are in some way
more fundamental than the whole and that, if you could describe them completely, you could predict everything about the
behaviour of the whole. In the end, for example, a complete
knowledge of atomic structure would explain completely the
behaviour of the DNA in genetic material, a complete knowledge of DNA would explain completely the behaviour of each
individual and this, in turn, would completely explain society.

The danger inherent in such arguments is apparent. For example, if the behaviour of the individual is determined by his or
her DNA, then misbehaviour can be dealt with by interfering
with the DNA (or some larger structure "determined" by it). For
thorough-going reductionists, things have to be the way they
are: racial, sexual and class oppression are all determined, in the
last analysis, by the properties of atoms, and so nothing can be
done about them.[22]

So the view is widespread, and the stakes are high.

This leads to the third – *methodological* – aspect of the paradigm,
which was considered inseparable from the ontological and epistemological aspects. The methodological aspect was prescriptive. More
or less enthusiastically, it was urged that people do experiments, but
whatever the varying views on this issue, it was agreed that scientific
conclusions should take the form of explaining all phenomena in terms
of matter and motion. This injunction is summarized in the preface to

Newton's *Principia*: 'All the difficulty of philosophy seems to consist in this – from the phenomena of motions to investigate the forces of nature, and then from these forces to demonstrate the other phenomena.'[23] In these days it is not worthwhile to claim that one knows exactly what Newton meant, as much attention is being devoted to what he was trying to hide by what he said. Therefore, perhaps, one can venture an eighteenth-century paraphrase and say that the injunction was taken to mean that no appeal should be made to secondary qualities in explaining the phenomena of the natural world.

It is the failure successfully to apply the programmatic aspect of this paradigm of explanation in the biological and behavioural sciences that provides the subject of the remainder of this paper. However, before moving to this, I should like to make my own position explicit. First, it seems clear to me that Cartesian dualism, and the doctrine of primary and secondary qualities, remain central to the philosophy of science. Secondly, recent discussions of the concept of 'action' and, *a fortiori*, the development of phenomenology on the Continent and in America, do not appear to me to succeed in transcending these problems. Rather, they confine the philosophy of nature to a realm that is a modern manifestation (*mutatis mutandis*) of the Cartesian world of subjectivity and thinking substances. Whatever progress may have been made in transcending the epistemological dualism of the subject–object distinction has been bought at the price of further separating the study of humanity from the categories of natural science. Briefly, phenomenology improves our conception of the person's relations with other persons and the external world by widening the gulf between subjectivity (mind) and the external world (body), and ontological dualism becomes more, not less, intractable.

Thirdly, we are in a position to examine the alternative interpretations of the status of secondary qualities as a result of a very lucid analysis and classification of theories by D.M. Armstrong.[24] Assuming only the independent existence of the external world, Armstrong shows that five different ontologies and nine different views of the status of secondary qualities are available and that each of these positions has been or is being held by one or more philosophers whom one must take seriously.

Finally, I should mention that I am satisfied that the paradigm of explanation, which has been characterized above, results in only one relatively consistent position, one that I find most convincingly argued in May Brodbeck's essay on 'Mental and Physical: Identity versus Sameness'.[25] Her interpretation will not be repeated here, although I am inclined to argue that some version of identity theory[26] is the only reasonable and consistent philosophy of physical science and the only valid application of the paradigm of explanation of modern science to biology, psychology and the social sciences. There is considerable evidence that these disciplines have not shown themselves to be adequately catered for by this paradigm. The persistent appeal to concepts that belong to the realm of mind or secondary qualities as a part of attempted explanation in these disciplines, reveals insubordination that I think should be taken seriously.

Some disobedient biologists

In the examples that follow, recourse can be had to at least four interpretations. First, one can argue that failure to reduce phenomena to explanation in terms of matter and motion simply reflects the limited scientific progress at the time in the subject. Thus, for example, Galen pointed out that 'so long as we are ignorant of the true essence of the cause which is operating, we call it a *faculty*'. Similarly, we could claim that Harvey lacked an adequate physiology of respiration and muscular contraction; Haller did not know about the electro-chemical transmission of the peripheral nervous impulse; Hartley was ignorant (as we largely are) of central neurohumours and the molecular basis of memory. Finally, Charles Darwin suffered from the lack of a particulate, genetic theory of inheritance and a molecular biology.

The second view that is available has been outlined by Ernest Nagel and hinted at by Gillispie.[27] Purposive and other non-reductionist accounts do not represent an *explanatory* alternative to mechanistic explanation but are only a matter of selective attention to certain features of biological processes. They are an alternative point of view, not an alternative explanation. Physicists can also adopt this approach but seldom do. However, a sustained attempt to do so can be found in Lawrence Henderson's essay, *The Fitness of the Environment*, where he draws many of his examples from nineteenth-century natural theology without adopting the associated question-begging explanatory scheme of design and vitalism.[28]

A third interpretation is related to the first two. That is, either because of the limited state of knowledge of micro-processes or because one wants to be brief (or provocative), one might speak in metaphorical or summary terms. Thus, Darwin thought of 'natural selection' as a metaphor and not as a mechanism itself.[29] Similarly, cyberneticists are prone to refer to the activities of their machines in mentalist terms in order to tease mentalists without implying that there are special vital laws at work in their semiconductors or a ghost in the machine.

A fourth interpretation has a strong and a weak form. The strong form is that special explanatory concepts are, after all, required in biology and the human sciences. This view is certainly held, albeit defensively, by some biologists and by even more students of animal and human behaviour. The weak form is a view I hold – that people, as a matter of fact, do continue to make some appeal to concepts drawn from the realm of secondary qualities and/or the concept of intention itself. They write papers in these terms that get published in reputable journals, get funds to support their research, and have distinguished careers. It remains to be seen whether or not this fact is philosophically interesting. I believe it is.

I should now like to refer to certain key episodes in the history of biology and psychology. In each case reference will be made to modern analogies in the use of question-begging, secondary qualities and mentalist concepts. One might caricature the accounts of historians who write about these episodes in terms of the advance of the official paradigm of explanation by saying that the persistence of the alternative modes of explanation reflects the disobedient, sorry, question-begging record of biology, psychology and the social sciences.

Let me begin with DESCARTES and HARVEY. When the paradigm of explanation of modern science was applied to living systems, it did not fare very well. Indeed, mechanistic physiology got off to a very bad start. Descartes used Harvey's discovery of the circulation of the blood as the key to all the rest of the investigation of living systems. He – and most historians of these developments – wrenched Harvey's discovery from its context in the Aristotelian philosophy of nature and interpreted it mechanistically.

Relying heavily on A.R. Hall, Gillispie gives the following picture of Harvey's achievement:

> His work was the first, if partial, breach opened by the scientific revolution in the life sciences. His subject is not the ineffability of life. It is a problem in fluid mechanics. The heart is a pump. ... The veins and arteries are pipes. The blood ... is simply a liquid, a lubricant to be passed periodically through the air filter of the lungs. No vital spirit, no principles of nourishment intrude into the analysis.[30]
>
> Galileo had excluded biological metaphor from physics. Harvey went further and introduced mechanistic thinking into organic studies. And by a simple though systematic extension, Descartes would find a machine in man.
>
> [Harvey's] hydraulics of the bloodstream destroyed a whole philosophy of the body in order to establish a single phenomenon of nature.[31]

The above quotations tell us something about historians who either have not read or have forgotten the books about which they are writing. To be slightly less rude, historians will falsify their sources in order to substantiate an oversimplified view of the onward movement of the 'edge of objectivity'. However, this observation would not justify recounting a bad account in such detail. The second reason for doing so is that Gillispie's version of the matter tells us something about Descartes and about the subsequent progress of biology. Gillispie is wholly inaccurate in fact and in judgement with respect to Harvey, but his account does accurately reflect the effect of Harvey's work on the development of physiology: that is, on mechanistic thinking. To put it another way, historians distort their sources to substantiate a particular view of the history of science – but so did Descartes. There are at least two versions of 'what somebody said'. The first is what we make of what they said, and the second is what their contemporaries and subsequent thinkers made of it. 'What they really said' is a will-o'-the-wisp – a noumenon.

John Passmore has reflected in a very interesting way on this matter.[32] The circulation of the blood might appear to be a simple fact, of no interest to philosophers. For example, the passage in Descartes' *Discourse on Method* that deals with Harvey is omitted from the Anscombe and Geach and the Smith editions of Harvey's works. Yet Harvey is the only Englishman mentioned in Hobbes' *Elements of Philosophy*, and Harvey and Galileo are the only persons named in Descartes' *Discourse on Method*, while Harvey is the only person mentioned in Descartes' *Passions of the Soul*. Copernicus and Galileo had applied mechanism to heavenly bodies, identifying them with terrestrial mechanics. Harvey allowed Descartes to extend mechanics to the living organism. Only the soul was left outside.

Descartes examined Harvey in detail, rejected Harvey's conclusions, and reverted to certain medieval views, even though he accepted the fact of circulation. Descartes supposed that the heart performed its functions by virtue of containing a peculiar source of heat. Are we to see this merely as a scientific difference? Descartes differed from Harvey's account because he wanted all of the body's functions to be explained in terms of concepts derivable from general mechanics – in this case, heat and expansion of the blood. Harvey, on the other hand, considered the contraction of the heart a fact, whose cause is either unknown or explained, for the time being, by a faculty. Descartes saw this as a pseudo-explanation, a reversion to the medieval mode of speaking: the 'faculty pulsifica', which Harvey actually mentioned several times. Thus, appeal to 'brute fact' is suspicious for Descartes.[33]

In his *La description du corps humain* (written in 1648–9), Descartes said of Harvey, 'If we suppose that the heart moves in the manner in which Harvey described it we shall have to imagine some faculty which causes this movement, the nature of which is much more difficult to conceive than everything he claims to explain by it.'[34] Descartes is not simply saying that Harvey resorts to faculties; he is saying that he must do so.

Facts were an irreducible foundation for Harvey, while for Descartes resorting to them was a confession of failure. They should be deducible from first principles. In the sixth part of the *Discourse*, Descartes did confess to the need for observations and experiments, but still, for him, this was a concession. He needed to connect the circulation with the general principles of mechanics. He would accept the explanation in terms of heat, as this was a valid mechanical principle of broad applicability. He rejected the special explanation of its dependence on the contractile properties of the heart.

Thus, Descartes would say of Harvey what he said of Galileo: without having considered the first causes of Nature, Galileo only looked for certain particular effects, and upon this he built without foundations. Harvey, on the other hand, was content with facts. He was also vigorously opposed to faculty explanations (see his 'Second Disquisition to Riolan'). For Descartes, not to explain was equivalent to appealing to occult faculties. The facts must be explained, because explanation is deducible from first principles, thereby demonstrating what the facts must be. For Harvey, demonstration was examination by the senses, ocular demonstration by experiment. In fact, he said of Descartes, rather laconically, that he had observed wrongly.

The point of this example is to show, as I shall re-emphasize below, that Harvey's discovery of the circulation of the blood can in no sense be considered a triumph of mechanism – the first great discovery in mechanistic physiology – even though that is what Descartes wished to make of it. Moreover, if we consult a modern textbook of physiology (the one I have to hand is the one I studied in medical school – Fulton's), it reads rather more like a text by Harvey than one that follows Descartes' strictures. The modern explanation of the heartbeat is put in terms of 'spontaneous automaticity', 'inherent irritability', 'intrinsic or autonomous rhythmicity', 'automaticity'. These are not faculties but biological properties (of which more below). The heartbeat is said to begin in the sino-auricular node near the termination of the great veins in the right atrium. The contractile property variously described above is characteristic of all heart tissue, but the heart is driven by the one with the highest rate. Electrical pacemakers work by taking over this role. We are dealing here with an essential physiological property of pacemaking miocardial cells. We find it localized in the muscle cells – it is 'myogenic'. It is localized but it is not 'explained' in the Cartesian sense. This is not to say that such explanations will not be forthcoming or that they have not appeared since the edition of Fulton that I studied.[35] My point is that biologists, including physiologists, feel quite at home with explanations that are not couched in terms of mechanistic first principles.

Returning to Harvey, Walter Pagel is emphatic about Harvey's vitalism and offers the following quotation from Harvey's writings *On Generation*:

> It is a common mistake with those who pursue philosophical studies in these times, to seek for the cause of diversity of parts in the diversity of the matter whence they arise. Thus medical men assert that the several parts of the body are both engendered and nourished by diverse matters, either the blood or similar fluid. . . . Nor are they correct who like Democritus, composed all things of atoms; wherewith Empedocles, of elements. As if generation were nothing more than a separation, or aggregation or disposition of things.[36]

The quotation goes on to appeal to Aristotle and to the divinity of nature, which is said to work as an efficient cause. Indeed, none of Harvey's contemporaries thought of him as a mechanical philosopher. *De Generatione* contains a more general natural philosophy. Harvey was unsympathetic to the mechanical approach of his contemporaries. He was emphatic in repeating Aristotle's criticism of atomism: it errs in ignoring formal and final causes.

Next, I should like to make a partial contrast between ROBERT BOYLE and JOHN RAY, two of the most effective exponents of the investigation of the phenomena of living organisms as a form of worship: natural theology. Although Boyle adhered to the mechanical and corpuscular philosophy, and sought explanations in terms of matter and motion,[37] he also wrote a very sober and restrained defense of the use of final causes in biological explanation. After making diffident gestures towards Galileo and Descartes' reasons for banishing final causes, he concluded 'that all consideration of final causes is not to be banished from Natural Philosophy', but that "tis rather allowable, and in some cases commendable, to observe and argue from the manifest uses of things, that the Author of Nature pre-ordained those ends and uses'.[38] However, Boyle cautioned 'that the Naturalist should not suffer the search of the discovery of final cause of Nature's Works, to make him undervalue or neglect the studious indagation of their efficient causes'.[39] '. . . The neglect of efficient causes would render physiology useless to me but the studious indagation of them, will [also] not prejudice the contemplation of final causes.'[40] In short, we must simply be cautious in the use of final causes and not employ them as a substitute for mechanistic explanation. Once again, there are abundant modern analogies in the descriptive teleology that is a commonplace among ethologists.

The views of Ray were in marked contrast with those of Boyle. His classic work in taxonomy and in establishing an unequivocal concept of species was conducted in the context of an explicitly anti-mechanistic natural philosophy and natural theology, which was spelled out in his essay *On the Wisdom of God*, which appeared in 1691, three years after Boyle's *Disquisition about Final Causes* and a year after Locke's *Essay Concerning Human Understanding*. Ray's philosophy of nature drew heavily on the ideas of the Cambridge Platonists. He opposed the mechanical philosophy on philosophical, theological and practical grounds. He felt that 'the Atomick Theists utterly evacuate that grand argument for a God, taken from the phaenomena of the artifical frame of things, ... the atheists are meanwhile laughing in their sleeves, and not a little triumphing, to see the cause of Theism thus betrayed by its professed friends and assertors', who do the atheists' work for them.[41] On the practical issue of the success of the mechanical philosophy, Ray remarked that its advocates are

in no way able to give an account [of the formation and organization of the bodies of animals] from the necessary motion of matter, unguided by mind for ends, and prudently therefore break off their system there when they should come to [the topic of] animals and so leave it altogether untouched.[42]

And those accounts which some of them have attempted to give of the formation of a few of the parts, are so excessively absurd and ridiculous, that they need no other confutation than *ha, ha he*.[43]

Ray rejected the concept of animal automatism. It could be said that no Englishman could pass the crucial test of loyalty to the Cartesian doctrine: guiltlessly kicking a dog. Ray appealed to a vitalistic force – the Plastick Nature – as God's purposive agent and medium in the natural world. Echoes of this view persisted in biological theory in Britain throughout the eighteenth century and it was being advocated well into the 1840s, in the writings of William Kirby and William Whewell, for example. There are modern analogies in emergentism, holism and gestalt.

My next example is concerned with the concept of a biological property and relates to the example of Harvey and Descartes. If Descartes can be said to have laid down the fundamental principles of mechanistic thinking in biology, ALBRECHT VON HALLER can lay claim to being the 'father of modern physiology'. His *Elementa* is recognized as the first modern handbook or systematic treatise in the field. It appeared between 1757 and 1766 in eight volumes. What interests me in Haller's thinking is the easy way in which he relied on biological concepts without feeling under any obligation to reduce them to mechanistic explanations. I shall focus on his concept of 'irritability'. This concept was first put forward by Francis Glisson in the seventeenth century and was an important step in providing a scientific – but not mechanistic – version of the phenomena that had formerly been explained by the dreaded faculties.[44]

Between Glisson and Haller, most investigators tried to explain irritability by either mechanical or vitalistic ideas. On the one hand, the iatromechanists tried and failed to account for everything in terms of matter and motion. On the other, the followers of Stahl insisted that everything depended on the soul, Ray invoked a Plastick Nature as a vitalistic principle, and so on. Haller's achievement was to ignore these alternatives and to characterize vital properties as phenomena in their own right. Irritability and sensibility were to be defined experimentally. He would not consider the question of mechanism. By explicitly and selfconsciously refusing to carry out the reductionist programme, he licensed his colleagues and those who came after him to use question-begging intermediate concepts under the general heading of 'biological property'. I see this as the thin edge of a wedge for separating biology and the sense of humanity from reductionism.

Haller based his views on experiments performed by himself and Dr Zimmerman in 1746 and 1751. He wrote that, since 1751:

> I have examined several ways, one hundred and ninety animals, a species of cruelty for which I felt such a reluctance, as could only be overcome by the desire of contributing to the benefit of mankind, and excused by that motive which induces persons of the most humane temper, to eat everyday the flesh of harmless animals without any scruple.[45]

He added, 'I am persuaded that the great source of error in physic has been owing to physicians, at least a great part of them, making few or no experiments, and substituting analogy instead of them.'[46]

Haller is very matter of fact: the purpose of his essay is to distinguish those parts of the body 'which are susceptible of Irritability and Sensibility, from those which are not':

> But the theory, why some parts of the human body are endowed with these properties, while others are not, I shall not meddle with. For I am persuaded that the source of both lies concealed beyond the reach of the knife and microscope, beyond which I do not chuse to hazard many conjectures, as I

have no desire to teach what I am ignorant of myself. For the vanity of attempting to guide others in paths where we find ourselves in the dark, shows, in my humble opinion, the last degree of arrogance and ignorance. . . .

I call that part of the human body irritable, which becomes shorter upon being touched; very irritable if it contracts upon slight touch, and the contrary if by a violent touch it contracts but little.

I call that a sensible part of the human body, which upon being touched transmits the impression of it to the soul; and in brutes, in whom the existence of the soul is not so clear, I call those parts sensible, the Irritation of which occasions evident signs of pain and disquiet in the animal. On the contrary, I call that insensible, which being burnt, tore, pricked, or cut till it is quite destroyed, occasions no such pain, nor convulsion, nor any sort of change in the situation of the body. For it is very well-known, that an animal, when it is in pain, endeavours to remove the part that suffers from the cause that hurts it; pulls back the leg if it is hurt, shakes the skin if it is pricked, and gives other evident signs by which we know that it suffers.

We see that experiments only can enable us to find what parts of the human body are sensible or irritable, and what the physiologists and physicians have said upon these qualities, without having made experiments, has been source of great many errors, both in this case and in a number of others.[47]

Haller's conclusions were based on a very large number of experiments for his time. He condemned a sensibility dependent on the nerves and continuous with the brain. Irritability, by contrast, was an inherent property of the muscles, independent of the nervous connection, because they contracted on stimulation after the nerves were detached. This was an important defeat for the spiritualists, as irritability persisted after connection with the organ of the soul was eliminated. Once again, I want to point out the modern analogy. We now consider contractility to be a specific property of muscles, while irritability is a general property of living matter to respond to stimuli. The spectre of faculty explanations has completely receded, and there is no fear, when we mention biological properties, that we are offering them as sufficient explanations.

DAVID HARTLEY would appear at first sight to be a poor candidate for question-begging because of appeals in his work to non-material causes in psychology. Locke and Newton are the two main sources for his doctrine, and his *Observations on Man, His Frame, His Duty and His Expectations* (1749) is a *tour de force* in the explanation of psychological and behavioural phenomena in terms of the vibrations of material, corpuscular particles in the nervous system. Hartley retained the doctrine of separate mental and bodily substances but abandoned the Cartesian concepts of the indivisibility of mind and free will, adopting psychological atomism and mental determinism as a parallel to corpuscular determinism. In the first chapter of his book he laid down a programme for physiological psychology that served as the basis for nineteenth-century associationist psychology and its integration with physiology, a programme that is still being pursued in brain and behaviour research.[48]

> The Doctrine of *Vibrations* may appear at first sight to have no connection with that of *Association*; however, if these Doctrines be found in fact to contain the Laws of the Bodily and Mental powers respectively, they must be related to each other, since the Body and Mind are. One may expect that Vibrations should infer Associations as their Effect, and Association point to Vibrations as its cause. I will endeavour . . . to trace out this mutual relation.[49]

For the next thousand pages Hartley did just that, using the concept of association by repetition as the mental analogy to the universal law of human nature, just as gravity was of physical nature (it was Hume who formulated this analogy). All mental phenomena are explained in terms of the vibrations (and 'vibratiuncles') of the corpuscular philosophy. In particular, the secondary qualities are discussed, one by one, in these reductionist terms.

Thus, Hartley's doctrine conforms perfectly to the paradigm of explanation of seventeenth-century science. However, as with Descartes, the uses to which it was put were very different. Beginning with Erasmus Darwin and, by a different route, in the doctrines of Lamarck (now mixing with the English associationist tradition), associationist psychology was used as a cloak for reintroducing purposive variables into biological theory. The fundamental significance of Hartley's hypothesis is that for the first time a mechanism had been worked out for the evaluative and teleological principle of utility. Put another way, adaptations can be acquired through experience. Perfect adaptation is obtained by the pleasures and pains resulting from the correlation of external phenomena, the vibrations that these cause in the nervous system, the sensations, ideas and motions that these build up by repetition, and the pleasures and pains that they engender. Adaptation is assured by experience. This became a powerful explanatory principle, not only in psychology, but also in biological – including evolutionary – theory. A massive dose of purposiveness got injected into living nature in the theories of

Erasmus Darwin, Lamarck and Herbert Spencer, as well as Charles Darwin. Striving is common to the theories of the first three. For them, evolution occurred as a result of perceived challenges and effort, the results of which were passed on to the next generation. As Spencer put it, evolution becomes a simple extension of sensationalist –associationist learning theory from the *tabula rasa* of the individual to that of the race. Biological evolution was thus explained by the paradigm of learning – not a very materialist reduction, I would say.[50]

But surely, it would be objected, we know that the so-called 'Lamarckian mechanism' involved appeals to progressive tendencies and more or less intentional striving. That is why we had no scientific theory of evolution before the appearance of Darwin and Wallace's theory of *natural* selection. The objector would continue to make sure that I do not argue that natural *selection* implies a selector, as we all know that neo-Darwinism put this problem to sleep in the 1930s, and if any lingering doubts remained, they were certainly solved by the late 1950s, with the establishment of the structure of DNA and the specification of the alphabet for amino acid sequences.

I do not feel shaken by these objections and will refer to current biological concepts below. For now, I want to remain in the historical mode and recall some of the features of Charles Darwin's thinking that did involve purposive and intentional variables. First, it should be recalled that Darwin's initial work on the mutability of species was done with domestic varieties and that he explicitly sought a natural analogue for the intentions of the breeder, which gave persistent directionality to random variation by means of purposive selection. Artificial selection was replaced by natural selection as a result of Darwin's use of Malthus' theory of population (which, it is worth noting in passing, was not reductionist but Hartleyan-utilitarian in its ancestry). Although Darwin felt that he had found a natural, non-teleological mechanism, he was plagued by doubts and criticisms. The title of his book reflects the risks: *On the Origin of Species by Means of Natural Selection, or the Preservation of Favoured Races in the Struggle for Life.* Selection? Preservation? Favoured? Struggle? Where do these explanatory concepts appear in a physicist's reductionist textbook? What place have they in the paradigm of explanation of modern science?

Darwin's response to criticisms of his theory of inheritance was to become increasingly Lamarckian, while Wallace eventually abandoned natural selection and appealed directly to the Will of the Creator to explain certain aspects of human evolution. *None* of the mainstream nineteenth-century evolutionists – with the partial exception of Robert Chambers – considered that evolutionary continuity overthrew the separation of mind and body, much less the doctrine of primary and secondary qualities. Vestiges of the Cartesian dualism and the separation of men and animals remained in the theories of Darwin's most ardent disciples. For example, Huxley, Wallace and Lyell were all troubled by the anthropomorphic aspect of Darwin's theory, and Tyndall confirmed the worry in his famous 'Belfast Address'. 'Can nature thus select?' he asked. 'Assuredly she can.'[51] Among friends, this kind of talk was permissible; others used the metaphor – which Darwin claimed to use only for brevity's sake but found indispensable – as an excuse for speaking of Designed Evolution. I will not dwell on this example, as I treat it at length elsewhere,[52] but before moving on, it is worth recalling that Spencer never abandoned his belief in use-inheritance and Progress, that the distinction between language users and non-language users was held by Huxley, and the distinction between savagery and culture was retained as a qualitative leap by most students of anthropology. These, it seems to me, are Cartesian vestiges.

My discussion of Hartley and utilitarian and intentional aspects of evolutionary theory has been designed to show the persistence of taboo concepts in biology. Other examples might be drawn from research in psychology and the study of the nervous system in the late nineteenth century up until the present, but there is an extensive literature on these topics, so I will not dwell on the matter.

Reintegration of teleology

I want to turn now to current concepts in biology, psychology and social science and suggest that they are very oddly related to mind–body dualism on the one hand and the doctrine of primary and secondary qualities on the other. In my view, the biological, psychological and social sciences are very defensive with respect to the reductionist paradigm. However, there is a large gap between what practitioners in these disciplines do when they are writing normal papers and what they do when they are reflecting as philosophers at prize-giving ceremonies and in presidential addresses. Among the key concepts in the biological, psychological and social sciences are adaptation, utility, function, property, goal, purpose and drive. Attempts systematically to reduce these to the phenomena of matter and motion have been spectacular failures. I am thinking of Hull's behaviourism, operationism, operant conditioning, and other forms of positivism and reductionism. Indeed, it has been impressively argued that in order to speak about living organisms, we find ourselves doing so by analogy to human intention. Charles Taylor's strike me as the best arguments for this, although I have attempted some myself.[53] Concepts such as function, adaptation and utility are common to disciplines extending from field biology through physiology and psychology to sociology and social anthropology. The use of such concepts in the human sciences has extended from phrenology through Comtean positivism and includes the powerful influence of Spencer on the growth of functionalist thinking in psychology, sociology, anthropology, political science and architecture. Concepts such as *milieu intérieur*, homeostasis, feedback and cybernetics are all ways of retaining purposive explanation in the biological and human sciences. The widespread use of such concepts and their pedigree – as outlined above – makes me think that the official reductionist paradigm of explanation in modern science never got a proper foothold in the biological and human sciences.[54]

In my view, scientists in these disciplines lead a split existence. They adhere to one paradigm of explanation in their philosophical reasonings, while happily practising another in their day-to-day work and published writings. Purposive, evaluative and teleological explanations have been as influential and have been as routinely extended down the line from the human towards the physical, as reductionist explanations have been used in successfully accounting for biological and human phenomena. Indeed, I think that the purposive has been more influential than the mechanistic.

The philosophical consequence of these historical observations would be that we should be more tolerant of conceptual hiatuses. Instead of the official reductionist programme, which I set out at the beginning of this essay, we might have a much looser (though no less structured) one. Here is a story that extends from the clinical to the material:

A patient is a person in a role.
A person is an organism.
Organisms are analysed in terms of functions.
Functions are about properties.
Properties are interpreted in terms of certain
    (secondary) qualities (colours, odours, tastes, temperature).
According to the rules of scientific explanation, these qualities, in turn, are to be interpreted in terms of primary qualities – extension, figure and motion, with number as the key concept. They are also supposed to be caused and explained by primary qualities.

The reason for the three dots in my title is that I do not think that we can pass smoothly across this divide. The reasons, as I indicated in my introduction, have to do with the limitations of the model of explanation seventeenth-century natural philosophers chose. Put another way, if a set of laws are enacted, and a whole section of the population fails to obey them, it may turn out that the laws (pun intended) are bad ones. Whitehead – to whom I have continued to return as a guide since I first read him as a second-year undergraduate – points out that during the seventeenth century there evolved a

> scheme of scientific ideas which has dominated thought ever since. It involves a fundamental duality, with *material* on the one hand, and on the other hand *mind*. In between them lie the concepts of life, organism, function, instantaneous reality, inter-action, order of nature, which collectively form the Achilles heel of the whole system.[55]

He says: 'The field is now open for the introduction of some new doctrine of organism which may take the place of the materialism with which, since the seventeenth-century, science has saddled phi-losophy.' His approach 'would lead to a system of thought basing nature upon a concept of organism, and not upon the concept of matter'; he calls his theory 'organic mechanism'.[56]

I believe this way of thinking has been implicit throughout the history of the biological and human sciences and that it is time to say so and to embark upon a radical metaphysical reconstruction in the light of this persistent way of thought. Moreover, I believe that if we avowedly (as opposed to surreptitiously) reintegrate purposes and values with material explanations, many forms of alienation will be more transparent and more amenable to being contested.

Notes

1. John C. Greene, *The Death of Adam: Evolution and Its Impact on Western Thought*, 1959 (New York: Mentor Books, 1961). See my *Mind, Brain and Adaptation: Cerebral Localization and Its Biological Context from Gall to Ferrier* (Oxford: Clarendon Press, 1970), esp. ch. 5.

2. John C. Greene, 'Biology and Social Theory in the Nineteenth Century: Auguste Comte and Herbert Spencer', in Marshall Clagett, ed., *Critical Problems in the History of Science* (Madison: University of Wisconsin Press, 1959), pp. 419–46.

3. John C. Greene, 'Darwin as a Social Evolutionist', in *idem, Science, Ideology and World View: Essays in the History of Evolutionary Ideas* (Berkeley: University of California Press, 1981), pp. 95–127. See my 'Darwinism *is* Social', in David Kohn, ed., *The Darwinian Heritage* (Princeton, N.J.: Princeton University Press, 1985), pp. 609–38.

4. Thomas S. Kuhn, *The Structure of Scientific Revolutions* (Chicago: University of Chicago Press, 1962), p. 15.

5. Charles C. Gillispie, *The Edge of Objectivity: An Essay in the History of*

Scientific Ideas (Princeton, N.J.: Princeton University Press, 1960).

6. Charles C. Gillispie, 'Lamarck and Darwin in the History of Science', in Bentley Glass, Owsei Temkin and William L. Straus, Jr, eds., *Forerunners of Darwin, 1745–1859* (Baltimore, Md.: Johns Hopkins Press, 1959), pp. 266–91 (279). See the almost identical wording in *idem, Edge of Objectivity*, p. 276.

7. The title of ch. 8 in *Edge of Objectivity*.

8. Gillispie, 'Lamarck and Darwin', p. 286.

9. Gillispie, *Edge of Objectivity*, p. 317.

10. Gillispie, 'Lamarck and Darwin', p. 282.

11. Gillispie, *Edge of Objectivity*, p. 303.

12. Robert M. Young, *Darwin's Metaphor: Nature's Place in Victorian Culture* (Cambridge: Cambridge University Press, 1985), ch. 2.

13. I have outlined these issues in 'The Mind–Body Problem', in J. Christie *et al.*, eds., *Companion to the History of Science* (London: Croom Helm, in press).

14. E.A. Burtt, *The Metaphysical Foundations of Modern Physical Science*, 2nd edn (London: Routledge and Kegan Paul, 1932), pp. 111–12.

15. E.J. Dijksterhuis, *The Mechanization of the World Picture* (Oxford: Clarendon Press, 1961), p. 431. See also A.C. Crombie, 'The Primary Properties and Secondary Qualities in Galileo Galilei's Natural Philosophy', in *Saggi su Galileo Galilei* (Florence: G. Barbera, 1967).

16. Dijksterhuis, *Mechanization of the World Picture*, p. 431.

17. A.N. Whitehead, *Science and the Modern World*, 1925 (London: Free Association Books, 1985), pp. 68–9.

18. Ibid., p. 70.

19. Burtt, *Metaphysical Foundations*, pp. 236–7.

20. Jonathan Bennett, 'Substance, Reality and Primary Qualities', *American Philosophical Quarterly*, 2 (1965), 1–17.

21. R.J. Hirst, 'Primary and Secondary Qualities', in Paul Edwards, ed., *The Encyclopedia of Philosophy*, 8 vols. (New York: Macmillan, 1967), V, 455–7 (456).

22. Felix Pirani, 'Little and Large', *New Statesman*, 19 Feb. 1988, pp. 33–4.

23. Quoted in Burtt, *Metaphysical Foundations*, p. 204.

24. D.M. Armstrong, 'The Secondary Qualities: An Essay on the Classification of Theories', *Australasian Journal of Philosophy*, 46 (1968), 225–41.

25. May Brodbeck, 'Mental and Physical: Identity versus Sameness', in P.K. Feyerabend and G. Maxwell, eds., *Mind, Matter, and Method: Essays in Philosophy and Science in Honor of Herbert Feigl* (Minneapolis: University of Minnesota Press, 1966), pp. 40–58.

26. What I have said here is too cryptic. Because I am not going to develop it at present, I will note some of the ambiguities in 'identity' theory with respect to my topic. There are at least four senses of identity in identity theory:

1. Identity theory is thought to refer to the *logical* concept of identity. In this sense, the theory is simply false.

2. Identity theory also refers to *explanation* of mental states in terms of physiological processes, thus conforming to the paradigm of explaining all phenomena in terms of matter, motion and number.

3. Identity theory alludes covertly to the *scientific* claim that 'the brain is the organ of the mind' in the sense that mental states are *caused by* physiological processes, not conversely.

4. [again related to (2)] We try to get out of the *epistemological* difficulties involved in identifying mental states by relating them to physiological processes of physical objects. This is an attempt to make mental events *public*.

On the relations between (4) and (2), see A.O. Lovejoy, 'Cartesian Dualism and Natural Dualism', in *idem, The Revolt against Dualism: An Enquiry concerning the Existence of Ideas* (LaSalle, Ill.: Open Court, 1969), pp. 1–46.

27. Ernest Nagel, *The Structure of Science: Problems in the Logic of Scientific Explanation* (London: Routledge and Kegan Paul), 1961, chs. 11, 12. See also Gillispie, *Edge of Objectivity*, p. 285.

28. L.J. Henderson, *The Fitness of the Environment: An Inquiry into the Biological Significance of the Properties of Matter* (New York: Macmillan, 1913).

29. This is the point of the title essay in my *Darwin's Metaphor*, ch. 4. I have spelled out the philosophical implications in a subsequent essay (see below, n. 52).

30. Gillispie, *Edge of Objectivity*, p. 73.

31. Ibid.

32. John Passmore, 'William Harvey and the Philosophy of Science', *Australasian Journal of Philosophy*, 36 (1958), 85–94. Much of what follows below is taken directly from Passmore's article.

33. Ibid., p. 90.

34. Quoted by Passmore from the Adam and Tannery edition, vol. 1, p. 243.

35. J.F. Fulton, ed., *A Textbook of Physiology*, 17th edn (Philadelphia: W.B. Saunders, 1955).

36. Walter Pagel, 'William Harvey and the Purpose of the Circulation', *Isis*, 42 (1951), 22–38.

37. Dijksterhuis, *Mechanization of the World Picture*, pp. 435–56.

38. Robert Boyle, *A Disquisition about the Final Causes of Natural Things* (London, 1688), p. 235.

39. Ibid., p. 229.

40. Ibid., p. 232.

41. John Ray, *The Wisdom of God manifested in the Works of Creation*, 5th edn (London: Benj. Walford, 1709), p. 47.

42. Ibid., p. 49.

43. Ibid., p. 341.

44. Owsei Temkin, 'The Classical Roots of Glisson's Doctrine of Irritation', *Bulletin of the History of Medicine and Allied Sciences*, 38 (1964), 297–328; Walter Pagel, 'Harvey and Glisson on Irritability with a Note on Van Helmont', *Bulletin of the History of Medicine and Allied Sciences*, 41 (1967), 497–514.

45. Albrecht von Haller, 'A Dissertation on the Sensible and Irritable Parts of Animals' (with an introduction by Owsei Temkin), *Bulletin of the History of Medicine and Allied Sciences*, 4 (1936), 651–99 (657).

46. Ibid., p. 658.

47. Ibid., pp. 657–8, 658–9.

48. Robert Young, 'Association of Ideas', in P.P. Wiener, ed., *Dictionary of the History of Ideas*, 4 vols. (New York: Charles Scribner's Sons, 1968), I, 111–18; *idem, Mind, Brain and Adaptation; idem*, 'David Hartley', in Charles Coulston Gillispie, ed., *Dictionary of Scientific Biography*, 16 vols. (New York: Charles Scribner's Sons, 1970–80), VI, 138–40; *idem, Darwin's Metaphor*, ch. 3.

49. David Hartley, *Observations on Man, His Frame, His Duty, and His Expectations*, 2 vols. (London: Leake and Frederick, 1749), I, 6.

50. All these issues are spelled out in greater detail in my two essays, 'Animal Soul', in Edwards, ed., *Encyclopedia of Philosophy*, I, 122–7, and *Darwin's Metaphor*, ch. 3.

51. John Tyndall, *Address delivered before the British Association assembled at Belfast* (London: Longmans, Green and Co., 1874), p. 40.

52. Young, *Darwin's Metaphor; idem*, 'Implications of Darwin's Metaphor for the Philosophy of Science' (talk delivered in Geneva, Switzerland, 1986); *idem*, 'Charles Darwin', in the BBC-2 television series, *Late Great Britons*, first broadcast 2 Aug. 1988.

53. Charles Taylor, *The Explanation of Behaviour* (London: Routledge and Kegan Paul, 1964); Young, 'Animal Soul'.

54. I have discussed these matters briefly in 'Why Are Figures so Significant? The Role and the Critique of Quantification', in J. Irvine *et al.*, eds., *Demystifying Social Statistics* (London: Pluto Press, 1979), pp. 63–74, and, at length, in 'The Naturalization of Value Systems in the Human Sciences', *Problems in the Biological and Human Sciences*, A381, 'Science and Belief: from Darwin to Einstein', block 6, unit 14 (Milton Keynes: Open University Press, 1981), pp. 64–118.

55. Whitehead, *Science and the Modern World*, p. 71.

56. Ibid., pp. 47, 93, 90.

Afterword

JOHN C. GREENE

The task of responding to a collection of essays offered as a tribute to oneself presents a considerable challenge. Let me begin by saying how much it means to me that scholars of substantial reputation should find what I have written sufficiently interesting and worthwhile that they should wish to contribute to this volume. Although I have devoted my life to studying the influence of ideas in Western culture, somehow it never occurred to me that my own ideas would have any particular effect. How gratifying, then, to discover that other historians have thought to draw some inspiration from my writings, and how interesting it is to observe to what extent the lessons they have drawn from reading my works are such as I might have expected or hoped for.

The essays in this volume fall under several categories. The largest number describe and analyse what I would call the ideological uses of scientific ideas. As the prestige of science has grown, partly because of its intellectual triumphs and partly because of its practical applications, various groups and individuals have tried to attach that prestige to programmes of social action and accompanying world views lying outside the domain of science. Just as, in earlier times, the sanction of revelation or of Nature was invoked in support of political, economic and social institutions and initiatives, so in modern times the sanction of science has been claimed on behalf of the *status quo* or, alternatively, in support of programmes challenging the established order.

In nineteenth-century Britain, as several of the essays in this volume make clear, the established order found its ideological justification in the Anglican Church. Consequently, aspiring professionals, middle-class reformers and champions of the proletariat, waging war on what Adrian Desmond characterizes as 'Tory–Anglican dynasties in their corporation "pest houses"', were driven to seek an alternative sanction in science and natural law. Evolutionary science, whether in the astronomical mode described by Simon Schaffer or in the Lamarckian modes outlined by Ludmilla Jordanova and James Secord, was especially attractive, for it seemed to suggest that change and progress were the inevitable outcome of the order of nature and hence of society. Scientific naturalism in various forms became the ideological battle cry of whatever groups were discontented with the *status quo*. George Combe's *Constitution of Man*, John Nichol's *Architecture of the Heavens*, Robert Chambers's *Vestiges of the Natural History of Creation*, and the works of Baden Powell, George Lewes and Herbert Spencer formed, says Secord, 'a series that exhibits an increasing intersection between an alternative scientific vision of progress ... and an elite science under the control of gentlemen'.

But the ranks of the so-called scientific naturalists were by no means unanimous in their aims and ideas, as the essays by Evelleen Richards and Bernard Lightman show. Struggles for power and prestige broke out among the gentlemen who controlled or sought to control the scientific establishment. Disputes erupted between those like Thomas Huxley and John Tyndall who considered themselves the anointed spokesmen for science and agnosticism and those who, like Charles Albert Watts, presumed to spread the gospel of scientific agnosticism and explain its social and political implications without proper scientific credentials.

The use of scientific ideas for social and political purposes ranging far beyond the purview of science was not confined to Great Britain nor to the nineteenth century, as Peter Bowler, Paul Weindling and John Durant have shown. Ernst Haeckel, Julian Huxley, Ralph Burhoe and Edward O. Wilson all typify the scientist-ideologue bent on saving society by promulgating a new ethics and a new religion claiming the sanction of science. The pattern is depressingly familiar. When will scientists and others learn that naturalism is a philosophical point of view with no more claim to the status of science than any other philosophical viewpoint, whether Marxian, Freudian, Russellian, Whiteheadian, or whatever. Scientists have as good a right to expound their philosophical, ethical and religious views as anyone else, but they have no right to palm these off as the findings of science. In so doing they hurt the cause of science by robbing it of such claims to objectivity as it may rightfully assert. It is no excuse, as Weindling and Durant seem to imply, that their ideals are noble, their influence beneficial in some respects, and their science relatively unimpaired by the ideological uses to which it has been put. In the pursuit of truth, as in other spiritual undertakings, the path to hell is paved with good intentions.

The distinction between science and ideology is important not only for scientists and for society but also for historians of science. There is a real danger, as I observed in my introductory conversation with the editor, that in our enthusiasm for depicting the ideological uses and abuses of science we may seem to imply that science is nothing but ideology. To say that 'Darwinism is social' is not equivalent to saying that the Darwin–Wallace theory of evolution by natural selection is pure ideology. Scientific theories may, and usually do, have ideological overtones and uses, but they are primarily efforts to grasp relationships in the actual order of things, to attain a limited but highly useful and satisfying comprehension of the way things work. To the historian, science may sometimes appear as 'a prolific and cruel mother, forever spawning scientisms and forever abandoning her illegitimate offspring',[1] but to the lover of truth, of knowledge for its own sake, it is a noble endeavour so long as it does not mistake its gradual advance in comprehending nature's ways for knowledge unto salvation, whether individual or social. That kind of knowledge has other sources, of which the scientist in his role as scientist knows nothing.

Of special interest to me as author of *The Death of Adam* are the studies of Erasmus Darwin and Jean Baptiste de Lamarck, and also Martin Rudwick's delightful essay on the origins of the pictorial representation of the epochs of earth history. As Roy Porter shows, Erasmus Darwin's evolutionism was no mere extension of the seventeenth-century system of matter in motion to the realm of life, but predominantly a biomedical speculation grounded in a naturalistic physiological psychology with strong vitalistic overtones. Likewise, Lamarck's version of what I have called the law-bound system of matter in motion, depicted by Ludmilla Jordanova, drew heavily on biomedical literature and diverged in important ways from the mechanical world view in defining concepts of nature, man, life and science appropriate to an evolutionary conception of reality. But, as Rudwick shows, the emerging radically altered view of human history in relation to earth history had to be presented pictorially, imaginatively, before it could be fully accepted conceptually. Ironically, it was the progressive creationists who set the stage for mankind's descent into the abyss of time.

James Moore's moving account of Charles Darwin's spiritual struggle with Christian doctrine in the crucible of personal bereavement, domestic love and growing intellectual doubt gives a much more human and believable picture of the man than the over-intellectualized descriptions of his religious development heretofore available. It is worth noting, moreover, that Darwin continued to be a theist through the publication of the *Origin of Species* and, with increasing misgivings, for many years thereafter. One cannot but feel the pathos of his last years as he oscillated between evolutionary theism and outright agnosticism: conscious that he could no longer respond to

Handel's *Messiah* as he had once responded, at times hating the passion for science which had robbed him of esthetic sensibility but which alone could take his mind off his 'accursed stomach', yet defiantly asserting that he had done rightly in devoting his life to science; confiding to Alfred Russel Wallace that life had become a burden to him even though he knew he had everything to make him happy, and taking what comfort he could in the hope that natural selection, aided by the inherited effects of mental and moral training, would eventually produce a 'more perfect creature', a race of hominids who would look back on him and Newton and Lyell as 'mere barbarians'; all the while little suspecting that not long after his bitter rejection of Christianity in his autobiography he would be buried with full Christian honours in Westminster Abbey. *Resquiescat in pace.*

What shall I say about John Durant's summary and critique of the central themes in my writings on evolutionary topics? First, I would like to thank him for undertaking a task that no one else has attempted, thereby presenting the issues more clearly than I myself have done. I must add, however, that he does not fully understand the role I have tried to play. I am not a philosopher of science concerned with the justification of scientific ideas. I am a historian of ideas interested in the interaction of science, ideology and world view in Western culture. In *The Death of Adam* I argued that the Cartesian idea of nature as a law-bound system of matter in motion, taken in conjunction with empirical discoveries in astronomy, geology, palaeontology, zoology and botany, undermined traditional static conceptions of nature and natural history, paving the way for dynamic and causal views of the same but giving no support to the idea that change was progressive, moving from lower to higher levels of being. That idea came into natural history from other sources, such as the 'great chain of being', the fossil record and the growing belief in human progress. Moreover, it was an idea totally incompatible with the mechanical view of nature and with the non-normative conception of science attached to that view.

From Lamarck onward, evolutionary theory attempted to reconcile mechanistic views of nature and natural science with notions of progression from lower to higher forms of being, from inanimate matter to conscious, thinking people. Darwin exacerbated the problem by giving chance a fundamental role in evolution and by sinking mankind, and hence mind, into nature. Later Darwinians, such as Ernst Mayr, have gone a step farther by making natural selection, a blind, statistical process, the source of the order and harmony evident in the organic world – a conception that would have horrified Newton and Descartes.

At the same time, evolutionary biologists, in their roles as citizens and ideologues, have done their best to spell out the implications of evolutionary biology for human duty and destiny, and these projections of world view have displayed the same tensions and contradictions manifest in their evolutionary science. On the one hand, they have rejected teleological views of nature as inconsistent with the positivistic, mechanistic conceptions of nature and science inherited from the seventeenth century. On the other, they have repudiated atomistic reductionism as incapable of accounting for levels of organization, the creativity of evolution, and the like. And they have couched their descriptions of evolutionary processes in language implying striving, purpose and achievement.

In my role as historian of ideas I have tried to call attention to these inconsistencies and tensions in the writings of biologists and to suggest the possibility that they indicate underlying transformations in Western thought. After all, if Newton's world view displayed an uneasy tension between the idea of nature as a law-bound system of matter in motion and the traditional static conception of nature as a framework of stable structures fitted as a stage for the activities of intelligent beings, might not the obvious tensions in the thought of the founders of the modern synthesis in evolutionary biology be symptomatic of an emerging transformation in our ideas of science and nature as profound as the transformation inaugurated in the seventeenth century? From this point of view, Darwin's writings may represent both the culmination of the mechanistic conception (leading from Boyle's *Origin of Forms and Qualities* to the *Origin of Species*) and the beginnings of the breakdown of that idea of nature and natural science.

The foregoing analysis is a partial answer to John Durant's query as to what significance I attach to the patent contradiction between the formal philosophy of nature and natural science of evolutionary biologists and the teleological, vitalistic and anthropomorphic figures of speech they employ in biological discourse. But there is more to be said on this subject. I have said repeatedly that I regard these figures of speech as ways of giving meaning and value to processes that would otherwise seem meaningless and devoid of value, and I have added that these metaphors and similes may in some cases express intuitions biologists have about nature but which they cannot acknowledge openly in the current state of ideas about science and nature. From both of these points of view I concur with Robert Young in finding metaphysical significance in these modes of expression.

But John Durant, like many other writers, finds no harm in these figures of speech so long as they can be 'translated into purely naturalistic terms', whatever that means. But why should a translation be necessary? Why don't biologists say what they mean? Does it save any words to say that the 'function' of mutation is to replenish the gene pool instead of saying that that is its effect? Or to describe reproductive isolation as 'a method guaranteeing evolutionary success' instead of saying that it may eventuate in a population capable of rapid increase? Or to say that biological species were 'invented' as a 'method' of preventing unsuccessful gene combinations from occurring instead of saying that reproductive isolation between biological populations has that effect? What contribution to clarity or conciseness of expression is made by saying that evolution is 'opportunistic', that it proceeds by 'trial and error', that it reaches 'dead ends' or accomplishes 'breakthroughs' by sneaking through 'loopholes', or that natural selection 'can remodel proteins in order to improve interactions'? Language of this kind is downright dishonest. If evolutionary biologists really believe that evolution is mechanistic, blind and purposeless, why don't they describe it in those terms? Perhaps because it would then be difficult to persuade the public, or themselves, that evolutionary science is a safe and sure guide to social action and to a knowledge of human duty and destiny.

As for John Durant's suggestion that I should identify 'genuine conceptual or theoretical difficulties', a perusal of the current literature on teleology and teleonomy, on adaptation, fitness and adaptedness, or on biological progress, will provide examples galore of conceptual difficulties. George C. Williams has described these in detail in his book *Adaptation and Natural Selection* (1966), criticizing his fellow biologists for mistaking effects for functions and for trying, like Julian Huxley, to derive cumulative evolutionary progress from the theory of natural selection.

'The concept of progress', Williams writes:

> must have arisen from an anthropocentric consideration of the data bearing on the history of life. There were undoubtedly some important, long-term cumulative trends in the early evolution of life. Some may have continued even after evolution became stylized by the establishment of precise chromosomal inheritance and sexual reproduction.... On the other hand, it seems certain that within any million year period since the Cambrian such trends were of very minor consequence. The important process in each such period was the maintenance of adaptation in every population.... Evolution, with whatever general trends it may have entailed, was a by-product of the maintenance of adaptation.... It is mainly when biologists become self-consciously philosophical ... that they begin to

stress such concepts as evolutionary progress. This situation is
unfortunate, because it implies that biology is not being
accurately represented to the public.[2]

The source of these misrepresentations of the theory of natural
selection, Williams speculates, is an uncritical reliance on analogies
between biological adaptations and humanly devised mechanisms,
combined with 'a desire . . . to find not only an order in Nature but a
moral order'. Even Darwin, he adds, claimed to find moral grandeur
in the work of natural selection, failing to realize that the 'ultimate
essence' of the theory of natural selection was compounded of 'a
cybernetic abstraction, the gene, and a statistical abstraction, mean
phenotypic fitness'. Perhaps, says Williams, biology might have
matured more rapidly in a culture not dominated by Judaeo-Christian
theology and the Romantic tradition – in a Buddhist culture, for
example.[3]

But, having delivered himself of this ahistorical fantasy, Williams
goes on to propose that the adaptations produced by the double
abstraction anthropomorphically styled 'natural selection' should be
described in language borrowed from human artifice!

> Whenever I believe that an effect is produced as the function
> of an adaptation perfected by natural selection to serve that
> function, I will use terms appropriate to human artifice. The
> designation of something as the *means* or *mechanism* for a certain
> *goal* or *function* or *purpose* will imply that the machinery involved
> was fashioned by natural selection for the goal attributed to it.
> This is a convention in use already.[4]

It is indeed the convention in use, a convention that enables biologists to have their cake and eat it. They can scoff at the idea of teleology
in nature and continue to describe the organic world in teleological
language, attributing to the double abstraction 'natural selection' the
providential role formerly assigned to the Creator and putting aside,
as a metaphysical question outside the domain of science, the question of how it happens that the terrestrial system is of such a character
as to maintain adaptation amid changing circumstances.

Oddly enough, Williams is as opposed as Aristotle was to the idea
that the marvellous adaptations of living beings could be the result of
chance. 'One should never imply that an effect is a function unless he
can show that it is produced by design and not by happenstance.'[5]
But, for Williams, the 'design' of organisms is a product, not of
Aristotelian final causes, but of a complicated interplay of material
processes devoid of intelligence and lumped together under the anthropomorphic label 'natural selection'. Design without any designing intelligence and selection without any selecting aim or agency are

the heart and soul of the 'science of teleonomy'. Translators, get ready to translate the language of this new science, with its 'evolutionary strategies designed to solve problems of living', into cybernetic and statistical abstractions, remembering always that the process adumbrated by these abstractions gave rise to the mind that conceived them!

My position with respect to these paradoxes and contradictions in the literature of evolutionary biology has always been that biologists should either abandon their vitalistic, teleological, anthropomorphic figures of speech or else, like Pierre Teilhard de Chardin, revise their philosophy of nature and natural science to make sense of them. No such general revision is under way, but many biologists have begun to question whether the conceptions of nature and of science associated with the physical sciences are adequate for the investigation of living beings. Julian Huxley, G.G. Simpson, Ernst Mayr and others have called for a new philosophy of science that recognizes the irreducibility of biological phenomena to physico-chemical models of explanation and makes room for levels of organization, variational evolution, means–ends relationships, goal-directed activity, and the 'incurably historical' character of biology. George Williams demands only that natural selection as conceived by Ronald Fisher, J.B.S. Haldane and Sewall Wright, be added to the list of explanatory principles recognized by science, but Mayr finds the 'atomistic reductionism' of 'Fisherism' incapable of explaining genetic homeostasis, the emergence of evolutionary novelties in macroevolution, and the like; he proposes instead a semi-holistic, emergentist approach to biological phenomena. Wright and Huxley, confronted with the presence of mind in nature, embraced panpsychism. More recently, Richard Lewontin and Richard Levins have resurrected Friedrich Engels's dialectic of nature, portraying Darwinian and neo-Darwinian theories of evolution as ideological reflexes of bourgeois capitalism and advocating 'the dialectical view of organism and environment as interpenetrating so that both are at the same time subjects and objects of the historical process'.[6]

All of which brings me to Robert Young's essay calling for a new philosophy of nature encompassing mind, purpose, intention and value, together with a new history and philosophy of science capable of transcending empiricist-positivist-mechanist conceptions in dealing with the biological and human sciences. In a counterpoint to my own analysis of the discrepancy between the formal philosophy of modern biologists and their metaphorical *vocabulary*, Young describes a similar disjunction between that philosophy and the *concepts* of purpose, function, utility, adaptation and the like, that form the mainstay of scientific analysis in biology and the behavioural disciplines.

With the general drift of Robert Young's argument I quite agree, but I think he did not foresee the extent to which the conception of science he attacks would be undermined not only in biology, but in the history and philosophy of science as well. Thomas Kuhn's paradigm of scientific development never caught on among historians of science and was never put to practical use by Kuhn himself. His book, *The Structure of Scientific Revolutions*, had a decided impact on philosophers of science, but not by way of reinforcing the conceptions of science and nature to which Young objects. Nor has Charles Gillispie's 'edge-of-objectivity' approach enlisted many followers. In recent years, the history of science has turned more in the directions suggested in Young's own stimulating essays on the social and political context of science than in those recommended by Kuhn and Gillispie.

As for biology, many evolutionary scientists, including Julian Huxley, Theodosius Dobzhansky, Ernst Mayr, Sewall Wright, and the Marxian biologists, have discarded mechanism and reductionism as incapable of making sense of organic phenomena. At the same time, however, they reject teleology as a threat to science, inventing terms like 'teleonomy' to describe the purposiveness of living beings. Thus they hang suspended between two worlds – the old one dying, the other still inchoate, struggling to be born.

The cure for these ambivalences, Robert Young suggests, lies in developing (1) an organicist philosophy of nature that can 'reintegrate purpose and value with material explanations' and (2) a philosophy of society and history that can 'shed light on the meaning of life – of life itself, of humanity, and the husbanding and enhancement of generous values'. With both of these objectives I feel sympathetic. The difficulty is that there are numerous conflicting philosophies of nature, society and history, ranging from Julian Huxley's evolutionary humanism to Edward O. Wilson's 'cosmic epic' to Young's 'humanistic marxism', all purporting to integrate values and purposes with material explanations and to shed light on the meaning of life and history. Many of these claim the sanction of Science with a capital 'S'.

As I said in the introductory conversation, the first need intellectually in the present state of Western civilization is to achieve a saner, better balanced view of the role of science in human culture. It should be plain by now that science, for all its intellectual triumphs in discovering nature's ways and nature's history, cannot provide the spiritual insight and moral direction essential to human welfare, both individual and social, nor can it resolve the age-old problems central to philosophical speculation and religious faith and doctrine. It should be equally clear that the application of science to practical concerns creates at least as many problems – social, political, environmental, moral – as it solves. The Baconian Dream, whether in its original form, its Enlightenment dress, or its Marxian transformation, shatters on the hard rock of human nature, a nature known in part to politicians and novelists but highly refractory to the tools of science.

Karl Marx and Friedrich Engels to the contrary notwithstanding, there is and can be no science of history, a science that will tell us not only what has happened and will happen but also why it must happen and what our duty and destiny are in the light of that science. Such a conception belongs more in the realm of philosophy and religion than in that of science. The historian must be content to study limited segments of the past in the hope of gaining some small insight into the tangled web of personalities, ideas, demographic and economic conditions, institutions, and events in a chosen period, in part by establishing as exactly as possible what happened and in part by trying empathetically to enter into the minds and hearts of the actors in the historical situation. The historian of science in particular will have to choose which of the many aspects of scientific activity and institutions to focus attention on, always keeping in mind that his historical vision and conclusions will be limited by that choice. If, in the end, he succeeds in shedding some light on the meaning of life, of humanity, of moral value, it will be because he has entered into the thoughts and feelings of the men and women he studies and has managed to connect them with his own thoughts and feelings and with those of his contemporaries.

In this undertaking science plays an important but limited role. Like the scientist, the historian must use care and ingenuity in establishing the facts relevant to the subject he studies. Like the scientist, he must propose some explanation as to why things happened as they did. But, unlike most scientists, he cannot experiment on his subjects or (like the palaeontologist) draw on established scientific laws and principles. Finally, unlike the scientist, he must eventually undertake the imaginative reconstruction of the past by entering as much as possible into the minds and hearts of the actors in the historical drama and, so to speak, re-enacting it for his readers, as James Moore has done in his essay in this volume. (The palaeontologist reconstructs the life of past epochs with scientifically controlled imagination, but he does not try to enter the mind of a dinosaur or an ammonite or a cycad.)

Different historians will choose different epochs to study or different aspects of the same epoch, and they will adopt or construct different explanatory frameworks and draw different conclusions, both intellectual and moral. There is no prescribed way to write the history of science, or any other history. George Sarton's approach to the history of science was bio-bibliographical; Alexandre Koyré and his followers concentrated on the interactions between science and philosophy; Robert Young and others on the social conditioning and political uses of science. The essays in this volume fall predominantly, although not exclusively, into the thir̃d category. I welcome them all and rejoice in the thought that my own writings may have exerted some small influence on those who have paid me this tribute.

Notes

1. John C. Greene, 'Biology and Social Theory in the Nineteenth Century: Auguste Comte and Herbert Spencer', *Science, Ideology, and World View: Essays in the History of Evolutionary Ideas* (Berkeley: University of California Press, 1981), p. 87.

2. George C. Williams, *Adaptation and Natural Selection: A Critique of Some Current Evolutionary Thought* (Princeton, N.J.: Princeton University Press, 1966), pp. 54–5.

3. Ibid., pp. 33, 254–5.

4. Ibid., p. 9.

5. Ibid., p. 261.

6. See John C. Greene. 'The History of Ideas Revisited', *Revue de synthèse*, 4th ser. (1986), 219–27. For Mayr's critique of 'Fisherism', see Ernst Mayr, *Toward a New Philosophy of Biology: Observations of an Evolutionist* (Cambridge, Mass.: Belknap Press of Harvard University Press, 1988), pp. 535–6.

Index

415